Comprehensive Biomarker Discovery and Validation for
Clinical Application

RSC Drug Discovery Series

Editor-in-Chief
Professor David E. Thurston, *King's College, London, UK*

Series Editors:
Dr David Fox, *Vulpine Science and Learning, UK*
Professor Ana Martinez, *Instituto de Quimica Medica-CSIC, Spain*
Professor David Rotella, *Montclair State University, USA*

Advisor to the Board:
Professor Robin Ganellin, *University College London, UK*

Titles in the Series:

How to obtain future titles on publication:
A standing order plan is available for this series. A standing order will bring delivery of each new volume immediately on publication.

For further information please contact:
Book Sales Department, Royal Society of Chemistry, Thomas Graham House, Science Park, Milton Road, Cambridge, CB4 0WF, UK
Telephone: +44 (0)1223 420066, Fax: +44 (0)1223 420247,
Email: booksales@rsc.org
Visit our website at www.rsc.org/books

Comprehensive Biomarker Discovery and Validation for Clinical Application

Edited by

Péter Horvatovich and Rainer Bischoff
University of Groningen, Groningen, The Netherlands
Email: p.l.horvatovich@rug.nl; r.p.h.bischoff@rug.nl

RSC Publishing

RSC Drug Discovery Series No. 33

ISBN: 978-1-84973-422-6
ISSN: 2041-3203

A catalogue record for this book is available from the British Library

Published by The Royal Society of Chemistry,
Thomas Graham House, Science Park, Milton Road,
Cambridge CB4 0WF, UK

Registered Charity Number 207890

For further information see our web site at www.rsc.org

Printed in the United Kingdom by CPI Group (UK) Ltd, Croydon, CR0 4YY

Introduction to Biomarker Discovery and Validation

PÉTER HORVATOVICH*[a,b,c] AND RAINER BISCHOFF[a,b,c]

[a] Analytical Biochemistry, Department of Pharmacy, University of Groningen, A. Deusinglaan 1, 9713 AV Groningen, The Netherlands; [b] Netherlands Bioinformatics Centre, Geert Grooteplein 28, 6525 GA Nijmegen, The Netherlands; [c] Netherlands Proteomics Centre, Padualaan 8, 3584 CH Utrecht, the Netherlands
*Email: p.l.horvatovich@rug.nl

In 2001, the biomarkers definitions working group of the National Institutes of Health in the United States defined a biomarker as "a characteristic that is objectively measured and evaluated as an indicator of normal biologic processes, pathogenic processes, or pharmacologic responses to a therapeutic intervention".[1] A large number of analytical techniques are available to discover and measure biomarkers in complex biological samples such as immunochemical assays, molecular imaging, molecular arrays and mass spectrometry. From these techniques, mass spectrometry excels to be the most versatile analytical instrument able to comprehensively determine the identity and absolute or relative quantity of biomolecules present in complex biological samples. Mass spectrometry, a fairly recent approach to discover and validate biomarkers, requires close collaboration between multiple disciplines such as biology, pathology, analytical chemistry and bioinformatics. This approach is currently gaining momentum in the clinical environment. Different types of biomarkers are needed to support early diagnosis of disease onset, to predict the outcome of therapy, to monitor the efficacy of treatment or to get an objective clinical measure of drug efficiency based on molecular signatures.

RSC Drug Discovery Series No. 33
Comprehensive Biomarker Discovery and Validation for Clinical Application
Edited by Péter Horvatovich and Rainer Bischoff
© The Royal Society of Chemistry 2013
Published by the Royal Society of Chemistry, www.rsc.org

To perform successful biomarker research ending with validated assays that are suitable for clinical application requires collaboration between academic centers and industry with considerable investments of time and money. More importantly it requires meticulous research based on an experimental design, well-characterized biospecimen collections, targeted and comprehensive profiling analytical platforms, and the processing and statistical analysis of enormous sets of data as well as determined leadership to arrive at go/no go decisions throughout the process from discovery to final implementation. Last but not least, biomarker-based assays must provide sufficient benefit for patients and the population at large to be eligible for reimbursement by health-care providers.

This book presents some of the state-of-the-art methodology used by leading academic and industrial research groups worldwide to advance biomarker research from planning sample collection and storage in biobanks, to the initial comprehensive discovery study all the way to the requirements for approval of biomarker assays by regulatory authorities. It provides a comprehensive view of mass-spectrometry-based biomarker discovery in 12 chapters covering all parts of this process.

Introductory Chapters

Chapter 1 covers strategic and practical aspects related to biomarkers for translational and personalized medicine related to pharmaceutical drug development.

Chapter 2 shows the complex and laborious process to take a biomarker product from discovery to commercialization. This chapter will familiarize the reader with regulatory issues and give a general understanding of quality system requirements. The chapter gives an overview over the current regulatory environment at the US Food and Drug Administration (FDA) and the European Medicines Agency (EMEA) for approving a biomarker test for clinical use.

Chapter 3 describes approaches used to design experiments for clinical biomarker discovery, verification and validation, discussing protocols and good practices for sample collection, storage, and quality control. It presents the cardinal role of biobanks and human biospecimen collections in biomarker discovery and validation, discusses the crucial points of patient and control selection to design longitudinal and cross-sectional studies. Finally, this chapter gives guidelines on how to select patient cohorts and to stratify patient populations.

Sample Preparation and Profiling

Chapter 4 deals with approaches of biomarker discovery in easily accessible body fluids such as serum, urine, epithelial lining fluid and cerebrospinal fluid, discussing sample-preparation methods and problems related to sample stability. Sample prefractionation methods that try to cope with the large dynamic concentration range of proteins in body fluids are described in this chapter.

Chapter 5 describes various profiling platforms that are based on liquid chromatography coupled to mass spectrometry (LC-MS) for proteomic biomarker discovery, verification and validation. Platforms used for biomarker discovery provide qualitative (identity of proteins) and quantitative information about the analyzed proteomics samples. A discussion of targeted proteomics platforms as applied in the postdiscovery verification and validation phases is part of this chapter. Targeted protein analysis by LC-MS/MS allows quantification of peptides directly in clinical samples and contributes to verification and validation of biomarker candidates from biological samples.

Chapter 6 deals with the use of affinity array-based technologies in biomarker discovery and validation. The chapter demonstrates the strength of this technology by the use of peptide arrays to profile kinase activity to characterize phosphorylation-mediated signal transduction. Several applications of peptide arrays for kinome analysis are discussed including specific examples in which arrays have contributed to the identification of disease-associated processes/ biomarkers, provided therapeutic targets, and offered novel insight into complex biological processes.

Bioinformatics and Statistics

Chapter 7 describes the structure of data preprocessing pipelines used to extract quantitative information from label-free LC-MS data. The chapter provides an overview of the main steps based on the example of a "Threshold Avoiding Proteomics Pipeline". This chapter further presents data visualization techniques and quality-control methods to assess the accuracy of automated label-free LC-MS data processing.

Chapter 8 describes important aspects of statistical analysis of the extremely complex biomarker discovery data leading to a selected set of highly discriminating peptides and proteins that can be followed-up in a subsequent validation phase. The main concepts of statistical validation to identify potential biomarker candidates are discussed. The presented statistical methods and concepts are described without the use of intricate mathematics, in a way that scientists involved in biomarker research with a general statistical or informatics background can follow. Techniques to avoid false positives are discussed and practical advice is given, including recognition of situations where no reliable biomarker candidate can be identified.

Chapter 9 describes a novel bioinformatics approaches to place biomarker candidates in biological context by exploring the molecular interaction networks and pathways that help in understanding the biological relevance of the discovered proteins. This chapter provides a detailed overview of bioinformatics and statistical approaches based on a case study related to plasma protein markers for breast cancer.

Discovery and Validation Case Studies, Recommendations

Chapter 10 presents a comprehensive study to discover, verify and validate urinary biomarkers for the early detection of bladder cancer. The chapter demonstrates the integration of a discovery study using stable isotope labeling on low numbers of pooled samples followed by the validation of a limited number of candidates with Multiple Reaction Monitoring (MRM) and ELISA approaches using a large set of individual samples. This chapter gives an example of the development of a biomarker panel rather than an individual biomarker.

Chapter 11 demonstrates the power of current mass-spectrometry-based biomarker discovery and validation approaches on the example of developing a biomarker panel to improve the prediction of near-term myocardial infarction. This case shows the power of a pooling strategy in combination with stable isotope labeling and 2-dimensional liquid chromatography coupled to mass spectrometry for biomarker discovery followed by validation with multiplexed immunoassays and MRM.

The last chapter, *Chapter 12*, provides a comprehensive overview of the bottlenecks and challenges of current protein biomarker research and provides guidance and recommendations to avoid common pitfalls. The challenges faced during protein biomarker development are multifactorial from the initial phase of sample collection and study design to final clinical validation comprising financial aspects and regulatory approval. This chapter helps researchers to recognize these potential challenges and help them to plan their study in a way that produces reliable results that can be critically interpreted.

Biomarker research underwent enormous technological and methodological developments during the last decade, a trend that continues today. Biomarker research covers many technological platforms and aspects that are not possible to include in one book. For example, this book does not include discussions on the bioinformatics of mass-spectrometry-based protein identification, analytical methods revealing protein isoforms or the tissue and cellular localization of proteins with optical- or mass-spectrometry-based imaging techniques. This book will aid practitioners ranging from specialists working at biobanks or at regulatory authorities, statisticians, bioinformaticians or researchers working in academic or industrial environments on biomarker discovery and validation projects.

Reference

1. Biomarkers Definitions Working Group, *Clinical Pharmacology and Therapeutics*, 2001, **69**, 89–95.

Contents

Introduction

RSC Drug Discovery Series No. 33
Comprehensive Biomarker Discovery and Validation for Clinical Application
Edited by Péter Horvatovich and Rainer Bischoff
© The Royal Society of Chemistry 2013
Published by the Royal Society of Chemistry, www.rsc.org

Chapter 3 **Introduction: The Cardinal Role of Biobanks and Human**
Biospecimen Collections in Biomarker Validation: Issues
Impeding Impact of Biomarker Research Outcomes 73
Pascal Puchois, Lisa B Miranda and Alain van Gool

Sample Preparation and Profiling

Discovery and Validation Case Studies, Recommendations

Chapter 10 Discovery and Validation Case Studies, Recommendations: A Pipeline that Integrates the Discovery and Verification Studies of Urinary Protein Biomarkers Reveals Candidate Markers for Bladder Cancer   **271**

Yi-Ting Chen, Carol E. Parker, Hsiao-Wei Chen, Chien-Lun Chen, Dominik Domanski, Derek S. Smith, Chih-Ching Wu, Ting Chung, Kung-Hao Liang, Min-Chi Chen, Yu-Sun Chang, Christoph H. Borchers and Jau-Song Yu

Introduction

Introduction: Biomarkers in Translational and Personalized Medicine

CHANCHAL KUMAR[a] AND ALAIN J. VAN GOOL*[b,c,d]

[a] Translational Medicine Research Centre, Merck Research Laboratories, MSD, 8 Biomedical Grove, Neuros #04-01, Singapore 138665, Singapore; [b] TNO Quality of Health, Metabolic Health Research, TNO, Zernikedreef 9, 2301 CE Leiden, The Netherlands; [c] Laboratory of Genetic Endocrine and Metabolic Diseases, Department of Laboratory Medicine, Radboud University Nijmegen Medical Center, The Netherlands; [d] Faculty of Physics, Mathematics and Informatics, Radboud University Nijmegen, The Netherlands
*Email: alain.vangool@tno.nl

Summary

This chapter covers strategic and practical aspects related to optimal ways in which biomarkers for translational and personalized medicine can be applied to innovate pharmaceutical drug development, and contribute to improved health and disease management.

1.1 Introduction

Biomarkers have been around since the beginning of medicine when the colour of skin, various characteristics of urine (exemplified by the diagnostic "urine

RSC Drug Discovery Series No. 33
Comprehensive Biomarker Discovery and Validation for Clinical Application
Edited by Péter Horvatovich and Rainer Bischoff
© The Royal Society of Chemistry 2013
Published by the Royal Society of Chemistry, www.rsc.org

wheel" published in 1506 by Ullrich Pinder, in his book *Epiphanie Medicorum*[1]), and other qualitative assessments were interpreted as biological markers of a person's well-being. For a long time, phenotypic analyses combined with a patient's self-assessment were the only tools for diagnosis of disease and monitoring of treatment effects. Recent breakthroughs in molecular technologies to identify, understand and measure biomarkers have strongly increased the possibilities towards a person-specific assessment of disease. These include accurate prediction of a person's risk to develop a specific disease, early detection of a prevalent disease, prediction of disease progression, and prediction and monitoring of the effects of disease treatment, all in a personalized manner.

Biomarkers can be diverse. Ten years ago a useful definition of a biomarker was drafted, being "*a characteristic that is objectively measured and evaluated as an indicator of normal biological processes, pathogenic processes, or pharmacologic responses to a therapeutic intervention*".[2] There are several important aspects in this definition, both in terms of what is described and what is missing. First, a biomarker can be an *indicator* of a normal process, of a derailed process related to disease or of the effects of a certain treatment thereon. Although this covers many of the applications, biomarker scientists argue it does not describe biomarkers that indicate disease risk, *e.g.* through genetic predisposition or brought about by a certain lifestyle. Secondly, the biomarker is a *characteristic*, meaning it can have multiple identities ranging from a single protein in serum to a complex three-dimensionally reconstructed image of the brain. This has caused several discussions in the field; indeed, would an established biochemical assay that has been operational long before the current biomarker hype, such as estadiol analysis, qualify as a biomarker? If so, how about a mechanical read-out such as a pressure meter in a pen, used by anxiety patients filling in a questionnaire? How about the questionnaire itself? Thirdly, the biomarker is to be *objectively measured and evaluated*, implying the biomarker assay read-out is trustable and actionable. This leaves room to decide exactly how objective a biomarker should be measured before enabling a clinical decision, fueling discussions on fit-for-purpose robustness of the assay. Despite these alternative views, the stated definition is still a useful one and used by many to focus their attentions to the output defined.

Because of their potential in clinical applications, biomarkers have received much interest in the biomedical field. Their main applications seem to reside in two areas. In *translational medicine*, knowledge from preclinical models is translated to clinical practice and back using biomarkers that can reflect various aspects of a biological system including molecular pathways, functional cell–cell interactions and tissue metabolism. Such studies are expected to greatly increase the molecular knowledge of the mechanisms of human disease and pathophysiology, leading to a better diagnosis and more effective clinical treatment. In *personalized medicine*, biomarkers are used to profile patients and to define which treatment should be given to which patient at what time and at what dose. Such stratification biomarkers are expected to strongly increase the chance of a successful clinical treatment by selecting patients that are most likely to respond to a drug and/or to deselect patients that are predicted to exhibit adverse effects.

The availability of the human genome sequence in the late 1900s prompted many to believe that by 2010 personalized medicine would be fully implemented as each person would have his/her genome on a chip to enable a physician to determine the best personalized care. Former president of the USA Bill Clinton phrased it in his 1998 State of the Union Address as: "Gene chips will offer a road map for prevention of illness through a lifetime". There are many shining examples where hard work has indeed resulted in good clinical utility of biomarkers, but there still is a long way to go, as discussed in this chapter.

Biomarkers have become part of our daily lives as illustrated by advertisement of the positive effects of nutrition based on biomarkers (*e.g.* cholesterol lowering), by media-supported general education about the molecular processes in a human body and how biomarkers represent those processes, by availability and acceptance of biomarker-based "health checks" that can be performed through dedicated providers or even main-street pharmacies, by smartphone apps that provide a health check based on biomarker data, and so on. This all leads to more aware and vocal patients who debate with their physician about their best treatment, rather than "following doctor's orders".

Regarding industrial implementation, biomarkers in the pharmaceutical drug development field have been leading the way, as they matured from explorative pharmacological parameters to essential tools to characterize a patient in molecular detail and to monitor drug action after dosing. In development of neutriceuticals (functional ingredients of food) biomarkers can have a similar role and potentially similar biomarkers representing biological mechanisms or metabolic physiological states can be used. Also, cosmetics can be an interesting biomarker application area, whereby the biomarker read-outs can demonstrate absence of side-effects of the cosmetics. Interestingly, biomarkers are receiving increasing interest to quantify health. Health is described to be "not merely the absence of disease but the ability to adapt to one's environment", also called resilience.[3] Health biomarkers thus indicate the risk of an individual to develop a disease and can be key drivers of prevention strategies, including timely correction by lifestyle change, neutriceuticals or pharmaceuticals.

Despite these positive developments, it was anticipated that progress in translational and personalized medicine would be more advanced than it is today. The discovery of a biomarker and its maturation to a clinically usable test has been shown to require a thorough and long-lasting research and development process.

In this chapter we will mainly focus on biomarkers in pharmaceutical drug development, as lessons learned there can be applied to biomarkers in other application areas. After outlining how biomarkers do play a role in decision making during development of drugs, and more specifically their role in translational and personalized medicine, we will review trends, challenges and opportunities related to biomarkers in biomedical science.

1.2   Biomarkers

Before discussing the role of biomarkers in pharmaceutical drug development, translational medicine and personalized medicine, we would first like to list useful definitions of biomarkers and their utilities in this field that will guide the thought process.[4]

    I. **Biomarker:** A characteristic that is objectively measured and evaluated as an indicator of normal biological processes, pathogenic processes, or pharmacologic responses to a therapeutic intervention.[2]

    II. **Disease biomarkers:** Biomarkers that are correlated with the disease where correlation is established *via* rigorous biological and clinical validation. Disease biomarkers are not necessarily causally associated with the mechanism of the disease. Correlation to important phenotypes of the disease, relationships to its initiation, propagation, regression or relapses, however, must be established. Disease biomarkers can serve as diagnostic biomarkers (distinguishing patients from nonpatients), as prognostic biomarkers (identifying "rapid *vs.* slow progressing" patients) or as disease-classification biomarkers (elucidating molecular mechanisms of the observed pathophysiology). All three functions are crucial parameters to drive the selection of subjects in clinical studies.[5]

    III. **Target Engagement Biomarker:** Biomarkers that represent the direct interaction of the drug (small molecule or biological) with the molecular target. These are highly important to guide drug exposure as they reflect distribution of the drug to the specific location of target, the residency time of the drug on the target and the extent of the drug target modulation by the bound drug.

    IV. **Pharmacokinetic Biomarkers:** Biomarkers that represent the level of the pharmaceutical drug in circulating body fluids and/or at the site of action, and that are important to calculate the dose needed to induce a certain pharmacological response.

    V. **Pharmacodynamic Biomarkers:** Biomarkers that represent the functional outcome of the interaction of a drug with its target (also called pharmacological biomarkers). These biomarkers can have various identities, can be analyzed by a variety of methodologies (including enzymology, omics, imaging), but generally represent a read-out of complex biology. Pharmacodynamic biomarkers are specifically used to rationalize clinical therapeutic efficacy and adverse effects, typically measured as a multiparameter panel of biomarkers representing distinct functional events.

    VI. **Predictive Biomarker:** Biomarkers that are used for the selection of patients for clinical studies. These biomarkers serve to predict which patients are likely to respond to a particular treatment or drug's specific mechanism of action, or potentially predict those patients who may experience adverse effects.

VII. **Validated Biomarkers:** Biomarkers that are measured in an analytical test system with well-established performance characteristics, and with established scientific framework or body of evidence that elucidates the physiologic, pharmacologic, toxicologic, or clinical significance of the test results.[6]

VIII. **Surrogate Endpoint:** A biomarker that is intended to substitute for a clinical endpoint. A surrogate endpoint is expected to predict clinical benefit (or harm or lack of benefit or harm) based on epidemiologic, therapeutic, pathophysiologic, or other scientific evidence.[2]

IX. **Therapeutic Biomarker:** A biomarker that indicates the effect of a therapeutic intervention and can be used to assess its efficacy and/or safety liability.[10]

1.3 Biomarkers in Pharmaceutical Drug Development

1.3.1 The Pharmaceutical Research and Development Process

The pharmaceutical drug-development process (Figure 1.1) is a multistep process that on average takes 14 years from initiation of research to marketing of the new medicine.

It starts with the discovery of a drug target, the molecular target that the drug will act upon. In this target discovery phase researchers investigate how a particular disease is caused and what factors play a key role. The inhibition or stimulation of those key factors will be the basis for the new pharmaceutical drug used to treat the selected disease. For example, postmenopausal complaints are caused by a decrease in endogenously produced estrogens and the objective is to find estrogen-like compounds that can supplement the natural pool of estrogens. The drug target in this case is the estrogen receptor, which is a nuclear protein present in cells of specific tissues and acts as an estrogen-activated transcription factor with specific effects on each specific tissue. Indeed, the activated estrogen receptor in osteoblasts mediates the synthesis of new bone, whereas the estrogen receptor in breast epithelial cells is a key player in cell growth.

The next step is the identification of compounds that have the desired effect on the drug target, for example by inhibiting or stimulating its activity. This discovery is done in the lead discovery phase, during which large numbers of chemical or biological compounds are tested for the desired effects in biochemical and cellular assays. Often mechanistic biomarkers or derivatives thereof are used as read-outs of the screening assays. In the estrogen receptor example, a screening assay to identify compounds with agonistic or antagonistic estrogenic activity may comprise of a cell line expressing the receptor and containing an estrogen-receptor sensitive luciferase reporter module.

Following selection of the most promising hits and limited optimization, the lead optimization phase starts where systematically up to a thousand variants of the original positive substances are synthesized and tested in various tests. A stringent selection process aims to select those compounds that display

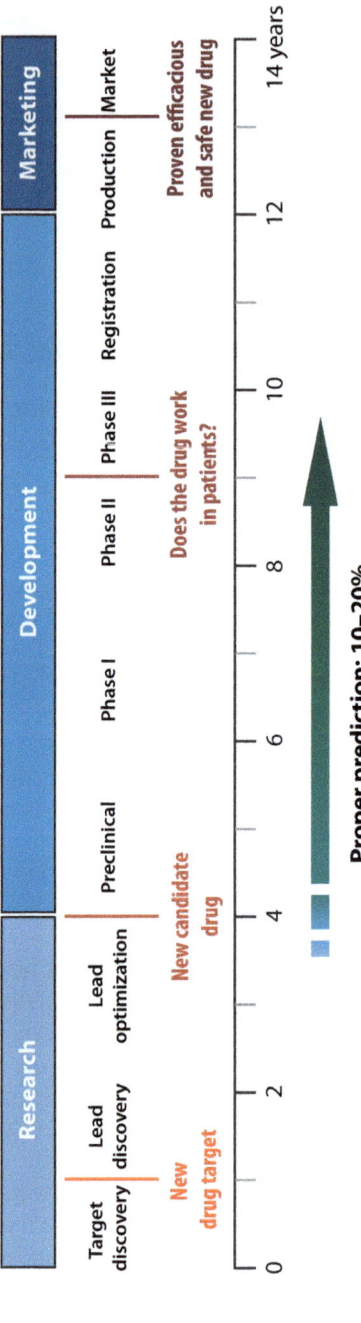

Figure 1.1   Schematic overview of the pharmaceutical drug discovery and development process, outlining the key decision points and associated time scale (adapted from[8]).

improved efficacy, specificity, safety, bioavailability and/or production efficiency (depending on the objectives of the project). Assays used during lead optimization include biochemical and cellular test systems, followed by *in vivo* assays to assess bioavailability, pharmacological and toxicological effects of the drug. Typically one or two of the best compounds are nominated to progress from research to development.

In preclinical development the substance is first investigated in animal models to test whether it is sufficiently bioavailable and safe. After successfully passing this phase, similar studies are performed in the phase 1 clinical trials, during which under strictly controlled conditions the compound is tested in human subjects. Typically, healthy human subjects participate in such trials; the exception being oncology trials whereby drugs are often tested directly in small numbers of patients. Subsequently, the compound is tested in patients, which is the first time the drug developer will determine whether the originally chosen approach of affecting the drug target has a positive effect on treatment of the disease. This occurs in a phase 2 clinical trial in which a relatively small group of patients are tested. A positive outcome of this trial, with an acceptable level of side effects, is very important as it is then proved that the approach chosen to treat the disease works, also known as the clinical proof of concept. After this milestone, the clinical phase 3 starts in which the effect of the substance is tested on large numbers of patients. Such a study can be very substantial. For instance, a phase 3 trial of testing estrogenic compounds in osteoporosis involves administration of the candidate drug in thousands of postmenopausal women per dose group for three years while recording how often a participant breaks a hip, a reduction of which is the currently accepted clinical endpoint of efficacy.[7]

The final stage of the drug-development process involves scaling up production, obtaining regulatory approvals, designing the drug label and packaging, and preparing the market launch. In addition, after introducing the drug on the market, often additional scientific studies are done to further elucidate the mechanism of action or to investigate whether the drug can be applied in other therapeutic indications.

Despite the strong underlying science and process-minded workflow of drug discovery and development projects, this process is highly inefficient. It takes on average 14 years to bring a new drug to the market starting from a new drug-discovery project, at the cost of 1.6 billion US dollars.[9] The main cause of this resides in the high attrition rate of 90% during clinical development, meaning that only one in ten projects being pursued through clinical testing in patients is successful[10,11] An analysis of various therapeutic areas has illustrated that there are differences in success rate in different therapeutic areas; drugs treating CNS, oncology and women's health are the most difficult to develop.[10,11] In particular, the proof-of-concept phase contributes to the high attrition because of a lack of efficacy and/or unacceptable safety liabilities. This leads one to surmise that preclinical studies in pharmaceutical research are insufficient to predict drug action in patients, resulting in a strong need for translational biomarkers to bridge research and development.

1.3.2   Biomarker-Based Decisions during Early-Phase Pharmaceutical Drug Development

Although the development of a drug is described above as a linear process, in reality it is often an intertwined process with many forward and backward translation loops. Indeed, to better define the objective of target discovery, a strong interaction is needed between the molecular biologists selecting the drug target and the clinical experts that define the patient needs. Also, a strong interaction is needed during lead optimization between pharmacologists and early clinical developers to ensure smooth transition of the compound from research to development and timely preparation of the clinical assays needed. Such translational medicine is supported by high-quality biomarkers that enable monitoring of drug action at all critical stages of drug development (Figure 1.2).

Most, if not all, pharmaceutical companies developed generic biomarker strategies to support their drug-development projects through a rational and consistent application of key decision-making biomarkers. Previously we published a question-based drug development biomarker strategy that is based on a set of translational questions.[12] In this strategy, answers are provided by translational biomarkers, enabling data-driven decisions throughout the development of a drug. The questions in this question-based drug development approach include:

- Does the compound get to the site of action?
- Does the compound cause its intended pharmacological and functional effects?
- Does the compound have beneficial effects on disease or clinical pathophysiology?
- What is the therapeutic window?
- How do sources of variability in drug response in target population affect efficacy and safety?

The emphasis on each question is different for each drug-development project and varies depending on therapeutic area, mechanism and drug target. In the example of the estrogen receptor and postmenopausal osteoporosis, biomarkers are mostly needed to reflect the tissue-specific modulation of the estrogenic pathway, particularly the pharmacological effects of the estrogen-like compounds on bone, and the absence of undesired extrogenic effects on breast and endometrium.[13] In some cases, a biomarker strategy and/or selected biomarkers can be used for other projects whereas similar approach is being followed, for instance when using the estrogen receptor α biomarkers for characterization of estrogen receptor β modulators.[14]

In all cases, however, a target engagement biomarker is essential, indicating the physical or mechanistic engagement of the drug target by the drug compound. Together with read-outs of clinical efficacy this will drive early decisions whether or not to proceed with a clinical development project, following the decision tree outlined in Figure 1.3. When the compound does not engage the drug target sufficiently to levels determined in preceding research, it makes no sense to test patients for drug efficacy. One needs to identify a more

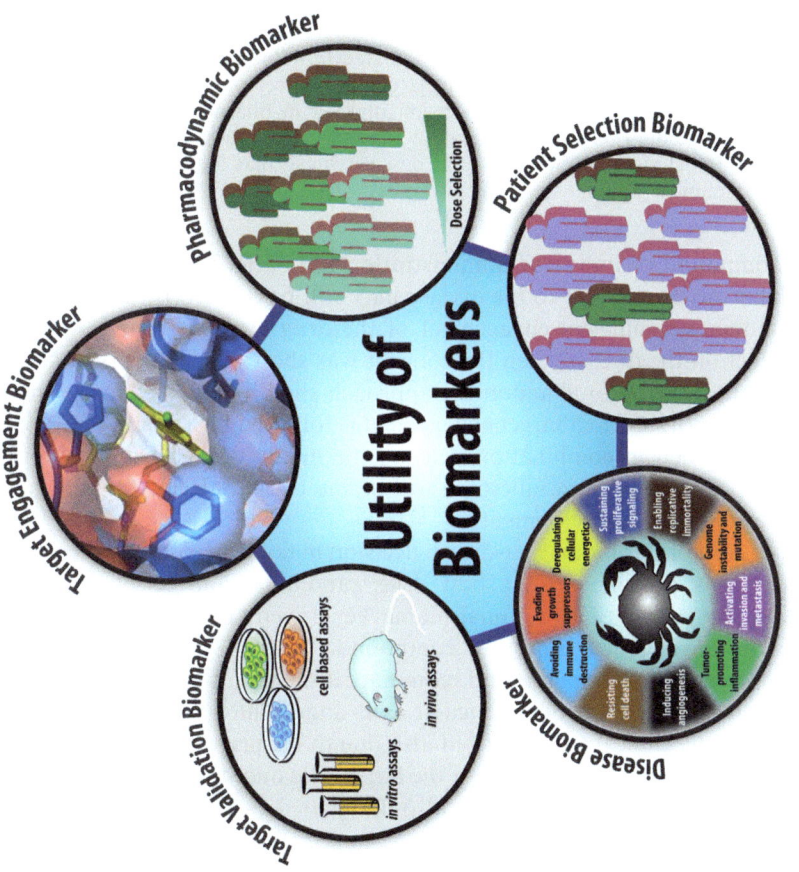

Figure 1.2 Schematic outlining key utilities of biomarkers in pharmaceutical drug development.

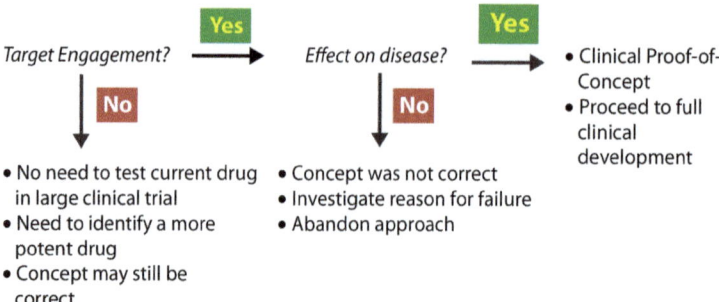

Figure 1.3   Simple schematic of the decision tree how biomarkers for drug-specific engagement of the drug target and for effect on the disease are being applied for data-driven decision making in pharmaceutical drug development. This strategy will allow for early selection of the best compounds for clinical development and deselect those that are unlikely to achieve clinical success.

potent compound or improved administration route, but the original concept that relates the drug target to the disease may still be correct. If, however, the compound shows sufficient engagement of the drug target but there is insufficient effect on the disease, then the concept was shown to be incorrect and a (data-driven) decision to abandon the approach should be taken. In the most positive scenario, sufficient target engagement is followed by a positive effect on the disease, which in combination delivers the highly desired clinical Proof-of-Concept. This important milestone will be the basis to decide for extensive clinical testing in full development.

The purpose of this biomarker-driven decision making is to decide as early in the process as possible on the best drug candidates and select the most successful approaches. This rational biomarker-driven decision making could shorten the development time per compound, but will surely improve the overall efficiency of the pipeline and reduce attrition later, providing an overall reduction in costs and time. In particular, late-stage failures in phase 3 are very costly due to the clinical trial itself but also due to the preparations to produce the new pharmaceutical ingredient on the large scale once the clinical tests were positively concluded. This strategy is also designated as "more shots on goal" and "stop early, stop cheap'.

For each project, it is of high importance to define early in the project the exact quantitative criteria that are to be used to evaluate the drug response using the biomarker data. Often it is not yet known what the thresholds of the biomarker are, for instance whether a change of 20% or 30% is needed to define a positive response. In some cases, this is particularly hard as the key disease-specific biomarkers can be below detection level in samples from healthy subjects. If so, one cannot define *a priori* to what extent a disease-specific biomarker should decrease following treatment to healthy levels. Clinical biomarker validation efforts should be strongly focused on this aspect, and the information should be translated back to preclinical studies where similar thresholds should be regarded. In particular, the relationship between drug exposure (pharmacokinetics, PK) and the changing levels of the biomarker (pharmacodynamics, PD) enable a forward translation of preclinical findings through modeling and simulation. Such PK/PD model can quantitatively predict the change to be expected in clinical testing and can lead to decisions whether the window between drug efficacy and safety is large enough for clinical success. In a truly translational medicine workflow, such quantitative analysis of key biomarkers should start at drug target validation, continue at lead optimization, and be extended to clinical trials.

1.3.3 The Use of Biomarkers in Late-Phase Pharmaceutical Drug Development

The majority of biomarker use in drug development is in the early phases for internal decision making on which drug to progress. Once pivotal trials are being run, typically phase 2b and phase 3 in late-stage development, biomarkers have a less-explorative role but are used as key decision points. Pivotal trials are used for registration of the drug for a defined indication, and the biomarkers that aid in building the filing dossier should be widely accepted by the scientific clinical communities and regulatory bodies. In some cases, biomarkers from research do progress to late clinical drug development as surrogate endpoints or companion diagnostics.

Surrogate endpoints are those biomarkers that substitute a clinical endpoint with the objective to predict the outcome of a clinical intervention where the primary endpoint is undesired (for instance mortality rate) or rare, requiring very large numbers of patients in the trial.[15] An example of the latter is the aforementioned phase 3 clinical trial in osteoporosis drug development. A surrogate endpoint that would replace such time-consuming clinical assessment would have added value, and a combination of noninvasive computer tomography imaging of bone structure combined with an algorithm to calculate bone strength based on 3D images[16] is a promising approach in this respect. The maturation of a biomarker to become a surrogate endpoint requires an extensive and lengthy clinical validation process that involves multiple clinical centers, intervention trials, key stakeholders and requires an intensive interaction with regulatory agencies such as FDA, EMA. An example of such surrogate endpoint biomarkers is cholesterol, that was used for registration of simvastatin.[17]

Companion diagnostics have received much attention in recent years, due to their intrinsic potential to enable personalized medicine. As discussed below, currently there are over one hundred FDA-approved drugs that contain pharmacogenomic information in their labels, and where the outcome of the biomarker test determines whether people should be prescribed this drug or not. Many of these genomic tests relate to metabolism, related to variants in the Cyp450 gene family, and the majority of applications reside in oncology and psychiatry.[18] To a lesser extent, but increasingly, companion diagnostics are being developed for other disease areas and much is to be expected for the years to come.

Lastly, biomarkers are being applied to the very last phase of drug development, the postmarket introduction phase 4, when the drug is being used in a certain therapeutic indication and additional studies are being performed to either learn more of its mechanism of action or to investigate potential applications in other disease areas. Here, the use of biomarkers is again of a more explorative nature, comparable to the early drug-development phases.

1.3.4   Fit-for-Purpose Biomarker Applications

The different applications of biomarkers to support the pharmaceutical drug-development process as described above already indicate the variable requirement of the biomarker assay robustness. When still in early research phases, such as the validation of a drug target in pharmacological studies, the biomarker merely serves to indicate functional effects and the corresponding assay does not necessarily have to be very robust. The test can yield qualitative data, *e.g.* condition 1 versus condition 2, although quantitative data is always preferred. In contrast, if the biomarker read-out is used to demonstrate clinical proof-of-principle or proof-of-concept, with decision-making implementations for the drug-development program, the assay needs to be robust, reproducible and quantitative. Importantly, the biomarker should have a solid body of evidence indicating its value as the intended read-out, and be accepted by the scientific clinical community. Thirdly, if the biomarker progresses to a surrogate endpoint or companion diagnostic, a rigorous process to generate a diagnostic test has to be followed. In this phase the focus is on thorough analytical validation and production of the biomarker test with a positive predicted health economic value. Clinical utility should be similar and acceptance of the biomarker test by clinical key opinion leaders and regulatory bodies is crucial. Useful guidelines to facilitate the assay development and support the definition of the fit-for-purpose robustness have been published.[19] In general, a robust quantitative biomarker assay is always preferred, as this facilitates translation of findings across different drug-development phases and data sharing among laboratories.

1.4   Biomarkers in Translational Medicine

1.4.1   Translational Medicine

The success rate of novel medical entities that are submitted for registration by the regulatory agencies and follow successful marketing has been stagnating for the past decade. Failure in efficacy and safety continue to be the prime hurdles and cause of failure and attritions in drug-discovery process,[20,21] particularly in phase 2 and phase 3 in the pharmaceutical pipelines.[22,23]

Translational medicine is an emerging paradigm within the pharmaceutical industry R&D organizations aimed to improve the predictability and success of drug discovery and development.[24] Translational Medicine is the integrated application of innovative pharmacology tools, biomarkers, clinical methods, clinical technologies and study designs to improve confidence in human drug targets, increase confidence in drug candidates, understand the therapeutic index in humans, enhance cost-effective decision making in exploratory development, all with the objective to increase success in clinical phase 2.

1.4.2 Translational Medicine in Drug Discovery and Development

The key objective of translational medicine in the pharmaceutical research is to improve the success rate of compounds and biologics identified in the discovery phase and chosen for clinical development. Towards this goal, translational medicine encompasses many different activities along the drug-discovery pipeline enumerated in Section 1.3.1. These include:

I. *Target validation* to establish that the molecular target selected for further development contributes significantly to a human disease and that manipulation of the target could provide desired benefits while minimizing potential adverse effect. The key goal of this activity is to establish the "therapeutic window" of safety and tolerability.

II. Better understanding of the *Pharmacokinetics (PK) and Pharmacodynamic (PD) relationships* of the drug with the molecular target. Understanding the pharmacological nature of compound interaction with the target (*e.g.* target occupancy and target engagement), extent of modulation needed to achieve desired therapeutic efficacy while reducing any "off target" effects.

III. Selection of *preclinical model systems* that provide insights into human disease. Such systems include *in vitro* biochemical systems, cell-based assays, *ex vivo* organ systems and integrated *in vivo* animal models. A key limitation in this aspect is the lack of sufficient representative cellular and animal models that approximate the human disease and their pathophysiology.

IV. *Patient selection/stratification* based on genetic variations for rational clinical trial designs. In recent years, substantial efforts have been made to carefully select patients that are likely to respond to a particular treatment regime. Cancer drug discovery and development is the prominent example of this paradigm shift. Modern molecular oncology seeks the discovery and development of drugs that specifically address the mechanism of the oncogenic transformation in particular cancer types so that the drugs are tailored towards them. In a very recent example FDA approved a new drug for cystic fibrosis on the basis of an expedited (3-month) review and smaller clinical studies of a compound VX-770 (Kalydeco) that targeted a rare mutation of CFTR gene (G551D).[25] Proper application of stratification biomarkers in development of preclinical models will ensure smooth translation of findings to clinical development.

1.4.3   Animal Models and Biomarkers in Translational Medicine

Animal models have been at the core of the drug-discovery process spanning nearly all phases of preclinical and clinical R&D phases. Rodent models have significantly contributed to our molecular and cellular understanding of various diseases, for instance various types of cancer, metabolic, immunological and neurological diseases. Genetically engineered animal models have enabled investigators to both observe and manipulate complex disease processes in a manner impossible to perform in patients[26] and have led to successful clinical translation of biological insights, for instance in the treatment of hitherto rare yet fatal acute promyelocytic leukemia (APL) that is now very effectively cured and managed in patients.[27,28]

However, animal models also have their own limitations due to species-specific genetic differences with regard to humans resulting in inaccurate and at times inadequate recapitulation of human disease pathophysiology. Mounting evidence also suggests that there are limitations on how closely some of the established models mimic human physiology leading to failure in accurately predicting the effects of drugs in human subjects, for instance in neurodegenerative and immunological diseases.[29,30]

In the context of immunology there are key molecular differences between mouse and human, which should be dealt with cautiously when using mouse as translational model for studying specific aspects of human immunology. For instance, important differences are seen in expression of key cellular proteins related to inflammation, for instance S100A12 (ENRAGE), which binds to RAGE (receptor of advanced glycation endproducts), and whose ligation has been implicated in various inflammatory-related diseases.[31] However, this mechanism cannot be studied in mouse as there is no evidence of functional murine S100A12 gene expression.[32] Similarly, a wide repertoire of chemokines have been identified in humans but not in mice, including IL-8 (CXCL8), neutrophil-activating peptide-2 (CXCL7), IFN-inducible T cell α-chemoattractant (CXCL11), monocyte chemoattractant protein (MCP)-4 (CCL13), HCC-1 (CCL14), hemofiltrate CC chemokines-2 (CCL15), pulmonary and activation-regulated chemokine (CCL18), myeloid progenitor inhibitory factor-1 (CCL23), and eotaxin-2/3 (CCL24/CCL26). Conversely, CCL6, CCL9, lungkine (CXCL15), and MCP-5 (CCL12) have been identified in mice but not humans.

A proper evaluation and definition of translational biomarkers could mitigate these limitations leading to improved translation of results from preclinical models to human. This approach requires components such as:

I.   Selection of the right species that allows monitoring biomarkers that can be monitored in preclinical models, healthy subjects and patients. For instance, nonhuman primates (NHP) to study key human diseases (metabolic, neurodegenerative and immunological) have for this reason gathered momentum in recent years.[33]

II. Biomarkers that enable translational PK/PD relationship in preclinical models and in human disease.

III. Conservation of the (patho)physiological process(es) studied in the animal with the human disease. whereby the functional role of biomarkers converge at spatial, temporal and functional levels across the experimental models and human studies. This principle in turn necessitates looking at cellular processes and their perturbations at systems levels. Humanized rodent models such as primary tumor transplant models may form such a hybrid translational test system.[34]

Rational applications of these principles have shown encouraging results in contemporary drug discovery. For instance, in strokes whereby studies in nonhuman primate (NHP) models have led to identification of PSD95 inhibitors that show promising results for stroke treatment, thus opening a new vista for future clinical validation[35,36] (Figure 1.4A). Similarly, clinical trials to treat a rare form of cancer, *i.e.* pancreatic neuroendocrine tumors (PNET) have been undertaken with encouraging phase III results, based on elaborate studies performed in mouse models[37,38] (Figure 1.4B). In metabolic disease space recently an engineered FGF21-mimetic monoclonal antibody "mimAb1" was tested in obese cynomolgus monkeys (NHP model) leading to decrease in body weight and body mass index (BMI).[39] Furthermore, mimAb1 treatment in NHP model led to decreases in fasting and fed plasma insulin levels, as well as a reduction in plasma triglyceride and glucose levels. While these results need to be replicated in human cohorts they nevertheless illuminate a beneficial therapeutic approach towards treating patients with diet-induced obesity and diabetes. In yet another study, a three-part therapy approach to cure type 1 diabetes was tested in a type 1 diabetes mouse model, which holds promises for multifaceted approach for curing this autoimmune disease.[40]

1.5 Biomarkers in Personalized Medicine

1.5.1 Personalized Medicine

Until recently the approach towards medicine (and healthcare) has been largely population-driven and reactive, whereby emphasis was placed on treating disease symptoms rather than on preventing disease. Predicated on this model, the drug-discovery process was also focused on developing drugs for masses that would have high return on R&D investments. The majority of the current drugs are approved and developed on the basis of their performance in a large population of people and each drug is prescribed to all patients with a certain diagnosis.

However, it is clear that this "blockbuster" model does not apply to all therapeutic areas. Indeed, hypertension drugs as ACE inhibitors are only efficacious in 10–30% of patients, heart failure drugs as beta blockers in only 15–25%, antidepressants in 20–50% of patients and so on.[41] With the advent of various molecular profiling sciences it has become feasible to investigate this in great detail at the genetic, transcriptomic, epigenetic, proteomic, metabolomic and other complementary "Omic" levels. The resultant knowledge and insights

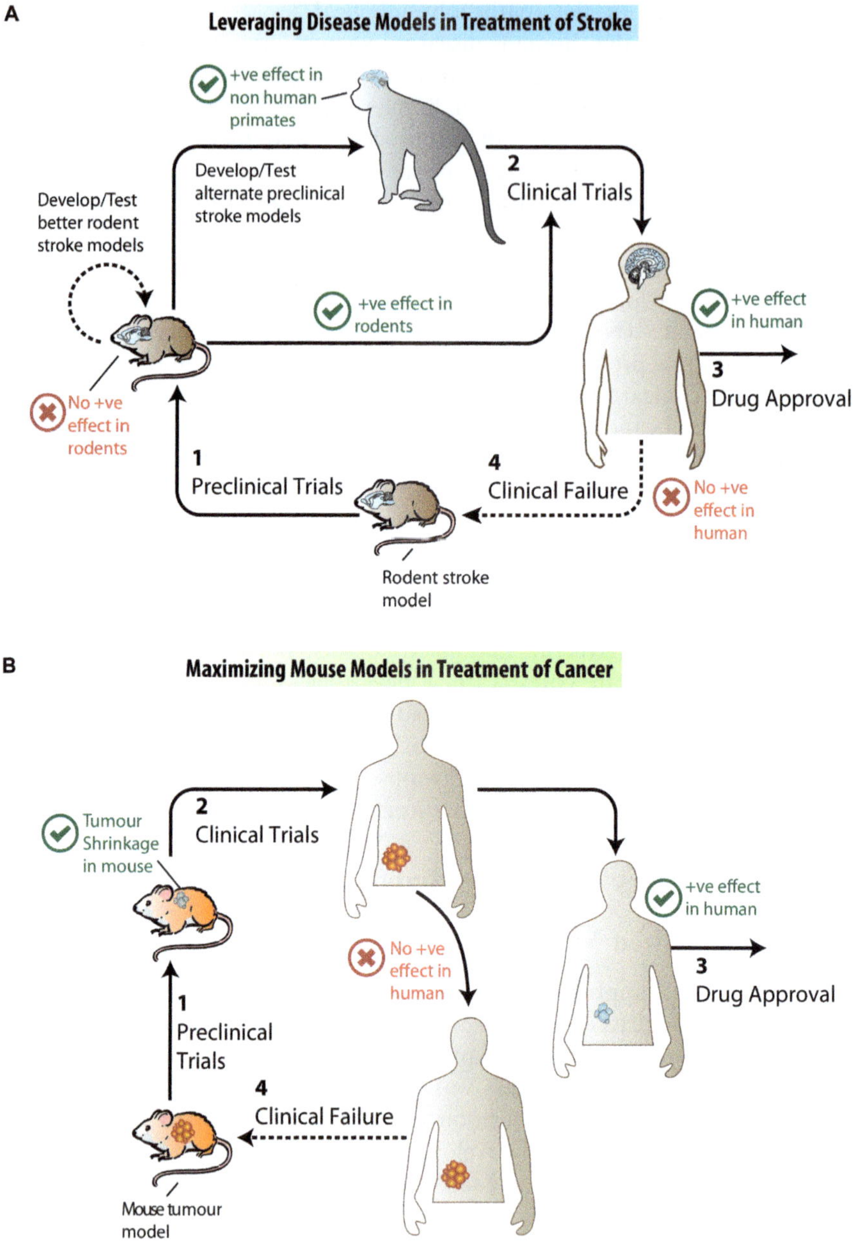

A

Leveraging Disease Models in Treatment of Stroke

B

Maximizing Mouse Models in Treatment of Cancer

Figure 1.4 Schematic outlining facets of translational medicine, depicting examples from stroke (A) and oncology (B) how animal and human studies can be used in forward and backward translation, thus increasing the chance of successful drug discovery and development.

have led to the appreciation that most of the diseases are extremely hetero-geneous and the "one-size fits all" approach to treatment is ineffective and unsustainable.[42] In the postgenomics era a new approach to medicine called *"Personalized Medicine"* is taking shape that is informed and driven by each person's unique clinical, genetic, genomic, and environmental information.[43] Personalized medicine is a paradigm shift in healthcare. Its success depends on multidisciplinary biomedicine, adaptive clinical practices and integrative tech-nologies to leverage our molecular understanding of disease towards optimized preventive health care strategies.[44,45] This new healthcare model predicates a central role of the emerging and empowered patient who actively participates in his health and disease management,[46] as part of an integrative framework defined by P4 (predictive, preventive, personalized, participatory) medicine.[47–49]

Personalized medicine is an attractive concept, however, as with biomarkers it is not new. A physician seeing a patient has always been making a personal judgement of the patient's status and best ways towards cure, based on visual observation, verbal interaction, physical examination and other personal impressions. What is new, however, is that the molecular analyses of patho-physiology that have recently become possible add to this judgement, enabling rational stratification of the patient and prescription of drugs with distinct mechanism of action to patient subgroups. Taking it one step further, if drug therapies are combined with individualized optimization based on the person's molecular phenotype, it may become possible to conduct $n = 1$ clinical trials and exercise truly individualized medicine.[50] Stratified medicine combined with the physician's personal advice to the patient, *e.g.* on adapting lifestyle, is already a major step towards that objective.

1.5.2 Impact of Personalized Medicine

Cancer treatment and therapies are leading the way in personalized medicine, where its impact is already being observed. With the accumulating molecular knowledge of cancer, it is clear that no two patients – even those with the same diagnosis – experience the same disease. Cancer in itself is a highly heterogeneous disease whereby cells within the same type of tumor may possess different genetic mutations and display different cellular activities and aberrations.[51,52] On the other hand, slight genetic variations in our bodies can alter how our immune systems fight cancer, how rapidly drugs are absorbed and metabolized, and the likelihood and severity of adverse effects. Thus, it is imperative that treatment be designed and prescribed to address the unique biology and molecular phenotype of each patient's own disease, thus to maximize the chance of successful outcome.

Personalized medicine aims to leverage our molecular understanding of disease to enhance preventive health care strategies while people are still well, thereby advocating preventive and pre-emptive medicine. The overarching goal of personalized medicine is to optimize medical care and outcomes for each individual, resulting in customization of patient care.[43] Towards this goal, the success of personalized medicine will depend on successful deployment of several practices in clinics:

I. **Pharmacogenomics:** whereby genomic information is used to study individual responses to drugs. When a gene variant (called an allele) is associated with a particular drug response in a patient there is the potential for making clinical decisions based on genetics by adjusting the dosage or choosing a more effective drug. Various genetic loci (and their alleles) associated with known drug responses have been identified, which can be tested in individuals whose response to a particular drug(class) is unknown. Modern approaches include multigene analysis or whole-genome single-nucleotide polymorphism (SNP) profiles, and these approaches are just coming into clinical use for drug discovery and development. For example, the enzyme CYP2D6, one of a class of drug-metabolizing enzymes known as cytochrome P450 gene(CYP) found in liver, metabolizes certain antidepressant, antiarrhythmic, and anti-psychotic drugs. Extensive molecular characterizations of this gene has led to identification of 70 variant alleles, which are known to exhibit substantial difference in drug metabolism.[53] Leveraging this information a diagnostic chip designed by Roche called AmpliChip P450 Array can determine genotypes for alleles of selected CYP genes, including CYP2D6, providing valuable information for appropriate drug administration.[54]

II. **Family Health History and Disease Risk Assessment:** Family health history (FHH) is an invaluable and till now rather underutilized information in the practice of personalized medicine. A robust and objective assessment of FHH, which reflect the complex combination of shared genetic, environmental, and lifestyle factors, can inform patients of potential risk factors for certain types of diseases (*e.g.* Type 2 diabetes, Coronary Artery Disease, Breast/Colon/Lung cancers).[55–57] Genetic testing can provide valuable information by profiling certain disease related genes – to enumerate particular changes or mutations that could render one susceptible to certain disease, for instance, BRCA1/2 for breast and ovarian cancer,[58] and APOE4 for Alzheimer's disease.[59] A timely FHH assessment would help identify persons at higher risk for disease, enabling pre-emptive and preventive steps, including lifestyle changes, health screenings, testing, and early treatment as appropriate.

III. **Digital Medicine:** While the "omics" revolution has changed the face of biomedical research, the "digital" revolution in the last decade has transformed our daily lives, engendering profound changes to the way knowledge is generated, shared and propagated. The impact of the digital revolution is being clearly felt in healthcare and it is predicted to become one of the key drivers of next-generation healthcare and personalized medicine. One of the components of the digital ensemble, noninvasive imaging technologies, has revolutionized clinical diag-nostics for cardiovascular diseases, various forms of cancers and neurodegenerative disorders. Recently, an imaging approach based on positron emission tomography (PET) with Pittsburg Compound B, called Amyvid, was approved by FDA as a test for diagnosis of Alzheimer's disease.[60] Digital medical devices coupled with powerful wireless technologies provide hitherto unavailable options to monitor various aspects of personal health and vital statistics (*e.g.* heart rate, glucose level, blood pressure) in a continual manner.[61] Electronic health records (EHR) intend to be the backbone of the proposed healthcare system of coming decades and till-date is a largely underutilized resource. However, with the growing amount of personal medical data being generated as part of modern clinical practices (clinical, omics, imaging, sensors), the role of EHR will be pivotal for the success of personalized and integrative healthcare.[44]

1.5.3 Molecular Profiling Toolbox and Personalized Medicine

The human genome project was a milestone in biomedical research that ushered in the era of molecular profiling (the "Omics") and opened a new vista for studying biological processes and systems at various levels of cellular organization and hierarchy.[62] In this section we briefly review the technology platforms that constitute the molecular profiling toolbox for the exploration of the role of the genome and its expressed products (transcriptome, proteome and metabolome) on health and disease states (Figure 1.5).

The foundations of personalized medical care would be built on rational clinical application of these technologies:

I. **Genome-Wide Variation:** Clinical characteristic of diseases such as cancer and certain individual response to drug treatment can be attributed to a person's genotype. Propelled by various technological breakthroughs particularly Next-Generation Sequencing, DNA sequencing has advanced our understanding of multiple diseases by elucidating underlying functional gene variants that are perhaps causal of the disease state, for example in cancer, rare genetic diseases, and microbial infection.[64–66] The use of whole-genome sequencing has matured to a level whereby it is has been successfully applied in clinical settings for definitive diagnoses and to guide treatment regimen.[67–70] On the other hand, various haplotyping projects have led to identification of SNPs that are the causative variants of disease,[71,72] thereby providing framework for various Genome-Wide Association Studies (GWAS). While many of these disease-related SNP associations obtained across over 1400 GWAS studies[73] have yet to show clinical relevance, there are a few exceptions. For instance, in a GWAS to see if the drug response to Hepatitis C treatment could be predicted, Ge *et al.*[74] identified a SNP at IL28B gene to be associated with a twofold change in treatment response. This finding was later confirmed in a clinical study that genotyped patients receiving treatment for the hepatitis C virus and found the same polymorphism to be a strong predictor of sustained virologic response.[75]

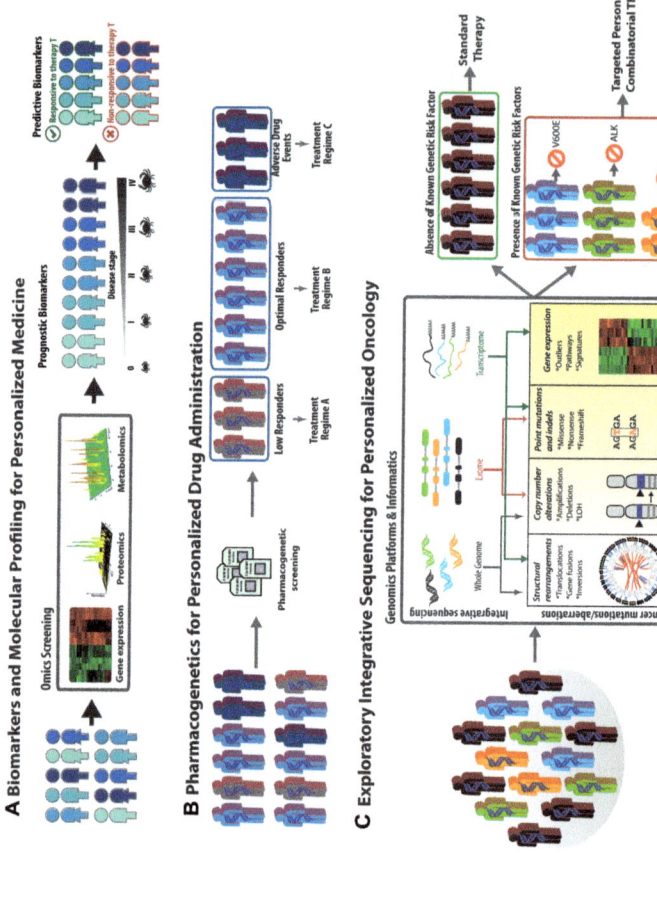

Figure 1.5   Schematic outline of the application of molecular profiling methodologies for personalized medicine. (A) Application of molecular profiling screening to identify prognostic biomarkers to predict response to a therapy. (B) Application of array-based pharmacogenetic screening to segregate subgroups of patients with predicted low response, optimal response and adverse drug events. (C) Application of deep sequencing methods to segregate patients with similar clinical phenotype into different genotype subgroups that require different therapies (adapted from Roychowdhury *et al.*[63])

II. **Transcriptomics:** For various diseases transcriptomic data generated using microarray platforms have been successfully used for diagnosis, prognosis, and in predicting response to therapy. In oncology, these data have uncovered distinguishable molecular classes for many cancers – including breast, lung, blood-based lymphomas, leukemia and melanoma[76–80] Similar disease classifications based on microarray data have been performed in other complex diseases, including cardiovascular disease, rheumatic diseases and their inflammatory pathways, and neurologic diseases such as multiple sclerosis.[81–83] Next-Generation Sequencing technologies such as RNASeq are providing insights into transcripts and their regulation at an unprecedented depth and precision.[84] For example, transcriptomes relating to cancer have been investigated through RNASeq, providing a new approach for understanding gene fusion and translocation events.[85,86] Owing to advances in genome-sequencing technologies *personalized oncology* is becoming a possibility.[87] In a recent pilot study, high-throughput sequencing involving whole-genome sequencing, exome sequencing and RNASeq were employed to guide clinical decisions for patients suffering from metastatic colorectal cancer and malignant melanoma.[63]

III. **Proteomics:** Advances in mass spectrometry (MS)-based proteomics has enabled high throughput, broader coverage, and accurate quantitation of proteins,[88] thereby opening newer vistas for diverse systems-wide proteomic investigations of clinical relevance.[89–91] Consequently, proteomic datasets are getting richer and being applied in understanding of human health and disease. In a recent example, global proteomic profiling was successfully employed to classify diffuse large B-cell lymphoma subtypes.[91] In another example, system-wide proteome analysis of breast cancer cell lines recapitulate the disease progression and provided novel prognostic markers for ER(–) tumors.[92] Recently, an innovative method of "proteome-wide analysis of SNPs" (PWAS) was reported. This method enables rapid screening of SNPs for differential transcription factor (TF) binding and was successfully applied to elucidate differential TF binding to type 1 diabetes- (T1D-) associated SNPs at the IL2RA or CD25 locus.[93] While still an emerging discipline, continued advances in mass spectrometry (MS)-based proteomics in combination with innovative experimental strategies, and advances in computational methods will surely play a profound role in personalized medicine.[94]

IV. **Metabolomics:** Study of changes in nonprotein small-molecules metabolites could reflect biological and (patho)physiological state associated with disease. Metabolomics has been applied to an array of diseases, including diabesity, cardiovascular disease, cancer, and mental disorders.[95–98] In drug discovery metabolomic profiling has led to identification of enzymatic drug targets and as a tool for assessing drug toxicity.[99]

1.5.4   Biomarkers in Personalized Medicine

Biomarkers are at the core of personalized medicine as they are the instruments by which clinicians decide which patients to give what drug at what dose and time. The impact of biomarkers on personalized medicine varies between therapeutic fields but encouraging results have started to emerge of patient selection for tailored therapies, for instance in cancer. Cancer is largely a genomic disease that manifests due to a spectrum of genomic abnormalities acquired by normal human cells, ranging from point mutation, structural variations (chromosomal rearrangement) and gene fusion and genomics biomarkers have shown attractive applications for patient selection.[100]

At present, 103 drugs are available whose label contains pharmacogenomic biomarker information, which drives the prescreening of patients to maximize drug efficacy or safety.[18] A total of 113 drug–biomarker combinations are characterized in multiple disease areas of which the main applications are in oncology (29%) and psychiatry (24%) (see Figure 1.6).

Roughly half of these drug–biomarker combinations relate to genetic Cyp450 family variants, mostly Cyp2C19 and Cyp2D6, that affect the metabolism of the drugs. On the one hand, this could result in too high (or too low) compound levels in poor (or ultrarapid) metabolizing patients, requiring dosing adjustment to have sufficient active compound present whilst not inducing safety liabilities. For psychiatry, multiple examples of this exist, reviewed by

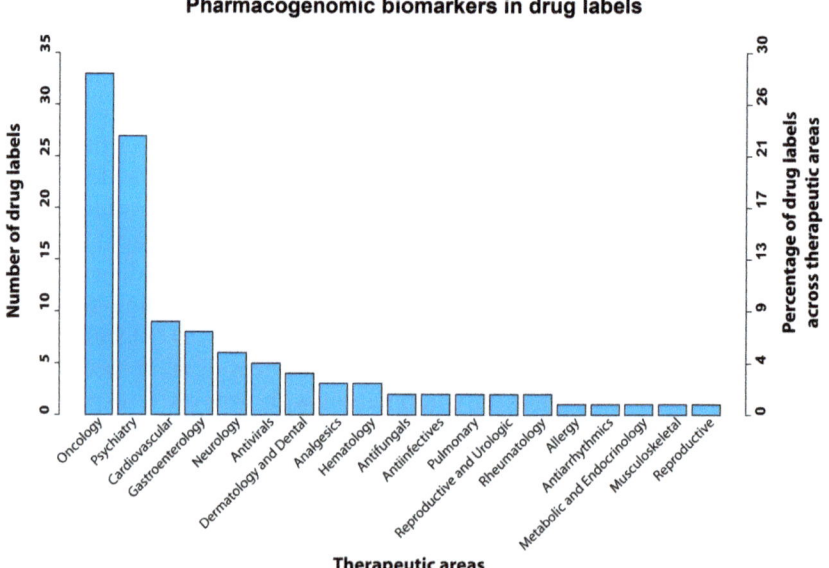

Figure 1.6   Schematic representation of the number of drug labels that contain companion diagnostic biomarker information, and the relative distribution of such drug labels over different disease areas.
Adapted from the table of Table of Pharmacogenomic Biomarkers in Drug Labels.[16]

Zhang *et al.*[101] On the other hand, Cyp450 metabolism may be required to convert the inactive parent drug to its active metabolite form, as is the case for the conversion of the antibreast cancer drug Tamoxifen to its active metabolite 4OH-Tamoxifen by Cyp2D6.[102]

The other part of the drug–biomarker combinations relates to biomarkers whose presence determines the efficacy of the drug in that patient. Particularly in oncology, there are striking examples. One of early examples of individualized cancer treatment based on genetic status of an individual was demonstrated in the case of Herceptin (Trastuzumab), a monoclonal antibody specifically used in treatment of early-stage and metastatic HER2-positive breast-cancer patients.[103] Based on data correlating efficacy to expression, FDA requires profiling of the tumor cells from breast-cancer biopsies to check for increased HER2 expression before Herceptin administration.

Alternatively, certain gene mutations are "driver mutations" that result in an overactive signalling pathway and strong tumor growth. An example is the $BRAF^{V600D/E}$ point mutation that is found in $\sim 60\%$ of patients with malignant melanoma.[105] This lethal form of cancer reacts poorly to conventional chemotherapy and radiation treatment, resulting in average survival rate of less than a year after diagnosis. Recently, an orally active BRAF-mutation-directed drug vemurafenib was developed with outstanding results, more than 80% of patients responded to this drug with tumor shrinkage in just 2 weeks. This effect was only seen in patients carrying the specific mutation and those without the mutation responded with worse outcomes upon treatment.[104,106] Other examples of drug–biomarker combinations include gefitinib–EGFR mutations[107] and crizotinib–ELM-ALK4 gene fusion[108] for treatment of nonsmall cell lung cancer.

1.6 Trends

In view of the potential roles of biomarkers as discussed above, it is interesting to reflect on the trends, challenges and opportunities regarding biomarkers in translational and personalized medicine in the pharmaceutical industry.

1.6.1 Importance of Biomarkers

Within pharmaceutical research and development, biomarkers have matured from being a hype to becoming an intricate part of the decision-making process. The low success rate and high costs associated with development of a novel drug has pushed pharmaceutical companies to try their utmost to be more efficient and cost effective. One of the strategies is to focus on biomarkers. Rather than trial-and-error, most if not all drug-development projects have key biomarker read-outs incorporated in their strategy to enable early derisking of the chosen approach and selection of the best compound moving forward. In parallel and in line with the desire for personalized medicine from society, there is a clear trend with oncology leading the way that new drugs have a companion diagnostic biomarker where possible.

1.6.2 Globalization

A clear trend in the pharmaceutical field has been the externalization of biomarker research. For both standardized and specialized activities it was found to be more cost-, resource- and time-effective to stop doing these activities inhouse and to outsource such tasks to dedicated service providers that perform high-quality contract research as their core business. As a direct consequence, many activities within companies and universities are currently being outsourced, creating interdependent functional networks. Such functional networks have a global nature and topology in which distance is not a key factor, but rather the quality, speed, costs and professional skills of the external partner. Many scientists nowadays have developed global connections and routinely send samples across continents for analysis by contract research organizations or academic collaborators.

The marketplace for pharmaceutical drugs has also globalized and with it, the research into their effectiveness and safety has taken a globalized view. Whereas earlier most, if not all, of the commercial and research focus of pharmaceutical and diagnostic industries was on Western markets in Europe and USA, recently this has shifted to the emerging markets, spearheaded by Brazil, Russia, India and China. However, distinct differences have been found in response to drugs originally developed for the Western markets, but also in the disease pathophysiology. For instance, the prevalence of certain cancers such as gastric and lung tumors is highly increased in Asia, caused by genetic and environmental risk factors that are different from those in western patients such as infection with hepatitis B or hepatitis C virus.[109] Also, South Asians from the Indian subcontinent have an increased risk of developing coronary heart disease as compared to Europeans, which is postulated to originate in different risk factors including insulin resistance, type 2 diabetes mellitus, lower physical activity, diet, in combination with a different coronary anatomy. Specific correlating biomarkers including blood levels of lipoprotein(a) and homocysteine, and candidate genetic markers were identified.[110] The global variation in disease prevalence and treatment success is due to variable exposure to risk factors, only some of which are known, making personalized medicine across the globe a real challenge but at the same time an opportunity. Without doubt, the different genetic backgrounds, dietary preferences and lifestyle of populations across the globe contribute to this, but clearly more research is needed to investigate these aspects.

Supported by their increased economic power, the emerging countries currently invest heavily in biomarker research, aiming to improve existing therapeutic treatments and to generate more personalized drugs for their markets. To facilitate such research, large science parks with impressive infrastructures have been set up, such as the Biopolis in Singapore and the Zhongguancun China Science Park in Beijing to name just two. These infrastructures bring together public and private partners to facilitate interactions towards applied research and development.

1.6.3 High-Content Biomarker Discovery

Methodological developments have strongly increased the potential for identification of new biomarkers. First, new data-rich biomarker discovery technologies have emerged that dramatically increased the amount of molecular information of preclinical and human systems. Particularly, next-generation sequencing technologies have revealed new possibilities to identify variants in the DNA and RNA species. Besides the genetic code itself, epigenetic regulation through methylation and acetylation, synthesis and composition of RNA species including messenger RNA and micro-RNA on nucleotide level can now be studied.[84,111] The costs of next-generation sequencing has dramatically decreased, enabling application of genome and whole exome sequencing in clinical diagnostic research to identify genetic causes of unknown diseases.[112] Also, mass-spectrometry methodologies have strongly developed and enabled the analysis of changes in peptide, protein and metabolite isoforms that reflect perturbations of biological systems. Complementing the sequencing analysis, mass spectrometry has yielded a wealth of observations on variants of the proteome and metabolome in healthy and diseased states.[113,114]

Biomarker discovery laboratories increasingly use multiple complementary molecular profiling approaches in parallel to identify a biomarker that fulfils the objective of a study.[12] This strategy not only increases the chance of identifying candidate biomarkers but it also generates independent data that provides additional mechanistic support of the best biomarker. As such, various molecular profiling combinations have been reported including of proteomics–metabolomics,[115] proteomics–lipidomics,[116] genetics–proteomics,[117] genetics–metabolomics,[118] and transcriptomics–proteomics.[90,119,120] Recently, considerable progress has been made in the development of novel chemometrics and bioinformatics methods to integrate biomarker-profiling data from the different platforms with phenotypic observations.[121]

This progress has dramatically changed the landscape of biomarker discovery and strongly increased the potential of identifying more specific and applicable biomarkers for human medicine.

1.6.4 Biomarker Combinations

The multitechnology biomarker-discovery approaches described above are likely to lead to biomarker panels rather than single biomarkers, as a panel of functional biomarkers read-outs is likely to have improved predictive value over single entities. However, it is not always practical to develop a biomarker panel test and the optimal condition for one test may be suboptimal for another in the same panel. Consequently, there is a need for simple tests that have an easy read-out, require minimal sample amounts and can be produced at low cost. Such assays can then be applied in single-plex format, but the data can be combined in multiplex format to have the desired multiparameter comprehensive view.

A related trend is to combine quantitative molecular biomarker data and clinical read-outs to provide a more detailed characterization of a disease state. This enables clinicians not only to diagnose diseases but also to specify the subtype or causal origin of the pathophysiology, potentially leading to a more tailored treatment. A nice example of this is the Disease Activity Score (DAS28) as applied in characterization of rheumatoid arthritis patients (see www.das28.nl/das28/en/). The DAS28 assessment is composed of a physical examination of the patient joints (the number of swollen and tender joints among 28 vulnerable joints), blood biomarker analysis (erythrocyte sedimentation rate) and a patient self-assessment. The DAS28 score will yield a number between 0 and 10, indicating the status of the rheumatoid arthritis disease at that moment.

1.7   Challenges

1.7.1   Interpretation of High-Content Biomarker Discovery Technologies

At present, biomarker scientists have become particularly efficient in collecting enormous amount of data at the whole-genome level. The translation of this wealth of data to knowledge and clinical insights, however, is still a big challenge. Improved statistical, bioinformatic and chemometric analysis methods are needed to smartly mine the high-content data and filter out the signals that can be used for identification of biomarkers and/or put them in a proper functional context.

Importantly, and a related challenge, many novel variants of genomic, transcriptomic, proteomic and metabolomic nature are identified, whereas knowledge on their biological function is scarce, resulting in a large number of potential new biomarkers that require further studies and validation. In this regard a new breed of clinical investigators and translational life scientists would need to be trained who are eager to embrace the breakthroughs in molecular-profiling sciences and amalgamate that with advances in medicine – to drive decisions for tailored treatment and therapies thereby advancing personalized medicine.[122,123]

1.7.2   Biomarker Validation

A major challenge in the biomarker field today resides in clinical biomarker validation. Even with the advances in molecular-profiling technologies we are still left with very few clinically accepted biomarkers. This discrepancy can be leveled by employing effective translational medicine approaches leading to expedited "bench to bedside" results. However, human diseases are mostly complex in nature and it requires several biomarkers to mechanistically describe the imbalance in the metabolic equilibrium responsible for the disease and the effect of treatment thereon. The number of published biomarkers with clear biological relevance needs to increase and, correspondingly, the clinical validation of such biomarkers needs to improve strongly. Without extensive validation, the outcome of a biomarker test cannot be used in driving important clinical decisions such as in selecting patients for personalized therapy and/or treatment. Clinical biomarker validation should optimally be performed by applying standardized procedures and protocols to test independent and well-characterized clinical samples across multiple independent laboratories. However, in reality, this is rarely carried out. The commitment of biomedical researchers to engage in long-lasting and expensive biomarker-validation projects is limited, which is at least partly due to the pressure to publish innovative findings in high impact journals and secure funding. Although clinical validation of biomarkers, including the development of quantitative biomarker assays with high accuracy, specificity and reproducibility is essential to enable translational and personalized medicine, it has limited new scientific value and as such is less prone to obtain funding and high-impact publication. Worryingly, the initial effect size of a biomarker in selected clinical populations can in the majority of cases not be confirmed in meta-analyses of subsequent studies.[124] Several reasons can be identified for this observation including differences in study designs, study sizes and subject inclusion, small variations in the biomarker analytical protocol used, differences in the isolation and preparation of the new clinical samples versus the ones on which the original finding was based, and potentially the biased selection of most differential data to publish the initial biomarker finding. This worrisome observation hampers the implementation of biomarkers by others and blocks the progression of the field at large.

1.7.3 Robust Biomarker Tests

To be able to mature to a molecular diagnostic test, candidate biomarkers need to be developed through a rigorous process. This starts with demonstration of the added value in clinical practice, followed by development of a prototype diagnostic test, its thorough analytical and clinical validation by several independent laboratories, demonstration of its positive health economic value, alignment with regulatory and insurance agencies to obtain registration and reimbursement of the test, and finally implementation by clinicians.[19,125] This lengthy process is aimed to optimize the specificity and selectivity of the diagnostic test, whilst minimizing its cost and demonstrating its usefulness in clinical practice. Whereas great progress has been made in the front end of this diagnostic development pipeline, little progress has been made in the latter part. In particular, major gaps remain in the diagnostic test development and multicenter validation of candidate biomarkers, as this type of work is generally not easily funded by grants. A few positive exceptions exist such as the recent multicenter testing of the key Alzheimer's disease biomarkers Aβ42, T-tau, P-Tau.[126]

1.7.4 Biospecimen

Despite constructive efforts by regional biobanks, there is an intrinsic lack globally of (pre)clinical samples that are highly characterized, are properly stored, have associated patient phenotype metadata and are being made available for multicenter biomarker validation. Consequently, lack of such biospecimen can lead the researchers to focus on objectives that are considered more achievable in the short-term rather than what is desired in the specific field in the long term. Hence, biomarker studies may be statistically underpowered, only rarely include reporting of robust clinical biomarker validation with sufficient clinical sample size, and hence may not be reproducible by other laboratories. It is imperative for high-quality biomarker data that high-quality biosamples are tested, but it will be a challenge to organize the global community so as to enable this.

1.7.5   Disconnected Biomarker Development Pipeline

Successful biomarker discovery, validation and development requires a smooth and cooperative interaction between key players in this field including patients, academia, method and instrument developers, the pharma industry, the diagnostic industry, clinicians, health-insurance providers, regulatory approval agencies, government policy makers, and many others. Although seemingly logical, such across-the-board collaborations only rarely exist, for various reasons including different interests and objectives, unawareness of the potential synergy of working together, lack of funding and others. In the current economic climate, securing funding is a major challenge. The consequence is that too few biomarkers progress from discovery to clinical implementation and are stalled at intermediate stages. The challenge here is to bring those key players together in a mutually beneficial working environment with sufficient funding of the joined research to enable the pipeline to be operational.

1.7.6   Translate Interindividual Findings in Personalized Biomarker Tests

An important component of application of biomarkers in real life, is to identify and understand the variation in biomarker levels across various population groups locally and in the world. Some of these can be related to variables in genetic background, lifestyle or nutritional intake, but many cannot. A greater understanding of this natural variation is imperative prior to the use of such biomarkers to mechanistically characterize disease and predict treatment effects.

1.7.7   Cost Reimbursement

Despite their potential in clinical settings to drive diagnosis and treatment strategies, unanswered questions still remain about the reimbursement of costs associated with expensive biomarker tests. Health-insurance companies generally do not want to pay for diagnostic tests that are not part of a hitherto established medical diagnostic toolkit. This has particularly become an issue with genomic biomarker screening that can be key to enable patient selection for rational treatment, as also recently exemplified in the case of Kalydeco to screen for a rare mutation of CFTR gene to maximize the success of cystic fibrosis treatment.[127] The cost of whole-genome analysis of an individual has dramatically decreased to about USD 1000, but it is yet unclear who should pay for these multibiomarker tests that will lead to one selected preferred treatment.

1.7.8 Handling of Personalized Data

The deluge of molecular biomarker and profiling data raises a key ethical question related to data privacy, access and use.[128,129] For any individual, having access to his/her own medical data through electronic health records could certainly boost their participation in disease management. However, the knowledge of having an identified genetic susceptibility to a certain disease may cause serious disruption of the person's peace of mind, aggravating the occurrence of disease symptoms. Even worse, if such data becomes available it may have serious implications for the individual's relation with employers and health-insurance companies. Who will have access to genetic information and how can we ensure that this information would not be wrongly used and interpreted? The challenge here is to define this properly so as to safeguard the individual's identity and interests.

1.8   Opportunities

Embracing an optimistic view, we feel that every challenge is an opportunity for progress and the challenges outlined above translate into specific objectives for improvement, both scientifically and process-wise. We would like to believe that the scientific improvements, such as those dedicated to improved data-analysis methods and improved understanding of the biological background of novel biomarkers, will be addressed by the innovation-driven research of academic and pharmaceutical researchers. The process-wise improvement in biomarker research and development, however, requires a different mind-set and working model. The desired pipeline of biomarker discovery–validation–development–implementation requires that multiple parties work closely together with the objective to apply known biomarkers rather than discover new ones.[48] Such an application-driven working model is ideally suited for public–private partnerships that bring together academic innovation, specialized technical skills, clinical expertise and industrial process focus.

The downsizing of internal biomarker activities in larger companies has already resulted in a considerable stimulation of new economic business opportunities. Many specialized spin-off companies have been formed by entrepreneurial scientists who excel in a specific part of the biomarker research and development process. Together with contract-research organizations, technology vendors and central analysis laboratories, they form a functional network with larger companies and universities to drive pharmaceutical and diagnostic biomarker development. Importantly, this functional network now also includes the emerging markets and their populations, driving biomarker research across the globe.

Although across-the-board funding opportunities have decreased, funding agencies generally prefer to support multicenter consortia rather than single laboratories. Good examples of public–private partnerships that we have recently been involved in include Netherlands programs such as Top Institute Pharma and Center of Translational Molecular Medicine, USA-based Critical Path Initiatives and European framework programs. These partnerships have brought together basic and applied biomarker researchers in academia and industry and have stimulated knowledge transfer and a focus towards applied biomarker research. Public–private consortia have been shown to lead to increased awareness of the value of crossdiscipline biomarker research, improved data-analysis workflows, and increased identification and development of biomarkers for specific diseases and mechanisms.

The opportunity now lies in the formation of a large public–private partnership that is fully focused on the validation and development of biomarkers, aiming at clinical implementation. We believe the mind-set of key stakholders is currently set to form and support such functional network. This should bring together pharmaceutical industry, academia, clinical experts, biobanks, analytical laboratories and diagnostic industries in a working model that creates a win-win situation for all. Pharmaceutical industries have a key interest in applying validated disease-related biomarkers and can contribute clinical samples from their intervention trials. Academic researchers can provide indepth knowledge on the mechanistic relationship of a biomarker with disease. Clinical experts are imperative to indicate the biggest clinical unmet need, thus focusing the network on the clinical application, and by making sure the right biosamples are included in the clinical validation projects. Biobanks have a platform to contribute their unique clinical samples to, thus obtaining compensation for their investments while maximizing the use of their samples to generate high-impact data. Analytical laboratories can bring in their specialized skills in biomarker-assay development and analytical validation and, where needed, contribute to the registration of the biomarker as a novel diagnostic product. Diagnostic companies can further develop the clinically and analytically validated biomarker test and convert it into a commercial product. The scale of such public–private networks should be large enough to enable independent cross-laboratory validation of a biomarker test, demonstrating its added value in a particular application or, perhaps even more importantly, its lack thereof due to insufficient robustness. Reporting of those positive and negative validation results are crucial to advance the field.

In parallel with well-defined biomarker development projects, academic researchers and specialized methodology scientists can contribute to this network by further improving biomarker technologies where needed. Through benchmarking of novel technologies versus established ones in biomarker development projects, the network can demonstrate their added value, thus facilitating acceptance of methodology improvements.

Looking ahead, focused biomarker development networks as outlined above are likely to boost the quality of published biomarkers and their application in translational and personalized medicine. Ultimately, this higher level of biomarker knowledge and tools will result in transformation of disease management into health management. At present, it may seem a utopia but with the impressive progress we have seen in recent years it may occur sooner than expected.

References

1. U. Pinder (Udalricus Binder). Epiphanie Medicorum. Speculum videndi urinas hominum. Clavis aperiendi portas pulsuum. Berillus discernendi causas & differentias febrium. Sodalitas Celtica: Nuremberg, 1506.
2. Biomarkers and surrogate endpoints: preferred definitions and conceptual framework. *Clin. Pharmacol. Ther.*, 2001, **69**(3), 89–95.
3. M. Huber M, *et al.*, How should we define health?, *BMJ*, 2011, **343**, d4163.
4. US FDA, Driving Biomedical Innovation: Initiatives for Improving Products for Patients, U.S.D.o.H.a.H. Services, Editor. 2011.
5. R. Katz, Biomarkers and surrogate markers: an FDA perspective, *NeuroRx*, 2004, **1**(2), 189–95.
6. US FDA, Guidance for Industry Pharmacogenomic Data Submissions, U.S.D.o.H.a.H. Services, Editor. 2005.
7. V. J. Swan, C. J. Hamilton and S. A. Jamal, Lasofoxifene in osteoporosis and its place in therapy, *Adv. Ther.*, 2010, **27**(12), 917–32.
8. A. J. van Gool, Via biomarkers naar betere medicijnen, *Inaugural speech Radboud University.*, 2010.
9. S. M. Paul, *et al.*, How to improve R&D productivity: the pharmaceutical industry's grand challenge, *Nat. Rev. Drug. Discov.*, 2010, **9**(3), 203–14.
10. I. Kola and J. Landis, Can the pharmaceutical industry reduce attrition rates?, *Nat. Rev. Drug Discov.*, 2004, **3**(8), 711–5.
11. I. Kola, The state of innovation in drug development, *Clin Pharmacol Ther*, 2008, **83**(2), 227–30.
12. A. J. van Gool, B. Henry and E. D. Sprengers, From biomarker strategies to biomarker activities and back, *Drug Discov. Today.*, 2010, **15**(3-4), 121–6.
13. B. Arun, M. Anthony and B. Dunn, The search for the ideal SERM, *Expert Opin. Pharmacother.*, 2002, **3**(6), 681–91.
14. K. Dechering, C. Boersma and S. Mosselman, Estrogen receptors alpha and beta: two receptors of a kind?, *Curr. Med. Chem.*, 2000, **7**(5), 561–76.
15. V. G. De Gruttola, *et al.*, Considerations in the evaluation of surrogate endpoints in clinical trials. summary of a National Institutes of Health workshop, *Control Clin. Trials*, 2001, **22**(5), 485–502.
16. J. F. Griffith and H. K. Genant, Bone mass and architecture determination: state of the art, *Best Pract. Res. Clin. Endocrinol. Metab.*, 2008, **22**(5), 737–64.

17. J. N. Cohn, *et al.*, Surrogate markers for cardiovascular disease: functional markers, *Circulation*, 2004, **109**(25 Suppl 1), IV31–46.
18. US FDA, Table of Pharmacogenomic Biomarkers in Drug Labels, Accessed 08 March 2012.
19. US FDA, Guidance for industry, bioanalytical method validation U.S.D.o.H.a.H. Services Editor, 2001.
20. J. Mervis, Productivity counts but the definition is key, *Science*, 2005, **309**(5735), 726.
21. J. A. DiMasi, *et al.*, Trends in risks associated with new drug development: success rates for investigational drugs, *Clin. Pharmacol. Ther.*, 2010, **87**(3), 272–7.
22. J. Arrowsmith, Trial watch: Phase II failures: 2008–2010, *Nat. Rev. Drug Discov.*, 2011, **10**(5), 328–9.
23. J. Arrowsmith, Trial watch: phase III and submission failures: 2007–2010, *Nat. Rev. Drug Discov.*, 2011, **10**(2), 87.
24. E. A. Zerhouni, Translational and clinical science—time for a new vision, *N. Engl. J. Med.*, 2005, **353**(15), 1621–3.
25. J. Kaiser, New Cystic Fibrosis Drug Offers Hope, at a Price, *Science*, 2012, **335**(6069), 645.
26. D. J. Cheon and S. Orsulic, Mouse models of cancer, *Annu. Rev. Pathol.*, 2011, **6**, 95–119.
27. V. Lallemand-Breitenbach, *et al.*, Opinion: how patients have benefited from mouse models of acute promyelocytic leukaemia, *Nat. Rev. Cancer*, 2005, **5**(10), 821–7.
28. M. S. Tallman, *et al.*, Acute promyelocytic leukemia: evolving therapeutic strategies, *Blood*, 2002, **99**(3), 759–67.
29. M. Jucker, The benefits and limitations of animal models for translational research in neurodegenerative diseases, *Nat. Med.*, 2010, **16**(11), 1210–4.
30. J. Mestas and C. C. Hughes, Of mice and not men: differences between mouse and human immunology, *J. Immunol.*, 2004, **172**(5), 2731–8.
31. D. Foell, *et al.*, Neutrophil derived human S100A12 (EN-RAGE) is strongly expressed during chronic active inflammatory bowel disease, *Gut*, 2003, **52**(6), 847–853.
32. G. Fuellen, *et al.*, Absence of S100A12 in mouse: implications for RAGE-S100A12 interaction, *Trends Immunol*, 2003, **24**(12), 622–624.
33. G. Yan, *et al.*, Genome sequencing and comparison of two nonhuman primate animal models, the cynomolgus and Chinese rhesus macaques, *Nat Biotechnol.*, 2011, **29**(11), 1019–23.
34. L. D. Shultz, F. Ishikawa and D. L. Greiner, Humanized mice in translational biomedical research, *Nat Rev Immunol*, 2007, **7**(2), 118–130.
35. D. J. Cook, L. Teves and M. Tymianski, Treatment of stroke with a PSD-95 inhibitor in the gyrencephalic primate brain, *Nature*, 2012, **483**(7388), 213–7.
36. D. J. Cook, L. Teves and M. Tymianski, A Translational Paradigm for the Preclinical Evaluation of the Stroke Neuroprotectant Tat-NR2B9c in Gyrencephalic Nonhuman Primates, *Sci. Transl. Med.*, 2012, **4**(154), 154ra133.

37. E. Raymond, *et al.*, Sunitinib malate for the treatment of pancreatic neuroendocrine tumors, *N. Engl. J. Med.*, 2011, **364**(6), 501–13.
38. J. C. Yao, *et al.*, Everolimus for advanced pancreatic neuroendocrine tumors, *N. Engl. J. Med.*, 2011, **364**(6), 514–23.
39. I. N. Foltz, *et al*, Treating Diabetes and Obesity with an FGF21-Mimetic Antibody Activating the β Klotho/FGFR1c Receptor Complex, *Sci. Transl. Med.*, 2012, **4**(162), 162ra153.
40. M. Wang, *et al.*, Mixed Chimerism and Growth Factors Augment β Cell Regeneration and Reverse Late-Stage Type 1 Diabetes, *Sci. Transl. Med.*, 2012, **4**(133), 133ra59.
41. E. Abrahams and M. Silver, The case for personalized medicine, *J. Diabetes. Sci. Technol.*, 2009, **3**(4), 680–4.
42. F. S. Collins, *The Language of Life: DNA and the Revolution in Personalized Medicine* First edition ed. 2010: Harper.
43. I. S. Chan and G. S. Ginsburg, Personalized medicine: progress and promise, *Annu. Rev. Genomics Hum. Genet.*, 2011, **12**, 217–44.
44. E. J. Topol, *The Creative Destruction of Medicine: How the Digital Revolution Will Create Better Health Care* First edition ed. 2012: Basic Books.
45. M. A. Hamburg and F. S. Collins, The path to personalized medicine, *N. Engl. J. Med*, 2010, **363**(4), 301–4.
46. T. Goetz, *The Decision Tree: Taking Control of Your Health in the New Era of Personalized Medicine*, First edition ed. 2010: Rodale Books.
47. L. Chouchane, *et al.*, Personalized medicine: a patient-centered paradigm, *J. Transl. Med.*, 2011, **9**, 206.
48. A. W. Evers, *et al.*, An integrated framework of personalized medicine: from individual genomes to participatory health care, *Croat. Med. J.*, 2012, **53**(4), 301–3.
49. L. Hood and S. H. Friend, Predictive, personalized, preventive, participatory (P4) cancer medicine, *Nat. Rev. Clin. Oncol.*, 2011, **8**(3), 184–7.
50. J. van der Greef, T. Hankemeier and R. N. McBurney, Metabolomics-based systems biology and personalized medicine: moving towards n = 1 clinical trials?, *Pharmacogenomics*, 2006, **7**(7), 1087–94.
51. B. S. Taylor, *et al.*, Integrative genomic profiling of human prostate cancer, *Cancer Cell.*, 2010, **18**(1), 11–22.
52. C. Curtis, *et al.*, The genomic and transcriptomic architecture of 2,000 breast tumours reveals novel subgroups, *Nature*, 2012, **486**(7403), 346–52.
53. U. A. Meyer, Pharmacogenetics and adverse drug reactions, *Lancet*, 2000, **356**(9242), 1667–71.
54. R. Weinshilboum and L. Wang, Pharmacogenomics: bench to bedside, *Nat. Rev. Drug Discov.*, 2004, **3**(9), 739–48.
55. S. Hariri, *et al.*, Family history of type 2 diabetes: a population-based screening tool for prevention?, *Genet. Med.*, 2006, **8**(2), 102–8.
56. K. Hemminki, J. Sundquist and J. Lorenzo Bermejo, Familial risks for cancer as the basis for evidence-based clinical referral and counseling, *Oncologist*, 2008, **13**(3), 239–47.

57. M. T. Scheuner, *et al.*, Expanding the definition of a positive family history for early-onset coronary heart disease, *Genet. Med.*, 2006, **8**(8), 491–501.

58. N. Petrucelli, M. B. Daly, G. L. Feldman, BRCA1 and BRCA2 Hereditary Breast and Ovarian Cancer. 1998 Sep 4 [Updated 2011 Jan 20]. In: R. A. Pagon, T. D. Bird, C. R. Dolan, *et al.*, eds. GeneReviews™ [Internet]. Seattle (WA): University of Washington, Seattle; 1993–. Available from: http://www.ncbi.nlm.nih.gov/books/NBK1247/.

59. J. Kim, J. M. Basak and D. M. Holtzman, The Role of Apolipoprotein E in Alzheimer's Disease, *Neuron*, 2009, **63**(3), 287–303.

60. R. J. Perrin, A. M. Fagan and D. M. Holtzman, Multimodal techniques for diagnosis and prognosis of Alzheimer's disease, *Nature*, 2009, **461**(7266), 916–22.

61. E. J. Topol, Transforming medicine via digital innovation, *Sci. Transl. Med.*, 2010, **2**(16), 16cm4.

62. E. S. Lander, *et al.*, Initial sequencing and analysis of the human genome, *Nature*, 2001, **409**(6822), 860–921.

63. S. Roychowdhury, *et al.*, Personalized oncology through integrative high-throughput sequencing: a pilot study, *Sci. Transl. Med.*, 2011, **3**(111), 111ra121.

64. M. Hornsey, *et al.*, Whole-genome comparison of two Acinetobacter baumannii isolates from a single patient, where resistance developed during tigecycline therapy, *J. Antimicrob. Chemother.*, 2011, **66**(7), 1499–503.

65. J. R. Lupski, *et al.*, Whole-genome sequencing in a patient with Charcot-Marie-Tooth neuropathy, *N. Engl. J. Med.*, 2010, **362**(13), 1181–91.

66. F. Dahl, *et al.*, Multigene amplification and massively parallel sequencing for cancer mutation discovery, *Proc. Natl. Acad. Sci. U S A*, 2007, **104**-(22), 9387–92.

67. J. C. Roach, *et al.*, Analysis of genetic inheritance in a family quartet by whole-genome sequencing, *Science*, 2010, **328**(5978), 636–9.

68. E. A. Worthey, *et al.*, Making a definitive diagnosis: successful clinical application of whole exome sequencing in a child with intractable inflammatory bowel disease, *Genet. Med.*, 2011, **13**(3), 255–62.

69. J. S. Welch, *et al.*, Use of whole-genome sequencing to diagnose a cryptic fusion oncogene, *Jama.*, 2011, **305**(15), 1577–84.

70. C. Gilissen, *et al.*, Disease gene identification strategies for exome sequencing, *Eur. J. Hum. Genet.*, 2012, **20**(5), 490–7.

71. P. Deloukas and D. Bentley, The HapMap project and its application to genetic studies of drug response, *Pharmacogenomics J.*, 2004, **4**(2), 88–90.

72. K. M. Weiss and A. G. Clark, Linkage disequilibrium and the mapping of complex human traits, *Trends Genet.*, 2002, **18**(1), 19–24.

73. L. A. Hindorff, *et al.*, *A Catalog of Published Genome-Wide Association Studies*. Available at: www.genome.gov/gwastudies. Accessed 26 June 2012.

74. D. Ge, *et al.*, Genetic variation in IL28B predicts hepatitis C treatment-induced viral clearance, *Nature*, 2009, **461**(7262), 399–401.

75. A. J. Thompson, *et al.*, Interleukin-28B polymorphism improves viral kinetics and is the strongest pretreatment predictor of sustained virologic response in genotype 1 hepatitis C virus, *Gastroenterology*, 2010, **139**(1), 120–9.

76. A. A. Alizadeh, *et al.*, Distinct types of diffuse large B-cell lymphoma identified by gene expression profiling, *Nature*, 2000, **403**(6769), 503–11.

77. A. Bhattacharjee, *et al.*, Classification of human lung carcinomas by mRNA expression profiling reveals distinct adenocarcinoma subclasses, *Proc. Natl. Acad. Sci. U S A*, 2001, **98**(24), 13790–5.

78. M. Bittner, *et al.*, Molecular classification of cutaneous malignant melanoma by gene expression profiling, *Nature*, 2000, **406**(6795), 536–40.

79. M. J. van de Vijver, *et al.*, A gene-expression signature as a predictor of survival in breast cancer, *N. Engl. J. Med.*, 2002, **347**(25), 1999–2009.

80. T. R. Golub, *et al.*, Molecular classification of cancer: class discovery and class prediction by gene expression monitoring, *Science*, 1999, **286**-(5439), 531–7.

81. M. M. Kittleson and J. M. Hare, Molecular signature analysis: the potential of gene-expression analysis in cardiomyopathy, *Future Cardiol.*, 2005, **1**(6), 793–808.

82. M. Comabella and R. Martin, Genomics in multiple sclerosis--current state and future directions, *J. Neuroimmunol.*, 2007, **187**(1-2), 1–8.

83. L. G. van Baarsen, *et al.*, Transcription profiling of rheumatic diseases, *Arthritis Res. Ther.*, 2009, **11**(1), 207.

84. Z. Wang, M. Gerstein and M. Snyder, RNA-Seq: a revolutionary tool for transcriptomics, *Nat. Rev. Genet.*, 2009, **10**(1), 57–63.

85. C. A. Maher, *et al.*, Transcriptome sequencing to detect gene fusions in cancer, *Nature*, 2009, **458**(7234), 97–101.

86. N. Palanisamy, *et al.*, Rearrangements of the RAF kinase pathway in prostate cancer, gastric cancer and melanoma, *Nat. Med.*, 2010, **16**(7), 793–8.

87. L. Chin, J. N. Andersen and P. A. Futreal, Cancer genomics: from discovery science to personalized medicine, *Nat. Med.*, 2011, **17**(3), 297–303.

88. R. Aebersold and M. Mann, Mass spectrometry-based proteomics, *Nature*, 2003, **422**(6928), 198–207.

89. T. Geiger, *et al.*, Super-SILAC mix for quantitative proteomics of human tumor tissue, *Nat. Methods*, 2010, **7**(5), 383–5.

90. N. Nagaraj, *et al.*, Deep proteome and transcriptome mapping of a human cancer cell line, *Mol. Syst. Biol.*, 2011, **7**, 548.

91. S. J. Deeb, *et al.*, Super-SILAC Allows Classification of Diffuse Large B-cell Lymphoma Subtypes by Their Protein Expression Profiles, *Mol. Cell Proteomics*, 2012, **11**(5), 77–89.

92. T. Geiger, *et al.*, Proteomic portrait of human breast cancer progression identifies novel prognostic markers, *Cancer Res.*, 2012, **72**(9), 2428–39.

93. F. Butter, *et al.*, Proteome-wide analysis of disease-associated SNPs that show allele-specific transcription factor binding, *PLoS Genet.*, 2012, **8**(9), e1002982.

94. M. Mann, Proteomics for biomedicine: a half-completed journey, *EMBO Mol. Med.*, 2012, **4**(2), 75–7.

95. J. R. Bain, *et al.*, Metabolomics applied to diabetes research: moving from information to knowledge, *Diabetes*, 2009, **58**(11), 2429–43.

96. J. L. Griffin and J. P. Shockcor, Metabolic profiles of cancer cells, *Nat. Rev. Cancer*, 2004, **4**(7), 551–61.

97. J. L. Griffin, *et al.*, Metabolomics as a tool for cardiac research, *Nat. Rev. Cardiol.*, 2011, **8**(11), 630–43.

98. R. Kaddurah-Daouk, *et al.*, Metabolomic mapping of atypical antipsychotic effects in schizophrenia, *Mol. Psychiatry*, 2007, **12**(10), 934–45.

99. H. C. Keun, Metabonomic modeling of drug toxicity, *Pharmacol. Ther.*, 2006, **109**(1-2), 92–106.

100. M. R. Stratton, P. J. Campbell and P. A. Futreal, The cancer genome, *Nature*, 2009, **458**(7239), 719–24.

101. J. P. Zhang and A. K. Malhotra, Pharmacogenetics and antipsychotics: therapeutic efficacy and side effects prediction, *Expert Opin. Drug Metab. Toxicol.*, 2011, **7**(1), 9–37.

102. H. Brauch, *et al.*, Pharmacogenomics of tamoxifen therapy, *Clin. Chem.*, 2009, **55**(10), 1770–82.

103. D. J. Slamon, *et al.*, Human breast cancer: correlation of relapse and survival with amplification of the HER-2/neu oncogene, *Science*, 1987, **235**(4785), 177–82.

104. K. T. Flaherty, *et al.*, Inhibition of mutated, activated BRAF in metastatic melanoma, *N. Engl. J. Med.*, 2010, **363**(9), 809–19.

105. H. Davies, *et al.*, Mutations of the BRAF gene in human cancer, *Nature*, 2002, **417**(6892), 949–54.

106. G. Bollag, *et al.*, Clinical efficacy of a RAF inhibitor needs broad target blockade in BRAF-mutant melanoma, *Nature*, 2010, **467**(7315), 596–9.

107. M. G. Kris, *et al.*, Efficacy of gefitinib, an inhibitor of the epidermal growth factor receptor tyrosine kinase, in symptomatic patients with non-small cell lung cancer: a randomized trial, *Jama*, 2003, **290**(16), 2149–58.

108. A. T. Shaw, *et al.*, Effect of crizotinib on overall survival in patients with advanced non-small-cell lung cancer harbouring ALK gene rearrangement: a retrospective analysis, *Lancet. Oncol.*, 2011, **12**(11), 1004–12.

109. A. P. Venook, *et al.*, The incidence and epidemiology of hepatocellular carcinoma: a global and regional perspective, *Oncologist.*, 2010, **15**(Suppl 4), 5–13.

110. K. Tziomalos, *et al.*, Vascular risk factors in South Asians, *Int. J. Cardiol.*, 2008, **128**(1), 5–16.

111. V. W. Zhou, A. Goren and B. E. Bernstein, Charting histone modifications and the functional organization of mammalian genomes, *Nat. Rev. Genet.*, 2011, **12**(1), 7–18.

112. M. Nelen and J. A. Veltman, Genome and exome sequencing in the clinic: unbiased genomic approaches with a high diagnostic yield, *Pharmacogenomics*, 2012, **13**(5), 511–4.
113. T. Nilsson, *et al.*, Mass spectrometry in high-throughput proteomics: ready for the big time, *Nat. Methods*, 2010, **7**(9), 681–5.
114. A. J. van Gool and R. C. Hendrickson, The proteomic toolbox for studying cerebrospinal fluid, *Expert Rev. Proteomics.*, 2012, **9**(2), 165–79.
115. M. P. Stoop, *et al.*, Quantitative proteomics and metabolomics analysis of normal human cerebrospinal fluid samples, *Mol. Cell Proteomics*, 2010, **9**(9), 2063–75.
116. A. Thomas, *et al.*, Mass spectrometry for the evaluation of cardiovascular diseases based on proteomics and lipidomics, *Thromb. Haemost.*, 2011, **106**(1), 20–33.
117. T. Geiger, J. Cox and M. Mann, Proteomic changes resulting from gene copy number variations in cancer cells, *PLoS Genet.*, 2010, **6**(9), e1001090.
118. C. Gieger, *et al*, Genetics meets metabolomics: a genome-wide association study of metabolite profiles in human serum, *PLoS Genet.*, 2008, **4**(11), e1000282.
119. J. N. Martin, *et al.*, Transcriptional and proteomic profiling in a cellular model of DYT1 dystonia, *Neuroscience*, 2009, **164**(2), 563–72.
120. T. Ideker, *et al.*, Integrated genomic and proteomic analyses of a systematically perturbed metabolic network, *Science*, 2001, **292**(5518), 929–34.
121. T. Ideker, J. Dutkowski and L. Hood, Boosting signal-to-noise in complex biology: prior knowledge is power, *Cell*, 2011, **144**(6), 860–3.
122. F. W. Frueh and D. Gurwitz, From pharmacogenetics to personalized medicine: a vital need for educating health professionals and the community, *Pharmacogenomics*, 2004, **5**(5), 571–9.
123. E. R. Edelman and K. LaMarco, Clinician-Investigators as Translational Bioscientists: Shaping a Seamless Identity, *Sci. Transl. Med.*, 2012, **4**(135), 135fs14.
124. J. P. Ioannidis and O. A. Panagiotou, Comparison of effect sizes associated with biomarkers reported in highly cited individual articles and in subsequent meta-analyses, *Jama*, 2011, **305**(21), 2200–10.
125. H. Mischak, *et al.*, Recommendations for biomarker identification and qualification in clinical proteomics, *Sci. Transl. Med.*, 2010, **2**(46), 46ps42.
126. N. Mattsson, *et al.*, The Alzheimer's Association external quality control program for cerebrospinal fluid biomarkers, *Alzheimers Dement.*, 2011, **7**(4), 386–395e6.
127. H. Ledford, Drug bests cystic-fibrosis mutation, *Nature*, 2012, **482**(7384), 145.
128. D. Greenbaum, *et al.*, Genomics and privacy: implications of the new reality of closed data for the field, *PLoS Comput. Biol.*, 2011, **7**(12), e1002278.
129. E. E. Schadt, The changing privacy landscape in the era of big data, *Mol. Syst. Biol.*, 2012, **8**, 612.

CHAPTER 2

Introduction: Regulatory Development Hurdles for Biomarker Commercialization: The Steps Required to Get a Product to Market

MELISSA A. THOMPSON

Matrix Clinical Research Management, 59 Pine Street, Peterborough, NH 03458
Email: melissaathompson@comcast.net

2.1 Introduction

Today's regulatory environment, for the development of biomarkers, genomics, personalized medicine, targeted therapy, and companion diagnostic products are under a great deal of scrutiny, change and new regulation for both the US Food and Drug Administration (FDA) and the EU. New guidelines to support the commercialization of these products have been released in draft form in 2011 by the FDA and are currently being revised under the EU directive. Despite the quickly changing regulatory environment, the basic principles for commercialization of any molecular diagnostic product, medical device, and pharmaceutical product are based on good business practices. Unfortunately, most academic scientists as well as many small/medium-size companies or startup companies are not familiar with the business practices for

RSC Drug Discovery Series No. 33
Comprehensive Biomarker Discovery and Validation for Clinical Application
Edited by Péter Horvatovich and Rainer Bischoff
© The Royal Society of Chemistry 2013
Published by the Royal Society of Chemistry, www.rsc.org

developing medical-related products. Without understanding these nuances, a great deal of time, money, and effort can be wasted in developing and validating biomarkers. This chapter will help to address and clarify some of the principles that need to be taken into consideration in order to commercialize a biomarker.

The basis of the draft regulations for both FDA and EU revolve around developing processes and procedures for developmental documentation to support a regulatory filing. This involves compliance with quality-system regulation for both US FDA, EU regulation and other international regulatory bodies. While this chapter will not incorporate the details of these regulations, references have been provided that help to commercialize a product to a regulatory agency. Biomarker assay developers may use these references to ensure that they have developed their product in compliance with in the law and international regulations.

The FDA is the regulating body for the United States and reviews and approves all new products and issues regulation for commercialization and marketing of these products. The European Medicines Agency for Europe operates as a decentralized scientific agency (as opposed to a regulatory authority) of the European Union and its main responsibility is the protection and promotion of public and animal health, through the evaluation and supervision of medicines for human and veterinary use. More specifically, it coordinates the evaluation and monitoring of centrally authorized products and national referrals, developing technical guidance and providing scientific advice to sponsors. Regulations are issued similar to the FDA for sake of this discussion under *In-vitro* Diagnostics Directive 98/78/EC. The scope of the IVDD covers all *in vitro* diagnostic devices and their accessories. An *in vitro* device is defined as any medical device that is a reagent, reagent product, calibrator, control material, kit, instrument, apparatus, equipment, or system to be used *in vitro* for the examination of blood or tissue specimens. The IVDD calls out specific design, manufacturing, packaging, and labeling requirements.

2.2 Regulatory Commercialization Path Options

No matter what type of product is developed it is essential that careful regulatory planning and implementation is performed. Regulatory planning is the foundation by which a product is developed. Knowing the regulatory path for the intended use of the product, and the intended market will help to establish a baseline for commercialization efforts. Biomarker development can assist in filling the gaps of uncertainty about disease targets, variability in drug response, animal and human discrepancy and new molecular selection. Biomarker or molecular selection is the identification of specific genes, gene segments, proteins, lipids, glycoproteins or other molecular signatures. Biomarker development can also assist in improving the success in future drug-development programs and improve clinical efficacy.

Each product needs to be developed with the intention of specific use, for example "the BRCA biomarker is intended to be used to determine a woman's risk of developing breast cancer". From this intended-use statement we know the therapeutic area is breast cancer, the target patient groups for use are women who are at risk of developing breast cancer and the result will generate a report indicating no increased risk or an increased risk is present. Evaluating the target application, the path of delivery or type of assay or instrumentation used for evaluation, and the intended commercialization (IVD diagnostic, LDT, Companion Diagnostic) will help to assess which regulatory path should be followed. Establishing these areas will help to determine any potential risks that the product may have to the patient. The risk associated with the intended use will help determine the correct regulatory path.

The FDA has issued two new guidances 17, 19 in 2011 affecting genomic/proteomic technologies and genetic testing, and includes details policies for making risk-based assessments of medical devices, and another covering the process for the submission of genomic biomarkers for qualification by regulatory authorities. Both can be found at the FDA website: http://www.FDA.gov.

Premarket approval (PMA) products are those that have been identified to have the highest risk associated with them. Submitting a PMA is mandatory if it is determined that the product is of high risk, and generally involves a face-to-face meeting with the FDA prior to submission. FDA developed a draft guidance for PMA and provides guidelines for industry and agency reviewers regarding factors that agency should consider when they assess the risks and benefits of medical devices. Medical devices, in this case, can be instruments, *in vitro* diagnostic assays, and molecular signatures such as biomarkers, with a specific intended use or targeted therapies developed as companion diagnostic products. The guidance is aimed to improve the predictability, consistency, and transparency of the PMA approval process.

The FDA provides a worksheet, within the guidance, to show how it to perform risk–benefit decisions and provides examples to illustrate this process.

When reviewing PMAs for medical devices, FDA uses safety and effectiveness data that address the potential risks and benefits of the devices, as well as the severity of any potential harm, and of the diseases or medical conditions that the products intend to diagnose. The data submitted in PMA approval request may consist of a combination of clinical trial data, analytical and preanalytical data, demographic data, chemistry and manufacturing control data and is reviewed by the appropriate division of the FDA or submitted as part of a technical dossier for the EU but will be a comprehensive package of material that provides significant overview of the design, development, and outcome of the studies required to bring the product to market.

All of this data is the foundation for a regulatory body to determine and assess the accuracy, safety and effectiveness of medical product requiring FDA premarket approval or EMEA/EU approval. As medical devices become increasingly complex, many factors may influence their benefit–risk determinations, especially for PMA devices. The guidances aim is to provide clear instructions to manufacturers to determine which factors they should consider when making and preparing an approval decision.

The FDA and EU, when assessing potential benefits and measuring the effectiveness of devices, looks at the types of benefits the devices offer, as well as the magnitude, probability, and the duration of the benefits. The FDA and EU, when assessing the possible risks or danger of these devices pose, considers the number, severity, and types of device-related serious adverse events, as well as the mild adverse events and procedure-related or indirect harms. The FDA and EU also weighs the probability of a dangerous events by determining the percent of the population that may face this danger, as well as the probability that a diagnostic test generate false-positive or false-negative results, and probability of subsequent consequences if incorrect diagnoses is provided, such as resulting in unnecessary procedures.

The FDA guidance includes recommendations covering the types of information that should be submitted in applications for review and approval that involves a judgment about whether a biomarker is accurate, repeatable and reliable to reflect a targeted biological process, response, or event. The guidance is aimed to address genomic, proteomic, and imaging-based biomarker categories. The FDA website at http://www.fda.gov provides numerous guidelines that can assist in planning the approval decision path forward.

According to the FDA guidance, the biomarker-qualification submission documents should include a section containing regional administrative information such as company name, location, contact and summary information covering the proposed context of use, data description, critical appraisal of the methods and data supporting the applications. These summaries could include analytical assay data, as well as nonclinical and clinical data related to the effectiveness of the biomarker to evaluate if it satisfies the intended use and or clinical utility claim.

The FDA guidance also provides recommendations for how to provide quality reports, clinical and nonclinical reports about assay development and validation, and clinical pharmacology and efficacy reports.

Biomarkers are typically classified in three categories:

- known valid: biomarkers accepted by a large part of the scientific community to predict accurately the clinical outcome;
- probably valid: where the biomarkers appeared to have a predictive value but not yet replicated or widely accepted;
- and exploratory: where the biomarkers have purely empirical value yet lay the groundwork for probable or known biomarkers.

2.3   Product Classification

The regulatory classification of a product must also be determined. The definition and the categories of classification are the following:

- Research Use Only (RUO);
- Investigational Use Only (IUO);
- Laboratory Developed Test (LDT);

- *In Vitro* Diagnostic (IVD);
- Companion Diagnostic (PMA);
- IVD Kit (IVD).

2.3.1   Research Use Only (RUO)

In the development process, it is possible to choose to market a product in a research-use application. The product could be a kit, or a service offering, for a client or organization seeking a research tool for multiple applications including but not limited to, identification, stratification or selection of patients or biomarkers based on chosen technology. While RUO products do not fall within the jurisdiction of the FDA, tools that support the test such as instruments, reagents, and arrays, do. In order to commercially distribute any product, basic standards of good quality control and documentation for product development must be in place. Following the General Quality System requirements and documentation this is an essential good business practice.

2.3.2   Investigational Use Products (IUO)

Investigational use products are generally used in collaboration with a pharmaceutical company in conjunction with the clinical trial. In collaboration with a pharmaceutical company specifically for a clinical trial application the product may be labeled as an IUO product. IUO products can be commercially available products, RUO products or products in development. IUO labeling is protocol specific as well as "sponsor" or pharmaceutical partner specific, and careful tracking mechanisms need to be in place to ensure accountability of all inventory associated with the clinical trial. Sometimes, as part of the development process, products are used in clinical trials to support developing the intended use of the product and, therefore, must be labeled appropriately. If products are used in a clinical trial application they must be labeled investigational use only or IUO. IUO is not a commercial label but rather a label that is used to support clinical trial products.

2.3.3   Laboratory Developed Tests (LDT)

"Laboratory-developed tests" or "LDTs" are used solely within the laboratory that developed the test(s) and are not distributed or sold to any other labs or health-care facilities. As long as LDTs are not marketed outside of the state in which the laboratory is located, they do not require approval for marketing from the FDA (as do commercially developed and marketed tests). If a company chooses to market an assay across the country, an FDA review and agreement would be required, acknowledging the test is in fact a laboratory developed test. Regardless, LDTs still must go through rigorous validation procedures and meet several criteria *before results* are used for decisions regarding patient care. Several governmental and nongovernmental entities regulate and guide the development and validation of LDTs. Nongovernmental entities such as the Centers for Medicare and Medicaid oversee the marketing of LDTs. This agency is also responsible for granting common procedural technical codes (CPT) for reimbursement of these tests provided by licensed laboratories. The FDA is attempting to regulate LDTs and has introduced several new guidance documents in attempt to regulate these products. These guidances, however, have been focused upon the manufacturers of reagents and instrumentation that are used in the industry. These guidances have set restrictions upon who these manufacturers can actually sell products to. If the product is not truly an LDT but rather a diagnostic product, then the manufacturer cannot sell to users RUO-labeled reagents or instrumentation that would be used in a commercial application.[19] The regulation of LDTs currently is not established.

In the US, the federal government, through the Centers for Medicare and Medicaid Services (CMS) and the Clinical Laboratory Improvement Amendments (CLIA) highly regulate development, evaluation, and use of lab-developed assays. CLIA states that laboratories must demonstrate how well an LDT test performs using certain performance standards. Additional information can be found at: http://www.cms.gov.

Examples of such criteria include:

Accuracy—the ability of a test to accurately measure the "true" value or measurement criteria of a substance.

Precision—the reproducibility of a test result.

Test sensitivity—the ability of a test to detect a substance especially at relatively low levels.

Test specificity—the test's ability to correctly detect or measure only the substance of interest and exclude other substances.

Although LDTs are not required to be FDA approved for marketing, some of the reagents, controls, and equipment used in these tests may be manufactured, and are FDA approved. Although the FDA is evolving their policy and guideline formation on LDTs, regulation of laboratory-developed tests, review of the 510(k) process, and promotion of "research-use only" devices and LDTs are actively on the agency's agenda. The 510(k) process is reviewed later in this chapter.

Historically, the FDA has practiced "enforcement discretion" over LDTs but is currently under pressure to change. According to Liz Mansfield, Director of Personalized Medicine at the Office of *In-Vitro* Diagnostic (OIVD), the FDA considers LDTs devices, not services, even though the agency has historically practiced enforcement discretion over the majority of such tests, leaving the CMS to regulate LDTs under the CLIA.

The FDA began reviewing the more complex assays run in a CLIA setting several years ago because they have been struggling with the potential regulation of LDTs. Tests that required multiple regulated components (instruments, reagents, test platform) and software or an algorithm to interpret results, were targeted for review and evaluation. These types of tests were referred to as

"*In-Vitro* Diagnostic Multivariate Index Assays" (IVDMIAs). An example of this type of test is the Agendia MammaPrint® gene expression test for breast cancer recurrence and Pathworks Tumor of Unknown origin assay.

The release of the IVDMIA guidance document by the FDA in July 2007 caused great controversy among device makers and assay developers. As a result, the FDA held a public hearing and issued two draft guidances with regard to IVDMIA regulation. In October 2009, the FDA met with all IVD manufacturers to review their opinions. Consequently, due to increasing objections from not only manufacturers but also independent laboratories, against these efforts, the agency has now said it plans to renounce the IVDMIA regulation. Instead, they are now proposing to regulate high-risk LDTs.

Despite the recent changes, it is believed the FDA will adhere to a well-defined path to market clearance. Until further guidance is issued by the FDA, we can only use examples of previously approved products to prepare ourselves for what could be eventual regulation. Using the Agendia Mammoprint at www.agendia.com for example, these tests are defined as tests run in a single lab, with a single serial numbered instrument using (whenever possible) current Good Manufacturing Practice (cGMP) manufactured reagents and 510(k) approved instruments. Good Manufacturing Practices are a set of policies and procedures that are implemented within a manufacturer to establish quality-system requirements such that, all processes and procedures for manufacturing follow consistent guidelines. cGMP manufacturing is directly linked to quality-system regulation and the expectations of the FDA. Similar directives such as ISO 9000 and, ISO 13485 are European directives that help to establish quality within the manufacturing environment. Not all reagent or instrument manufacturers currently offer cGMP manufacturing. In those instances, the FDA has stated that it is the obligation of the commercializing entity to bring these products under compliance through their own Quality System. These requirements will be addressed later in this chapter. In addition, software and/or algorithms that are used for interpretation of results must be validated in accordance with FDA software regulation. Many *off-the-shelf* products *e.g.*, Partek, www.partek.com are not validated systems and pose a similar quality management challenge for companies aiming to commercialize their products.

The FDA will need to provide leverage to companies, who are willing to commercialize under the new directives, knowing that there are still flaws within the abilities of the manufacturers to comply. However, the FDA has been very specific about what their expectations are for companies willing to commercialize these products as well as manufacturers' requirements. As a result, many companies are in the process of upgrading their manufacturing and quality systems to address these needs. A recent directive from the FDA mandates that companies manufacturing reagents labeled RUO only, is sold to companies that are truly doing research. If the manufacturer knows that the reagents will be used for diagnostic purposes for a product labeled as an LDT they are restricted from selling that product to the end-user. This has placed a great deal of responsibility on the manufacturer to comply with GMP or risk losing market share.

2.3.4 *In Vitro* Diagnostics (IVD)

An *in vitro* diagnostic product is defined as any medical device that is a reagent, reagent product, chemical, calibrator, control material, kit, instrument, apparatus, equipment, or system, whether used alone or in combination, and intended by the manufacturer to be used *in vitro* for the examination of specimens. There are multiple paths that can be taken for *In Vitro* Diagnostic (IVD) products, predicated again by their intended use. Most biomarker products developed for commercialization are classified as an *IVD*. These products are generally biomarker assays that require the combination of reagents, instrumentation and some sort of algorithm or software to interpret the results. All *IVD* products are regulated (and cleared, approved or denied) by the FDA. A specific intended use statement must be provided. In addition, an IVD application must include data supporting the achievement of analytical validation and clinical validation (in most cases) in an intended patient population. The clinical validation in many cases may be accumulation of retrospective data versus prospective data. The FDA defines very specific requirements for infrastructure and "system" development, documentation, validation and maintenance and this will be a rigorous path for any company to accomplish. For the sake of this discussion, a "system" is defined as the assay, instrumentation, reagents, software and or algorithms required for interpretation.

2.3.4.1 *Device Class and Regulatory Controls*

The FDA looks at each product and assigns a device classification in accordance with the risk associated with the product. Class I products are the lowest risk and do not impose any potential harm to the patient or have any direct contact with the patient and therefore are only required to be registered with the FDA. Class II products generally either come in contact with the patient or provide a result that can determine how a patient is treated or followed up. Since this class of products provides information to the physician on how to treat the patient this product may require FDA review and approval. Class III products offer the most significant risk to the patient and require the greatest oversight and data to be submitted as part of a premarket approval or PMA. Controls, as they relate to quality assurance, are defined as General Controls and Special Controls. Details of the requirements of each of these categories can be found on the FDA website at http://www.FDA.gov and once again link back to overall compliance with Quality-System Regulation and in the case of manufacturers Good Manufacturing Practices.

- Class I General Controls
  - generally exempt.
- Class II General Controls and Special Controls
  - require premarket notification and or 510(k) clearance.

- Class III General Controls and Premarket Approval
  - require premarket approval;
  - clinical data to support clinical utility.

Most new biomarker products that are currently under development that do not contain a predicate device, would undergo strict review by the FDA and may require a PMA based on the intended use, indications for use in the classification of risk to the patient.

Manufacturing IVD diagnostic products is not a simple process and must also incorporate expertise in design control, manufacturing practices in regulation, and quality system management. We will briefly touch upon these areas in this chapter. Due to the complexity of regulations it is recommended to consult an expert in-house or by external consultants.

2.3.5 IVD Kit

IVD kits are classified as products that combine reagents, controls, or other assay required products in combination with technology, process, and or product such as microarrays to run a diagnostic assay in a laboratory. IVD kits are generally commercially distributed and labeled with instructions for use and also proficiency samples for quality assurance. Products for diagnostic use should use the regent's suppliers with cGMP qualification. Recent regulation of FDA has put the burden on manufacturers of reagents that were traditionally labeled as RUO and restricted their sales to laboratories that were traditionally running assays as laboratory developed tests (LDT). If a manufacturer knows that the laboratory is using their products in a diagnostic capacity they are restricted from selling them the product.

Identifying manufacturers and suppliers who meet the qualifications of Good Manufacturing Practices is quickly becoming a requirement instead of an option. With an IVD kit, careful consideration must be taken in choosing the companies who provide products that are used to manufacture the product. Proper labeling, documentation, certificates of analysis and compliance are required and compliance with the FDA regulation is recommended. As mentioned above these restrictions are outlined in a new guidance document released by the FDA in 2011.[19] A summary of the main properties of CLIA/LDT and IVD diagnostic assay is provided in Table 2.1.

2.3.6 510(k) Clearance

The 510(k) process is required for most Class II products as well as some Class I products. There are two potential paths for the 510(k) process: one where a pre-existing or predicate device exists and submission is prepared that shows that the product is substantially equivalent to the already 510(k) cleared product. The second way is through what is call a *de novo* 510(k) where no predicate device exists. A *de novo* 510(k) may sometimes be reviewed and placed into a PMA category depending upon the FDA's review and the FDA risk

Table 2.1 Description of CLIA/LDT and IVD diagnostics.

Section	CLIA/LDT	IVD
Scope	Clinical Laboratory patient testing	Clinical Laboratory patient testing IVDMIA or IVD diagnostic kits
Definitions	Patient testing, reporting results for patient management	Patient testing, reporting results for patient management
Inspections	Centers for Medicare and Medicaid Services CMS or equivalent	FDA
Personnel	Licensed staff for testing patient samples, reports reviewed by licensed staff. Licensed staff must have constant and direct observation of unlicensed testing personnel and highly complex testing cannot be performed	Licensed staff for testing patient samples, reports reviewed by licensed staff. Licensed staff must have constant and direct observation of unlicensed testing personnel and highly complex testing cannot be performed
QA	Complex structure including project management, material management, testing, assessments, accessioning, reporting, QA meeting, process improvement, *etc.*	Complex structure including project management, material management, testing, assessments, accessioning, reporting, QA meeting, process improvement, *etc.* Full compliance with Quality System Regulation 21 CFR part 820
Director	Licensed physician understanding testing procedures and interpreting patient results for reporting. Responsible for the QA system implementation. Must be available to correct interpretation of results if a requesting physician needs help with interpretation of results.	Licensed physician understanding testing procedures and interpreting patient results for reporting. Responsible for the QA system implementation. Must be available to correct interpretation of results if a requesting physician needs help with interpretation of results.
Standard operating procedures	Making sure all SOPs are written and followed	Making sure all SOPs are written and followed
Compliance	CLIA inspections every two years	Full compliance with Quality System Regulation 21 CFR part 820 (I)
Maintenance Storage and record retrieval	Records maintained and kept for inspection Three years	Design History File/ Records maintained and kept for inspection Seven Years
Regular inspection of records	All records are reviewed on a regular schedule. All reports and systems are regularly monitored	All records are reviewed on a regular schedule. All reports and systems are regularly monitored
Disqualification	Removal of license must close doors	FDA deficiency citation or blacklist

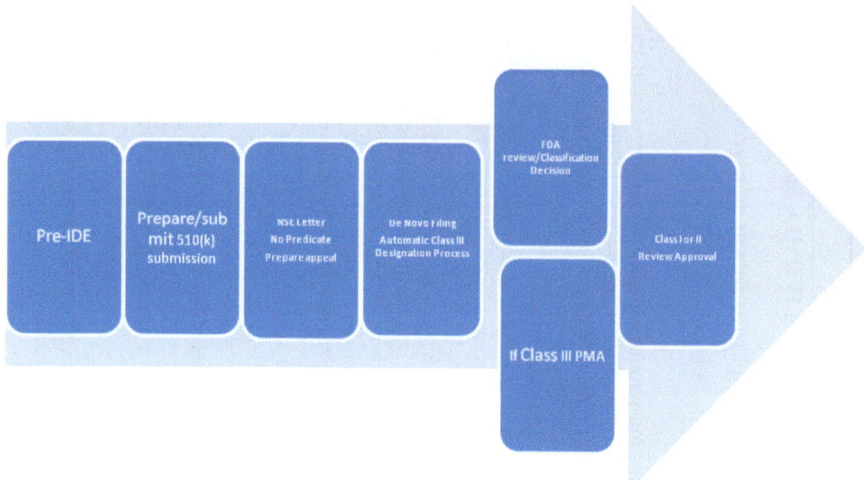

Figure 2.1 Flow chart of main steps required to obtain 510(k) clearance for an assay at FDA.

classification of the device. In other words if submission of a *de novo* 510(k) as a class II product is performed and the FDA believes that the product should be categorized as a class III product then the product would have to go through a PMA review and approval.

While the process may seem difficult and time consuming, it is essential to bring a product to market that has been determined to be regulated by the FDA. Figure 2.1 presents the outline of the steps required to get a product through to the FDA for a 510(k) clearance. Unless the manufacturer is absolutely certain that the manufacturer has a predicate device then it is possible to compare the product to predicate device and demonstrate the substantial equivalence. For this procedure it is recommended to have a pre-IDE meeting with the FDA. This is a free meeting whereby the FDA will review what manufacturer intend to market and to provide the manufacturer with guidance and insight as to whether or not manufacturer is following the correct path. These meetings typically last one hour, therefore it is important that the manufacturer is well prepared and that the manufacturer seeks advice on the appropriate method for conducting these meetings.

2.3.7 PMA

A PMA or premarket approval is required for class III devices. The PMA review process generally takes longer than the 510(k) process because of the level of scrutiny, the amount of detail that must be provided to the FDA, as well as audits that must take place not only for the manufacturer but also for suppliers. Many of the new biomarker products that are being brought to market have to undergo PMA approval since there is no predicate device

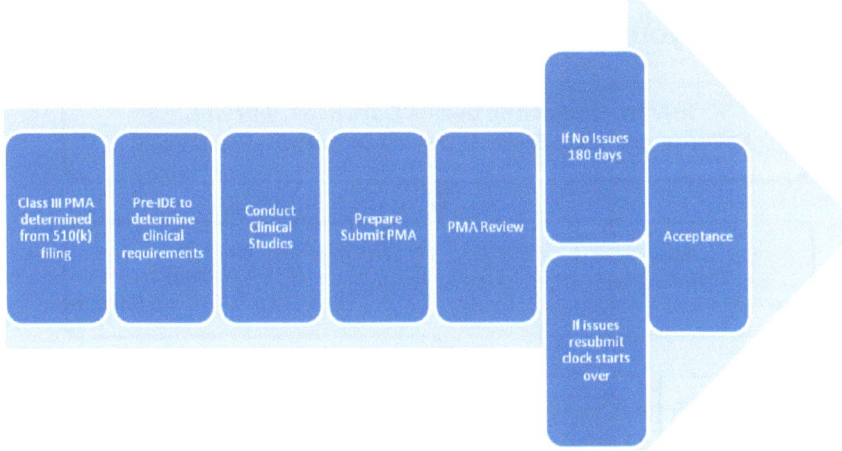

Figure 2.2   Summary and milestones of the PMA process.

available for a standard 510(k) and the fact that many of these are being developed as diagnostic and or prognostic medical devices. Companion diagnostic tests can also be expected to be reviewed as PMAs and must be reviewed in conjunction with the drug that the diagnostic test is indicated for. Figure 2.2 shows the summary and milestones of the PMA process.

2.3.8   CE Marking

Countries in the European Union or EMEA (Europe, Africa and Middle East) require CE marking as part of their review and approval process. CE Marking (CE refers to "Conformité Européenne" meaning "European Conformity") is an alternate or parallel path for consideration to launch a biomarker product in Europe. CE marking is the FDA equivalent for Europe and is required for *in vitro* diagnostics products. CE marking is necessary for commercial distribution of tests in Europe. The European Union does not recognize FDA approvals nor does the FDA recognize CE marking approvals for commercial distribution. The reimbursement structure in Europe is also significantly more complex and that will be discussed in detail later. The path from a documentation and validation perspective is as complex as any other path we have defined; however, the approval process has traditionally been less restrictive in the European Union for this type of assay. The *In-Vitro* Diagnostic Device Directive (IVDD) that relates specifically to CE marking of IVD diagnostic products currently does not have the depth to cover these products and is slated for revision in the near term. Figure 2.3 provides the summary of the current CE Marking review process at EMEA.

Determining the class of product is the starting point for assessing what level of quality assurance and registration is required. The higher the risk of the

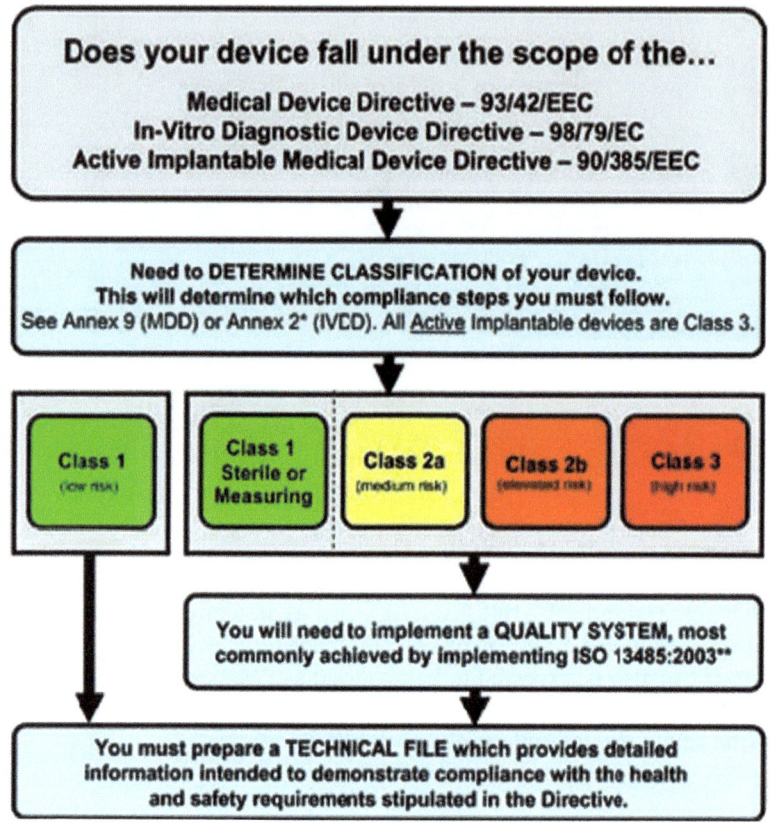

Does your device fall under the scope of the...

Medical Device Directive – 93/42/EEC
In-Vitro Diagnostic Device Directive – 98/79/EC
Active Implantable Medical Device Directive – 90/385/EEC

Need to DETERMINE CLASSIFICATION of your device.
This will determine which compliance steps you must follow.
See Annex 9 (MDD) or Annex 2* (IVDD). All <u>Active</u> Implantable devices are Class 3.

Class 1
(low risk)

Class 1
Sterile or
Measuring

Class 2a
(medium risk)

Class 2b
(elevated risk)

Class 3
(high risk)

You will need to implement a QUALITY SYSTEM, most
commonly achieved by implementing ISO 13485:2003**

You must prepare a TECHNICAL FILE which provides detailed
information intended to demonstrate compliance with the health
and safety requirements stipulated in the Directive.

Figure 2.3   Summary of the current CE Marking review process at EMEA.

product, the more stringent the quality-system requirements should apply. ISO 13485 is essentially equivalent to cGMP in the United States and is required for a company to manufacture a product as a medical device.

Generally, there are only two types of procedures in the CE Marking process flow for biomarker products:

1. "self-certification" by the manufacturer; or
2. obligatory certification by a third-party certification body (called a "Notified Body" or "Competent Body").

European Union regulatory filing is similar to the FDA *filing* requirements and requires companies to create a technical file for conformity testing. The technical file is equivalent to a Device History File for FDA submissions that outline the design, development, testing and validation of the product. Conformity assessment may involve product testing, validation, visual inspection, risk analysis, as well as a review of the product label and instructions. Every CE Marking directive requires the manufacturer or importer to compile and keep available for inspection a Technical File. The purpose of the Technical File is to provide documented evidence of the product's compliance. The Technical File must be kept available for at least 10 years after the date of the last manufacture of the product. The exact documentation to be provided in the Technical File is stipulated in the text of the directives. However, change control and traceability must be provided, from concept to function. As a rule, the Technical File should cover the design, manufacture, and operation of the product.

The Technical File should contain at least all information about the product that allows the authorities to:

- identify the product;
- determine the configuration of the product that was approved;
- determine which standards were applied;
- determine which directives were applied;
- determine how the product meets the requirements of the relevant directives and standards;
- identify the manufacturer and manufacturing process;
- determine how the manufacturer ensures consistent quality of production and continuous product compliance.

Most CE Marking directives require a full copy of the Technical File to be kept in one of the member or associate states of European Union. Each country outside of the European Union has its own regulatory requirements and classification structure but generally requires a Technical File submission and authorized representative. A good solution is to appoint an "Authorized Representative" that keeps the Technical File. An Authorized Representative is strictly a regulatory representative, and is not involved in the design, manufacture, or sale of the products. The "manufacturer" would be required to state the fulfillment of the requirements of the relevant directives and standards. The declaration of conformity contains the references to the directives and standards that have been applied, a product specification and the contact details of the manufacturer. The Declaration of Conformity must be signed by a person authorized to legally bind the company. The Declaration of Conformity is kept available in the Technical File.

Affixing the CE Marking to a product may only be done by the manufacturer or his authorized representative. The CE mark confirms that the product meets the applicable CE requirements. The CE Marking must be affixed to the product visibly, legibly and indelibly on the product itself and its packaging and labeling.

Examples of CE marked biomarker-based tests in Europe include Aptima HPV for cervical cancer, Ipsogen cancer profiler, AdnaGen Circulating Tumor profile and others. Once the CE mark has been approved the following label is applied to the product: ce. CE mark approval in most cases requires the submission of a technical file (similar to a device history file) and review by a notified body before being submitted to the agency for approval. With the CE marking on a product the manufacturer ensures that the product conforms to the essential requirements of the applicable European Community directives. Once again, if a company does not have this level of expertise, consultants are available for assistance. Additionally, registration of the company may also be required under ISO 9000, ISO 13485 or other regulation depending on the product.

As part of the development process it should be to determined what will be the market of the product. If the product is intended to market in European Union then CE marking, ISO 13485 and or ISO 9000 are the requirements for commercial distribution. ISO 13485 is an International Standards Organization (ISO) standard that defines the requirements for a comprehensive management system for the design and manufacture of medical devices. ISO9000 is a family of standards related to quality-management systems and is designed to help organizations to ensure that they meet the needs of customers and other stakeholders also developed by ISO. If the product is intended to market in the US then FDA review and or approval may be required in the regulations defined by the United States Government. Consideration for marketing in other countries including Canada, Australia, Japan, China, and non-European Union entities all have their own specific regulations that must be adhered to for commercial distribution of products in their countries. Country-specific requirements beyond the US or European Union requirement are not addressed in this chapter.

2.3.9 Companion Diagnostics

The US FDA recently approved Roche's personalized melanoma drug Zelboraf alongside a companion genetic test, highlighting the process as a model for future drug/diagnostic (Rx/Dx) codevelopment strategies.

An IVD companion diagnostic device is an *in vitro* diagnostic device that provides information that is essential for the safe and effective use of a corresponding therapeutic product. The use of an IVD companion diagnostic device with a particular therapeutic product is stipulated in the instructions for use in the labeling of both the diagnostic device and the corresponding therapeutic product, as well as the labeling of any generic equivalents of the therapeutic product. The development of companion diagnostic products is an important commercialization opportunity and is high on the priority list of pharmaceutical companies and the FDA. Several initiatives from the FDA, including the recently released *In Vitro* Companion Diagnostic Devices Guidance (July 12, 2011) as well as the Drug-Diagnostic Co-Development Concept Paper and the Critical Path Initiative, seek to promote biomarker usage in an attempt to speed the development process and create safer compounds (see www.FDA.gov).

The framework for combination product submissions requires a sustainable business model to account for the differences in timing between clinical development studies and diagnostic device trials. The need to establish biomarkers that can assist in targeting therapies to specific patients within a given population, has proven its necessity and importance with products such as Herceptin (HER2) and Tarceva (EGFR) and recently several others. Herceptin is the drug and HER2 is the biomarker that determines whether or not a patient will respond to the drug. Similarly, Tarceva is the drug and EGFR is the biomarker to determine if the patient will respond to the drug.

The regulation of companion diagnostics, in general, and pharmacogenomic tests in particular, remains the most critical issue in the development of diagnostic business models to support new therapies. The FDA regulates drugs and diagnostics differently.

The regulatory path for companion diagnostics requires a close collaboration between drug and diagnostic developers. Sharing of clinical and analytical data is important for both organizations. Establishing strict compliance with design control, quality systems, and product life-cycle development will be important for all developed products. Developing and understanding these requirements provide and important assistance in marketing efforts of diagnostic test and to collaborate with pharmaceutical partners, as all drugs require strict regulatory oversight. The level of oversight for devices and in particular LDTs does not align itself with the requirements of pharmaceutical developers. Establishing a highly compliant quality system early in the infrastructure of the manufacturer organization will provide added value to potential pharmaceutical partners as well.

The packaging and documentation of this product should be performed in collaboration with the pharmaceutical partner company and in some cases labeling of the IVD product is included as part of the package design and documentation of the drug. These types of products require concise timing for submissions of an New Drug Application (NDA) for the drug and IVD application for the *in vitro* diagnostic, all under the guise of a PMA. The PMA process can take up to a year once submitted. Approval submissions for these products in the European Union fall under the regulation of EMEA in combination with *In Vitro* Diagnostics Division IVDD.

2.4   Supporting Clinical Utility

Clinical Utility defines what medical outcome is expected by using a particular product. It can establish diagnosis, prognosis or provide a test measurement or assessment that can be used in combination with other standard of care assessments. Establishing a strong evidence base for the clinical utility of an assay will be essential to drive adoption and reimbursement – regardless of the regulatory pathway ultimately selected.

If a biomarker falling into a class II or class III (higher risk) category requiring FDA oversight is being developed, it may be required to conduct a clinical trial to support not only the intended use and indications for use but also to support the clinical utility of the product. Clinical trials can be retrospective or prospective in nature depending on indications for use. However, clinical trials need to be statistically powered in order to support an FDA filing and must be conducted under good clinical practice guidelines established by the FDA. Coordinating a clinical plan in conjunction with the product's regulatory plan is essential in the early phase of development to ensure that the developer does not need to go back and repeat steps that are required. Clinical trials, whether retrospective or prospective can be quite costly. Obtaining specimens from a reputable biobank can be difficult, however, companies such as Trans-Hit Biomarkers are working to provide a resource for high-quality biospecimen acquisition.

Just recently the FDA released this guidance, *Design Considerations for Pivotal Clinical Investigations for Medical Devices*, which outlines critical points to consider during the clinical trial planning process and in particular as it relates to products and would be submitted to the FDA for approval or clearance. Details of this guidance can be found at (www.fda.gov).

FDA compliance under good clinical practice guidelines (GCP) must be taken into consideration when planning or conducting clinical trials. If the developer company does not have internal expertise conducting clinical trials there are many consulting and or clinical research organizations available for assistance.

2.5   Infrastructure and Other Considerations for Commercialization

2.5.1   Instrument Platform/Reagent Manufacturer Selection Challenges

Choosing the right platform for commercial development is very important. Typically in a research setting, instruments are manufactured as research use only or RUO instruments. Identifying if the manufacturer or other instrument platforms that have been 510(k) cleared are available and should be taken into consideration in the case of manufactured or developed product is ultimately used as a diagnostic product and requires FDA clearance. If an assay is developed that will only be used in either a research setting or specifically as a LDT then the utilization of RUO instruments is acceptable.

While we have mentioned the FDA having met with instrument and reagent manufacturers and the need to have products manufactured under cGMP, choosing the right platform can significantly impact the commercialization timeline and resource requirements for development. Once again this is an ever-changing environment and the FDA has placed recommendations on instrument manufacturers to file for 510(k) submissions where instruments are used in clinical diagnostics. Reviewing marketing plans with instrument manufacturers prior to the beginning of commercial development can play an important role in saving multiple development efforts. If a class II or class III (higher risk) product is developed that will require FDA review, the user should discusses with instrument manufacturers to get a clear understanding of how they intend to market their product in the future. If instrument manufacturers are not developing an instrument to be submitted or reviewed by the FDA, then filing under a single instrument serial number, single lab, and single-use application could be the only alternative.

The recently issued draft guidance by the FDA entitled; *Commercially Distributed In Vitro Diagnostic Products Labeled for Research Use Only or Investigational Use Only*, (www.fda.gov) places restrictions on reagent manufacturers for distribution of products labeled RUO or IUO to be sold *only* to those companies who are using them for nondiagnostic purposes. Manufacturers of reagents are struggling to work within these guidelines and this should be considered when negotiating or exploring opportunities with reagent manufacturers. In addition, the FDA has stated that whenever possible reagents manufactured under cGMP should be used for *in vitro* Diagnostic Products. Not all manufacturers currently manufacture under cGMP and again careful attention should be made in selecting qualified vendors. Many manufacturers are changing their manufacturing facilities to become cGMP compliant. Companies like Agilent, Illumina and Qiagen have made enormous efforts to improve their manufacturing operations. It is important to note, however, that not all their products are manufactured in the cGMP facilities so care must be take to evaluate the product labeling as well as the manufacturers.

In addition, auditing the selected vendors for commercial development is part of the Product Life Cycle Development Process. It is critical to review and identify the *best* IVD instrument/reagent company partners.

2.5.2   Software

Instrumentation software as well as software that is either off-the-shelf or integrated into a laboratory information management system (LIMS) system has become an important aspect of product development. Software is often required for the interpretation of the results of the diagnostic test. Before the assays can be cleared for sale by the FDA, the software must be validated according to FDA guidelines: General Principles of Software Validation; Final Guidance for Industry (www.fda.gov). This document was issued on January 11, 2002 as well as 21 CFR Part 11 for protection of human subjects. The integration of software has been a major challenge for many companies looking to commercialize their tests, which involve an algorithm and software for interpretation. Much like the development of an IVD diagnostic test, the creation of design history files and documentation must be developed and maintained, to support the development of the software. Many of the systems developed today are "home-grown" systems designed specifically to support the genetic signatures. Having a locked and final design as well as the documentation to support the validation of the design and the ability of the software to interpret the result accurately and repeatedly is critical.

Developing bioinformatics infrastructure for data analysis is essential and ensuring that compliance with FDA 21 CFR part 11 (www.fda.gov) as well as compliance with validation requirements are in place as well as documentation for validation of the software and its ability to interpret the results accurately and repeatedly. Utilization of off-the-shelf packages and onboard software applications can be acceptable with proper documentation.

Some basic considerations for utilizing any software package or developing a custom software solution are:

- Title and Manufacturer of the Software.
- The Version Level, Release Date, and the Patch Number or Upgrade Designation as appropriate.
- A description of software documentation that will be provided to the end user.
- A statement defining why this Software is appropriate for this medical device.
- A statement defining the expected design limitations of the Software.
- The specific computer system configurations that will be used to validate the software are:
  - Hardware specifications: processor (manufacturer, speed, and features), RAM (memory size), hard disk size, other storage, communications, display, *etc.*
  - Software specifications: for example the operating system (Windows 7, Mac OSX or Vista) with specific version levels (*e.g.*, 4.1, 5.0, *etc.*), Internet browsers and any other software required to run/use the target software solution.
- How will the use of the software be controlled by the End User?
  - Aspects of other the software and system component, which should be installed/configured.
  - Steps permitted (or that must be taken) to install and/or configure the software.
  - How often the configuration will need to be changed.
  - Education and training suggested or required for the user of the software.
  - How use of nonspecified software will be prevented (system design, preventive measures, or labeling). For example if the customer adds a software package that has not been tested with the product, what kind of fail-safe mechanism is in place to prevent that software package from modifying the original data.
- Description of the Software's functions
  - What is the software intended to do? The design of the documentation should specify exactly which components of the software are included in the design of the medical device. Components not included in the product should also be defined. This could be something as simple as an Excel sheet but it must be specified to what extent the software is involved in areas such as error control and software messaging in the final software product. Being able to demonstrate that the software performs as intended over and over again and that safety measures are incorporated to ensure consistency is important.

- ○ What are the links/relationship with other software including software outside the medical device (not reviewed as part of this or another application) such as off-the-shelf software?
- ○ The links to outside software should be completely defined for each medical device/module. The design of the documentation should include a complete description of the linkage between the medical device software and any outside software (*e.g.*, networks, websites, *etc.*). For example, if Excel is used as part of a calculation sheet, this software does not comply with the same level of regulation as one that is custom designed for the product, however, it must be verified as part of the validation process to ensure it performs as intended.
- ○ In addition, for validation of software the following considerations must be made: A Verification & Validation Plan, outlining the software validation requirements as defined in user requirements, design input and output requirements and in accordance with any regulatory framework to ensure the software functions as intended and can produce reliable, repeatable results on a consistent basis.
- The software verification and validation plans must identify the exact software (title and version) that is to be used.
- When the software is tested it should be integrated and tested using the specific software packages and computer systems that will be specified for use by the end users.
- If the medical device is allowed to be used with different versions of the software then the medical device should be validated for each software version.
- A current list of known software anomalies (bugs) should be provided.

Software validation requires very specific skill sets and many consultants are available for assistance in this area.

2.6 Quality Systems

The FDA has been explicitly clear in stating that the commercialization of any IVD product requires compliance with 21 CRF part 820 Quality System Regulation (QSR). They have also been explicitly clear that if the selected manufacturers to work with are not currently cGMP compliant or ISO13485 complaint, it is the obligation of the manufacturer or company, to bring those products-in under the Quality System of the assay manufacturer company for acceptance, approval, and validation.

Most academic institutions as well as many large and small companies, do not have complete documentation to support the development of their biomarker assays, software, or other system obligations in accordance with the requirements set forth by the FDA under QSR. Requirements for early-stage development, namely activities performed prior to validation steps, such as good laboratory records should be maintained. However, once the assay developer get to a point where the development phase is close to the final form of the assay, compliance with quality-system regulations will be required. Often, this is the point at which academic institutions license out the technology or transfer the technology to a medical device manufacturer company. It is important to realize that good source documentation on the development of the assay is required to enhance the value of the assay. It is the obligation of the assay manufacturer to integrate the necessary development and infrastructure in-house to support these requirements. The Device History File (DHF) should be developed to support the product development and validation.

However, the assay manufacturer company management should be aware of the requirements for Design Control and have an understanding of the requirements for compliance with the FDA.

2.6.1   What is Involved?

A quality-management system (QMS) can be expressed as the organizational structure, procedures, processes, and resources needed to implement quality management. Figure 2.4. provides an overview of the structure of QMS.

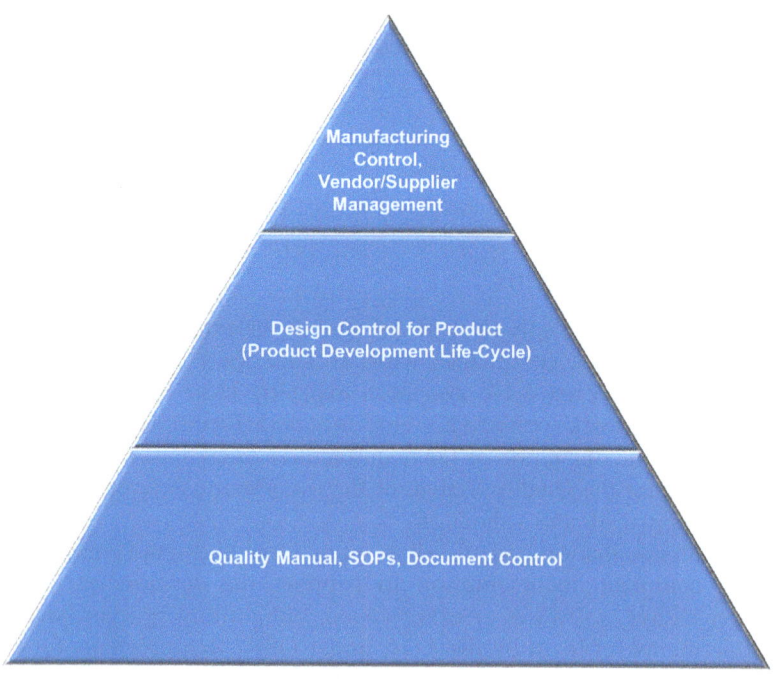

Figure 2.4   Structure of quality-management system.

2.6.2   Elements of a Quality-Management System

1. Organizational Structure;
2. Management Responsibilities;
3. Procedures;
4. Processes;
5. Resources;
6. Customer Satisfaction;
7. Continuous Improvement.

Quality-System requirements for medical devices have been internationally recognized as a way to assure product safety and efficacy and customer satisfaction. It is the responsibility of each manufacturer to establish requirements for each type or family of devices. These requirements will result in devices that are safe and effective, and establish methods and procedures to design, produce, and distribute devices that meet the quality-system requirements. This includes not only the manufacturers of the biomarker assays, reagents, and instrumentation, but also companies developing products as well.

The FDA has identified in the QSR the essential elements that a quality system shall embody for design, production, and distribution, without prescribing specific ways to implement these elements. These elements include:

2.6.3   Quality System General Requirements

1. personnel training and qualification;
2. controlling the product design;
3. controlling documentation;
4. controlling purchasing;
5. product identification and traceability at all stages of production;
6. controlling and defining production and process;
7. defining and controlling inspection, measuring and test equipment;
8. validating processes;
9. product acceptance;
10. controlling nonconforming product;
11. instituting corrective and preventive action when errors occur;
12. labeling and packaging controls;
13. handling, storage, distribution and installation;
14. records;
15. servicing;
16. statistical techniques.

It is the responsibility of the assay manufacturer company's Management and Quality Assurance leadership to oversee all of these elements.

It is the responsibility of the medical device establishment to determine the most applicable parts of CFR 21 Part 820 that relate to the products being offered, based on previous discussions with the FDA. The following list of specific Quality System Requirements of the FDA is used to establish commercialization of IVD diagnostic devices of this nature:

- Design Controls;
- Risk Management;
- Corrective and Preventative Actions;
- Process Controls;
- Good Validation Practices;
- Good Documentation Practices;
- Complaint Handling;
- Medical Device Reporting;
- Product Recalls;
- Product Labeling, Advertising and Promotions;
- Change Control;
- Purchasing Controls/Supplier Management.

This will entail the creation of the following:

- Quality Policy – Beliefs of the company;
- Quality Manual – Overview of the company's intent to maintain control in each identified element;
- Quality Procedures (QP) – May begin with the procedure detailing the program overview;
- Standard Operating Procedures (SOP) – Detail of the requirements to execute the specific tasks;
  - Forms/Templates – Are associated with the (SOP) and control entry of data;
  - Work Instructions (WI) – control tasks by a "step by step" approach; used for precise execution (laboratory, manufacturing, or any task requiring precision).

In addition, consideration for document control software (such as Documentum (www.emc.com), SharePoint(www.microsoft.com), Master Control (www.mastercontrol.com) or other commercially available document control systems) should be taken into consideration early on in the assay developer companies' infrastructure. Implementing these types of programs in the infancy of the company can prove to be beneficial for growth and expansion of the corporate infrastructure.

Documents that will be an integral part of product development and will be required and defined in a detailed Product Development Life-Cycle analysis. Product life-cycle management (or PLCM) is the succession of strategies used by business management as a product goes through its life cycle. The conditions in which a product is developed, manufactured and eventually sold, changes over time and must be managed as it moves through its succession of stages. As part of this management cycle, documentation must be maintained in an auditable form for regulatory authorities to review. From a high-level overview, the following controls are needed for identifying, collecting, indexing, filing, storing, and maintaining quality records to ensure integrity and accessibility:

2.6.3.1 Quality Record

- Provides objective evidence of conformance of products and systems to specifications/process requirements.
- Controlled document – transferred to Document Control when completed
- Retention period defined.

2.3.6.2 Design History File (DHF)

- Compilation of documents that describe the design history of a finished device.
- Design activities used to develop the device, accessories, major components, labelling, packaging and production processes.
- A DHF is required for each type of device or family of devices that are manufactured according to one DMR.

2.6.3.3 Device Master Record (DMR)

- Compilation of documents needed to manufacture a device.
- Product specifications: drawings, raw materials, software, components.
- Production process specifications: equipment, process, methods, environment, labeling.
- Quality Assurance procedures: acceptance criteria.
- Installation, servicing, and maintenance procedures.

2.6.3.4 Device History Record (DHR)

- Compilations of production history records of a finished device (*e.g.* batch record, *etc.*).
- Contains signatures of individuals performing the operations (entries are legible, no erasing).
- Reviewed by Quality Assurance for completeness, adequacy, and release.

2.7 Design Control

Design Control is critical as the backbone of the supporting documentation that will be used in an eventual package for submission to the FDA and EMEA. These deliverables are created in the execution of the Product Development Life Cycle program. All of these documents are critical in defining the product; how it will work, proving it works reliably and repeatedly, how the manufacturer organization will take it to market, *etc.* They are also the foundation for the files required for submission to the FDA or European Authorities for CE Marking.

Below are listed key highlights of Product Life-Cycle Development and Design Control:

Purpose of Design Control
Establish the requirements for product realization from planning to postlaunch (when the product is actually being sold on the market).

Design and Development
Product Life-Cycle Development offers a set of product development guidelines

Design and Development Planning
Defines key elements to provide evidence processes/resulting product meet requirements (*e.g.* quality objective, verification/validation, monitoring, inspection tests, risk analysis, *etc.*)

Design and Development Inputs
Design Input Requirements (DIR) are established and documented to address functional, performance and safety requirements.

Design Outputs
Documentation that enable verification against the DIR.

Design Review
Systematic reviews of the design and development at appropriate stages.

Design Verification
Confirmation that the Design Outputs meet the DIR (*e.g.* test reports).

Design Validation
Assures conformance to DIR given expected variations in components, raw materials, manufacturing processes, environment, *etc.*

Control of Design and Development Design Changes
Changes are documented, evaluated, approved, and validated.

Clinical Evaluation Report
Summary of conducted clinical trials are included as required and becomes part of Trial Master File. Figure 2.5 shows the flow chart of clinical trials approval at the FDA.

2.7.1   Product Development Life-Cycle Process

The Product Development Life-Cycle Process will help to integrate these design components for each phase of commercialization. The phases for commercialization are shown in Figure 2.6.

Within each of these phases of development are documents that are created to identify the design control requirements to take the product from one phase to the next. These documents are what constitute the Device History File and

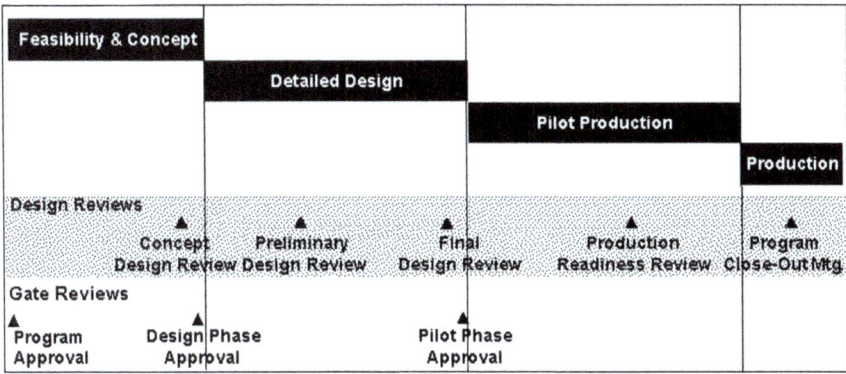

Figure 2.5   Flow chart of clinical trials approval at FDA.

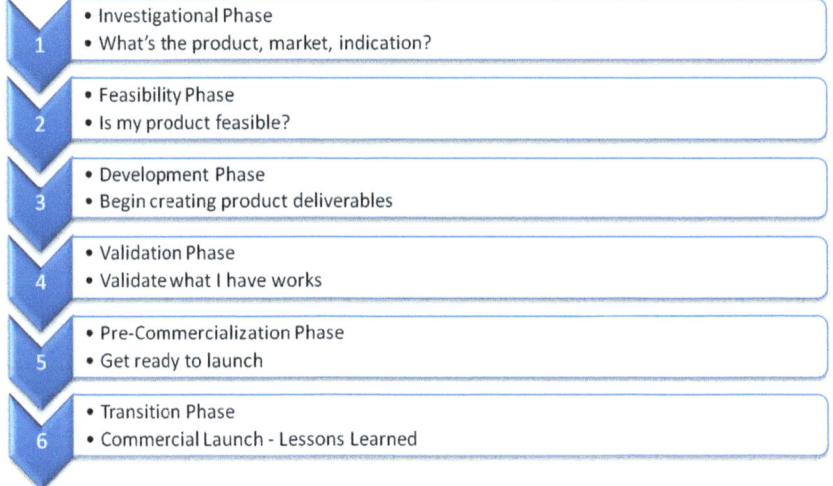

Figure 2.6   Phases of Product Development Life-Cycle Process for commercialization.

are critical for any product potentially being submitted to FDA as well as a general rule of good business practice. Each phase requires significant and important contributions of a team of developer and manufacturer companies. Table 2.2 summarize the most important aspects that should be defined during the investigational phase of assay marketing.

As an example, the second phase, the *Feasibility Phase* of design control, is required as part of the Product Development Life-Cycle Process. Table 2.3 shows the documents that should be created during such process.

Some of these documents are the building blocks for the foundation of Design Control, which is a key requirement for product development. This level of detail is required for *each* phase of development and therefore careful

Table 2.2   Most important aspects that should be defined during the investigational phase of assay marketing.

Marketing	Project Management
Marketing Requirements Document	Product Development Plan/Project Plan
	What is to be developed?
	What is intended market?
	Organizational Responsibilities?
	Major tasks to be undertaken?
	Deliverables of each task?
	Scheduling of each task?
	Product Requirements?
	Signature Confirmation?
	Feasibility of Product Success?
	Output Requirements?
	Define Operations Requirements
R&D	**Regulatory**
Technical Feasibility Plan	Regulatory Plan
Resource Requirements	General Standards
Testing Requirements	Define Regulatory Options
Stability Studies Plan	Intended Use
	Intended Market
	Regulatory Strategy Submissions
	Country-Specific Requirements
	Challenges Concerns Mitigation
	Labeling Requirements
Quality	
Quality Plan	
Quality-System Requirements	
Corporate Compliance	
Manufacturing Requirements	

planning of time and resources is required. These examples only represent some of the phases of development. Complete development plans can be found in a number of resources either in printed format or on the Internet. The basic concept is, however, to create document for each product, describing what the product is doing, how the product should be used, the obtained results to achieve high reliability, repeatability, and accountability for every aspect of the developed product.

2.7.2   Who is Involved?

Product life-cycle development and design control is a team effort and requires individuals form all aspects of the business to participate (Figure 2.7). Identifying these resources and responsible parties within the organization is an important first step in establishing the company. Teamwork is needed to be emphasized to ensure a unified and coherent path to market. The successful "division and ownership of responsibilities" between these team members is

Table 2.3 Most important documents that should be created during Product Development Life-Cycle Process.

MARKETING:	R&D:
Marketing Plan	Technical Feasibility Report
Financial Analysis	Technical Transfer Plan
Update MRD	Design Verification Plan
	Master Validation Plan
	Method Validation Plan
	Process Validation Plan
	IP and or Licenses/ requirements
	Instrument Validation Plan
	Software Validation plan
PROJECT MANAGEMENT:	**REGULATORY:**
Update Project Plan	Update Regulatory Plan
Outline Design Input Requirement (DIR)	Technical Risk Analysis
Begin Design History File	Risk Management Plan
	Regulatory Analysis
QUALITY:	**OPERATIONS/MANUFACTURING:**
Update Quality Plan	Identify Subcontractors
Quality Audit Plan	Manufacturing Plan
Escalation Plan	Operation Validation Plan
Training Plan	

Figure 2.7 Main participants of product life-cycle development.

critical in satisfying the requirements of the FDA approval. The product-development process is complex but one that is required. It is usually led by either the quality department or regulatory department or through coordination with a project manager.

2.7.3 Reimbursement

The regulation and requirements to get a product reimbursed once on the market is a long and arduous path. The Centers for Medicare and Medicaid Services (CMS) in the USA are responsible for the assignment of Common Procedural Technical codes (CPT). This process can take years since CMS only reviews new products twice a year. Many assays that are developed using similar technology such as FISH, ELISA, methods for DNA/RNA extraction, *etc.* have CPT codes that currently exist and that can be utilized for reimbursement for new molecular assays. Similar to FDA regulation, regulation over CPT codes is changing. The process is extremely complex and involves not only the assignment of a CPT code but also negotiation with payers for reimbursement. European regulation for reimbursement is as complex as, if not more complex than, the US and reimbursement for molecular assays is difficult to achieve. Hiring a technical consultant with expertise in the area of reimbursement is recommended. Numerous consultants can be found on the Internet.

2.8 Summary

Developing biomarkers for commercialization is a daunting task and one that cannot be taken without appropriate funding, dedication, and infrastructure support to manufacture product accordingly. While the regulatory landscape for commercialization is changing rapidly, gaining a better understanding of the basic concepts and principles is essential. In this chapter, many of these concepts have been outlined that are needed to take into consideration prior to or during developing a biomarker product. Each company and manufacturer have their own strengths and weaknesses, but having a good plan in place and a good understanding of the requirements will help to achieve a successful product development strategy.

Many external resources are available to assist the developer in this process. This can come from web-based material or from consultants. Numerous Web-seminars offered *via* the Internet are a good source for general knowledge and information for relatively low cost. Signing up to participate in LinkedIn groups is a valuable resource for participating in these webinars and most are free. Consultants will require cash resources to support long-term use, and spending the money up front for short-term advice may be a good investment. Companies like Trans-hit Biomarkers Inc., Matrix Clinical Research Management, Covance and others can provide assistance for the identification of biomarker specimens required for retrospective analysis and/or can provide the assay developer with consultative services for regulatory affairs, quality systems and/or product development.

Self-education and educating the product developer management team who build the foundation of the product is essential for successful commercialization. The old adage, "you don't know what you don't know" will not lend itself well in the eyes of the FDA and EMEA, therefore empowering the developer organization to be prepared for full-scale commercialization under all regulatory requirements is essential.

Resources

Applicable Domestic and International Regulations

1. Quality System Regulation, 21 CFR Parts 820.3 www.accessdata. fda.gov/scripts/cdrh/cfdocs/cfCFR/CFRSearch.cfm?fr = 820.3.
2. 21 CFR Part 809.10 Labeling for In Vitro Diagnostic Products www.accessdata.fda.gov/scripts/cdrh/cfdocs/cfCFR/CFRSearch.cfm?fr = 809.10.
3. FDA Guidance "Use of Symbols on Labels and in Labeling of In Vitro Diagnostic Devices Intended for Professional Use" www.fda.gov/ MedicalDevices/DeviceRegulationandGuidance/GuidanceDocuments/ ucm085404.htm.
4. 98/79/EC Directive http://www.hpa.org.uk/ProductsServices/ MicrobiologyPathology/UKStandardsForMicrobiologyInvestigations/ TermsOfUseForSMIs/AccessToUKSMIs/SMIQualityRelatedGuidance/ smiQ03EuropeanDirectiveonInVitroDiagnosticmedical/
5. ISO   13485:2003   http://www.iso.org/iso/catalogue_detail?csnumber = 36786.
6. Council Directive 93/42/EEC http://eur-lex.europa.eu/LexUriServ/ LexUriServ.do?uri = CELEX:31993L0042:en:HTML
7. MEDDEV.2.14/2 rev.1 http://ec.europa.eu/health/medical-devices/files/ meddev/2_14_2_research_only_product_en.pdf.
8. EN ISO 14971:2000 (for Risk analysis) http://www.iso.org/iso/home/store/ catalogue_ics/catalogue_detail_ics.htm?csnumber = 31550.
9. EN 980:2008 (for labels) http://www.techstreet.com/products/1589298.
10. EN 375:2001 (for pack insert) http://www.techstreet.com/products/ 1300765.
11. FDA guidance: www.FDA.gov.
12. Guidance for Industry, FDA Reviewers and Compliance on Off-The-Shelf Software Use in Medical Devices Sept 9, 1999. www.fda.gov/ downloads/MedicalDevices/DeviceRegulationandGuidance/Guidance Documents/UCM073779.pdf.
13. General Principles of Software Validation; Final Guidance for Industry and   FDA   Staff   www.fda.gov/MedicalDevices/DeviceRegulationand Guidance/GuidanceDocuments/ucm085281.htm.
14. Guidance for the Content of Premarket Submissions for Software Contained in Medical Devices www.fda.gov/downloads/MedicalDevices/ DeviceRegulationandGuidance/GuidanceDocuments/UCM089593.pdf.
15. Drug-Diagnostic Co-Development Concept Paper http://www.fda.gov/ downloads/drugs/scienceresearch/researchareas/pharmacogenetics/ ucm116689.pdf.

16. 21 CFR Part 11 Electronic Records; Electronic Signatures — Scope and Application http://www.fda.gov/RegulatoryInformation/Guidances/ucm125067.htm.
17. Commercially Distributed In Vitro Diagnostic Products Labeled for Research Use Only or Investigational Use Only http://www.fda.gov/downloads/MedicalDevices/DeviceRegulationandGuidance/GuidanceDocuments/UCM257460.pdf.
18. Design Considerations for Pivotal Clinical Investigations for Medical Devices www.fda.gov/MedicalDevices/DeviceRegulationandGuidance/GuidanceDocuments/ucm265553.htm.
19. Factors to Consider When Making Benefit Risk Determinations in Medical Device Premarket Approvals and De Novo Classifications http://www.fda.gov/MedicalDevices/DeviceRegulationandGuidance/GuidanceDocuments/ucm267829.htm.
20. Biomarkers Related to Drug or Biotechnology Product Development: Context, Structure, and Format of Qualification Submissions http://www.fda.gov/downloads/drugs/guidancecomplianceregulatoryinformation/guidances/ucm267449.pdf.

List of Important Internet Resources

www.trans-hit.com
www.matrixclinicalresearchmanagement.com
www.covnace.com
www.agendia.com
www.partek.com
www.fda.gov
www.emc.com
www.mastercontrol.com
www.microsoft.com
www.cms.gov

Documents for EMEA Submission

Each country has its own specific requirements for filing regulatory submissions. Countries that require submission packages and have specific rules for filing are as follows: Argentina, Australia, Belarus, Bolivia, Bosnia, Brazil, Canada, Chile, China, Colombia, Croatia, Ecuador, Egypt, Europe, Honduras, Hong Kong, India, Indonesia, Israel, Japan, Malaysia, Mexico, Peru, Russia, Saudi Arabia, Singapore, South Korea, Taiwan, USA, and Venezuela.

While it is impossible to cover each countries requirements in this book, the following list of resources may be of value and identifying and providing guidance for creating the right regulatory package for submissions at EMEA.

MDD 93/42/EEC
- CONSOLIDATED Medical Device Directive (93/42/EEC) (Interactive Web page. Downloadable .pdf document of the MDD 93/42/EEC also available.)

In-Vitro Diagnostic Devices Directive (98/79/EC)
- Directive 98/79/EC on In Vitro Diagnostic Medical Devices (Interactive Web page. Downloadable .pdf document of the In-Vitro Diagnostic Devices Directive (98/79/EC) also available.)

Vigilance Contact Points
- Vigilance Contact Points Within The National Competent Authorities (opens in a new window - link to EU Commission Website)

Major Regulations
- CONSOLIDATED Active Implantable Medical Devices Directive (90/385/EEC)
- CONSOLIDATED Cosmetic Products Directive (76/768/EEC)
- CONSOLIDATED Medical Device Directive (93/42/EEC) | View as web page
- In-Vitro Diagnostic Devices Directive (98/79/EC) | View as web page
- Personal Protective Equipment Directive (89/686/EEC)

General Guidance Documents
- MEDDEV 2.1/1 - Defines Medical Devices, Accessories and Manufacturer
- MEDDEV 2.1/2 REV 2 - Application of the Active Implantable Device Directive
- MEDDEV 2.1/2.1 - Computers used to program implantable pulse generators
- MEDDEV 2.1/3 REV 2 - Demarcation between MDD and Medicinal Products Directive
- MEDDEV 2.1/4 - Discusses demarcation between the EMC and PPE Directives
- MEDDEV 2.1/5 - Addresses medical devices with a measuring function
- MEDDEV 2.1/6 - Qualification and Classification of Stand Alone Software
- MEDDEV 2.2/1 REV 1 - Addresses EMC requirements
- MEDDEV 2.2/3 REV 3 - Discusses the "Use By" date
- MEDDEV 2.2/4 - Conformity Assessment of IVF and ART Products
- MEDDEV 2.4/1 REV 9 - Classification of medical devices
- MEDDEV 2.5/3 REV 2 - When a quality-related inspection of a subcontractor is needed
- MEDDEV 2.5/5 REV 3 - Clarifies translation procedures
- MEDDEV 2.5/6 REV 1 - Defines homogeneity of production batches
- MEDDEV 2.5-7 REV 1 - Discusses conformity assessment for breast implants
- MEDDEV 2.5/9 REV 1 - Medical devices containing natural rubber latex

- MEDDEV 2.5/10 - Guideline For Authorised Representatives
- MEDDEV 2.7.1 REV 3 - Clinical Evaluation: A Guide For Manufacturers And Notified Bodies
- MEDDEV 2.7/4 - Guidelines on Clinical Investigation: A guide for manufacturers and notified bodies
- MEDDEV 2.10-2 REV 1 - Designation and monitoring of Notified Bodies
- MEDDEV 2.11/1 REV 2 - Risk management in animal tissue products - TSE
- MEDDEV 2.12-1 REV 8 - Guidelines on a medical devices vigilance system
- MEDDEV 2.12-2 rev 2 - Post market clinical follow up for medical devices
- MEDDEV 2.14/1 REV 2 - Borderline issues between the IVD and Medical Device Directives
- MEDDEV 2.14/2 REV 1 - Dealing with IVD products for research use only
- MEDDEV 2.14/3 REV 1 - Requirements for e-labeling of IVDs
- MEDDEV 2.14/4 - CE marking of blood based IVD medical devices for vCJD
- MEDDEV 2.15/2 REV 3 - Committees and organizations related to medical devices
- COMMISSION REGULATION (EU) No 207/2012 - E-labeling regulations
- GUIDANCE Document on Interpreting Directive 2007/47/EC
- Manual on Borderline and Classification in the Community Regulatory Framework for Medical Devices - VER 1.13
- EU - GHTF Guidelines Regulatory Auditing of QMS Parts 4
- EU - GHTF Guidelines Regulatory Auditing of QMS Parts 5
- NBOG - Guidance for Notified Bodies Auditing Suppliers 2010-1
- NBOG - Guidance on Audit Report Content 2010-2
- NBOG - Certificates Issued by Notified Bodies Reference to Council Directives 2010-3

Forms and Applications
- NBOG - Checklist for audit or Notified Body review of clinical data/evaluation

CHAPTER 3

Introduction: The Cardinal Role of Biobanks and Human Biospecimen Collections in Biomarker Validation: Issues Impeding Impact of Biomarker Research Outcomes

PASCAL PUCHOIS,*[a] LISA B MIRANDA*[b] AND ALAIN VAN GOOL*[c,d,e]

[a] Trans-Hit Biomarkers Inc., Montreal, Canada; [b] Biobusiness Consulting Inc., Greater Boston Area, USA; [c] TNO Quality of Health, Metabolic Health Research, Leiden, the Netherlands; [d] Laboratory of Genetic Endocrine and Metabolic Diseases, Department of Laboratory Medicine, Radboud University Nijmegen Medical Center, The Netherlands; [e] Faculty of Physics, Mathematics and Informatics, Radboud University Nijmegen, The Netherlands
*Email: pascal.puchois@trans-hit.com;
lisabmiranda@biobusinessconsulting.com; alain.vangool@tno.nl

3.1 Introduction

Biobanking is a burgeoning market sector. Hundreds of biobanks exist around the world, with many others in prospective creation regularly. It is estimated

RSC Drug Discovery Series No. 33
Comprehensive Biomarker Discovery and Validation for Clinical Application
Edited by Péter Horvatovich and Rainer Bischoff
© The Royal Society of Chemistry 2013
Published by the Royal Society of Chemistry, www.rsc.org

that more than 1500 biobanks are currently in operation globally; the greatest prominence of activity aggregated across North America and Europe. The majority of biobanks appear to be cancer oriented, comprising as high as 70% of active biobanks. Other biobanks include a research focus that is disease based, *i.e.* CNS, cardiovascular, metabolic disorders, infectious, immune or rare diseases based. Biobanks collect, manage and store human or animal biological specimens such as blood or tissue from diseased and healthy participants. Historically, collective biological samples or biosamples have proven invaluable in establishing the natural evolution of disease and its transmission. In addition to increasing understanding of disease in general, biospecimens offer a precious resource to support advancement of translational research related to biomarker development and validation.

In the last decade, biospecimen research has also contributed to development and validation of novel diagnostic tests. Diagnostic tests can be utilized to diagnose and monitor disease progression as well as predict drug efficacy and risk for adverse events. Such tests can also facilitate stratification of patients based on response to therapy, thereby optimizing clinical treatment options and patient outcomes alike. One case study is the human epidermal growth factor receptor-2 (HER2) gene. Predictive testing for HER2 protein led to development of trastuzumab drug. Predictive tests for other genetic biomarkers (*e.g.*, BCR/ABL tyrosine kinase, EGFR/KRAS) have been similarly code-veloped for other targeted anticancer therapies. Utilization of biospecimen collections has also bolstered development of prognostic tests that stratify patients by disease outcome to guide clinical decision making. Recent examples include Oncotype DX 21-gene breast cancer and MammaPrint assays, both of which predict the risk of disease recurrence in certain women with early-stage, node-negative breast cancer.[1] Based on these and other successful proof-of-concepts, one may surmise that biospecimen research if harnessed efficaciously may considerably contribute to the discovery and validation of novel biomarkers and related clinical therapies.

To date, utilization of biomarker-disease based and molecular-medicine-related research potential has been at best somewhat diminutive. Only a minority of candidate biomarkers progress beyond initial discovery. The few biomarkers that do proceed beyond the initial phase of development tend to have a high degree of unreliability. To further complicate matters, the ability to progress candidate biomarkers into large-scale validation studies poses further challenges. Quality and relevance of biospecimen and data collections remain key limiting factors for biomarker research in general. In the present chapter, we commence with an overview of the biobanking–biomarker collaborative landscape in an effort to elucidate, fundamental planning considerations for optimizing biomarker research and validation. We will next present prevailing challenges in biospecimen cultivation and utilization that compound biomarker research outcomes, and then follow with recommendations as to how these challenges may be ameliorated. In tandem, we offer guidance as to how biospecimen and biomedical research practices may be augmented to advance biomarker discovery and amplify development outcomes towards maximal impact, emphasizing that cardinal collaborations are imperative to achieve such impact.

3.2 Navigating the Biobanking–Biomarker Collaborative Landscape

3.2.1 Crucial Considerations in Planning Biobank–Biomarker Research Collaborations

Historically, biomarker research practices and biobanking collection priorities have been developed prior to systematic surveying of the therapeutic and/or drug discovery market landscape. The majority of specimens collected have been interdependent and largely driven by internal stakeholder research biospecimen supply and demand. The past decade's surge in biobanking formalization sheds light on acute issues that require early incorporation into technical and scientific planning processes if the short- and long-term impact of biomarker research outcomes is to improve. Biomarker scientists may not be cognizant of the requisite due diligence affiliated requirements of the collaborative value chain, that is how to strategically plan for and successfully navigate biomarker–biobanking research pathways. Extramural and interoperational research activities tend to be frequently impeded by limited transboundary education, feasibility review and prospectively proactive logistical planning. To ensure "fit for purpose" and impactful research and clinical outcomes alike, it is especially important that biomarker scientists acquire education on quality issues biobanks face in operations. Scrutiny is not restricted to biospecimen quality review; it permeates the continuous research pathway. It is crucial that biomarker scientists optimize approach and evaluation of biobanks during preinvestigative feasibility planning. It is a false assumption that biobanks and their respective collections are equivalent to each other. Biobanking collection practices differ dramatically; the relevance and quality of collections vary in tandem. Historically, the majority of biospecimen and data collections are not prospectively designed with biomarker research requirements in mind. Clarity of understanding is a prerequisite to more aptly understand how best to define and cultivate relevant biomarker-ready biospecimen and data collections.

Biobanks range in infrastructure, operational models and extent of expertise. Biobank infrastructures permutate from informal collections in a researcher's freezer to dedicated facilities.[2] Historically informal collections were referred to as "tissue banks" or "biobanks" when inventory included biofluids and other source materials, in additional to tissue specimens. The term "biobank" is often the vernacular. The following terms are also relatively synonymous but tend to encompass a significantly higher degree of formal infrastructure: biospecimen resources, biorepositories, tissue banks, core bioresource facility, sample management laboratory, biolibraries and biological resource centers. Some bioresources identify with the term biological resource centers (BRCs). BRCs tend to be broader in scope of operations and services, and are designed as an applied research complex that processes source biological materials into cells and cellular components, and function in manufacturing environment conditions.

Trans-Hit Biomarkers (internal database) has identified over 1500 biobanks worldwide; it is unknown how many human biospecimen banks exist in total. It is known that biobanks are prominent in both nonprofit and commercial sectors across the market. In the nonprofit sector, biobanks function in government research laboratories, nonprofit research institutes, academia, and community-based hospitals, cooperative entities (United States based examples *e.g.* Eastern Cooperative Oncology Group (ECOG), Southwest Oncology Group (SWOG), and Gyneacologic Oncology Group (GOG)) and nonprofit charitable/community-based organizations (US-based examples *e.g.* Accelerated Cure Project for Multiple Sclerosis, National Cancer Society (NCS), Juvenile Diabetes Research Foundation (JDRF)). In the profit sector biobanks operate as services providers, as commercial laboratories and sometimes as Clinical Research Organizations (CROs), but may also function as pharmaceutical and biotech-based sample management facilities. As the precompetitive domain amplifies, market expansion of biobanks appears to be increasing. The resulting variance in market activity has a causative effect on further evolving biobanking model form and function.

The most widely known models in human biobanking are biosample-based bench collections, clinical trial repositories and pathology core resources. Recent efforts have focused on centralization of biobanking resources in concert with development and harmonization of networks. These activities as well as network to network linkages will continue to burgeon; it is not possible to discern if such efforts will be sustainable long term without timely integration of formal business models and cost recovery. Aggregation of sample and data collections is also an increasing trend. The shift towards aggregation appears largely due to the urgency to increase sample collection size aimed at ameliorating statistical significance of downstream research results. As a result, some scientists have begun to search for retrospective compatible specimens for their research efforts from other sources, both internal and external. Others have begun to amalagamate prospective banking of particular samples to acquire a statistically significant relevant sample size. Biobanking research objectives also differ. While some biobanks focus solely on collecting specimens for a key research purpose, other biobanks may collect for a wide range of research needs. Disease-specific banks are becoming more popular; it is speculated that disease-specific banks may have a greater propensity of producing quality collections due to their concentrated focus. Moreover, population-based banking which ranges in prominence geographically, appears to be a renewed consideration for those with access to disease-specific registries. Population-based banking entails procuring samples from both diseased and healthy individuals in communities towards alleviating a specific clinical outcome. Population-based banking has occurred for decades in the context of public and private epidemiologic research but has rarely been harmonized with central biobanking efforts. Historic collection of specimens has also occurred in vast quantities *via* drug clinical trials in which many biosample and data collections have been collected and stored for private research efforts. All of these collections may hold potential value for biomarker research providing evidence-based quality parameters can be defined and achieved.

There is often confusion in the nomenclature in regards to what denotes a "commercial" biobank *vs.* a noncommercial biobank and how operationally these biobanks may contrast. The domain specific term refers to what are in reality, commercial tissue procurement organizations. Commercial tissue procurement organizations appear to be a growing trend with at least fifty commercial tissue procurement organizational biobanks operating in the for-profit sector alone. While these resources may bear physical resemblance to academic bioresources in general, what differs is that these organizations typically procure specimens and data secondhand from primary sources, *e.g.* academic medical centers and community-based hospital partners. These partners tend to be located in a variety of geographic areas ranging from North America, Western and Eastern Europe, as well as Asia and South America. It is meaningful to understand the downstream issues resulting from access and availability of primary versus secondary source biological material and data. Issues tend to pertain to quality, *i.e.* the extent and authentication of annotation. It is difficult to conclude which type of bioresource offers the highest quality of biospecimens and data collections. However, it does appear that commercial vendors tend to be more aware of their inventory and have shorter lead times related to feasibility studies and to provision of retrospective samples. The type of biospecimens available may also differ, *e.g.* primary materials (blood, urine, tissue) or bioproduced materials, *e.g.* derivatives (extracted RNA, DNA, tissue array). One may therefore presume that biospecimen and data collections are likely to vary in quality and content based on the source origin and type of organization collecting them.

Harmonization of standardized sample management methodologies and research practice is thought to be a logical solution to defining and improving quality of biosample collections and related biobanking activities. However, the wide geographic span and vast diversity of biobanking activities creates challenges to expeditious standardization. In addition to disparities in infrastructure, research expertise and type of collections, biobanks differ in terms of types of sample management protocols, bioprocessing techniques, and biobanking workflows. Variation also exists in volume of biosample throughput, along with access to and utilization of manual and automated and benchtop sample management technology and instrumentation. There are some who intimate that greater implementation of automated technologies may alleviate prevailing issues related to standardization that impact quality. Degree of cohesion of oversight may also vary, which affects the extent of quality biobanking and sample management practices. Tissue banks are typically governed by internal parties, *i.e.* coordinating Directors, who provide primary oversight aided by key adjunct advisors and/or principal investigators, Secondary oversight is typically offered by founding stakeholders, *i.e.* department heads, and administrators. Core facilities tend to have scientific advisory committees composed of interdisciplinary parties both internal and external to the direct bioresource. Oversight may also be offered *via* bio-specimen "utilization" committees. A biospecimen utilization committee's degree of contribution, expertise and effort provided tends to vary from informal gestalt and "weigh in" on organizational planning and operations to formal review of upcoming projects. Current oversight practices, *i.e.* scientific and technical advisory tend to range in frequency (1–4 times a year) and mode (*e.g.* face to face meetings, conference calls, and/or emails). Further information on bioresource oversight may be referenced in Section B.1.2.2 of the 2011 NIH/NCI BBRB best practices focusing on guidance related to bioresource operation and management.[2]

As briefly referred to above, biobanking experts have contributed to the development of experiental best-practice-based technical and operational guidelines with the aim of improving quality and harmonizing standardization of specimen and data collections. Currently, there are four established primary best-practice reference documents:

- "National Institutes of Health/National Cancer Institute's Biorepositories and Biospecimen Research Branch (BBRB) Best Practices for Bio-specimen Resources."[2]
- "International Society of Biological and Environmental Repositories (ISBER) Best Practices for Repositories: Collection, Retrieval, and Distribution of Biological Materials for Research."[3]
- "Organization for Economic Co-operation and Development (OECD). Best Practice Guidelines for Biological Resource Centers".[4]
- "World Health Organization International Agency for Research on Cancer (IARC). Common Minimum Standards and Protocols for Biological Resource Centers Dedicated to Cancer Research").[5]

Scientific working groups have published reporting guidelines that dispense advice as how to report research methods and findings. While originally developed to increase the quality of data reported and/or presented in publi-cation, guidelines *e.g.* REMARK (prognostic tumor markers)[6] (STARD (diagnostic accuracy),[7] STROBE (observational studies),[8] and BRISQ (bio-specimen reporting)[9] can be effective tools when incorporated prior to execution of research, particularly in enhancing prospective biomarker research protocols and biobank collection procedure design. Further information including a comprehensive list of reporting guidelines may be obtained *via* the Equator Network website (http://www.equator-network.org). If used efficaciously best practice and reporting guidelines may then serve as tools to enhance research parameters, define benchmarks and increase statistical relevance. The long-term goal for both best practice and reporting guidelines is to incorporate empirical data and evidence-based lessons learned from biospecimen science research over time.[10] However, prevalence, rate of adoption as well as frequency and manner of utilization of best practice and reporting guidance remains unconfirmed. One could envision that statistics on these variables may function as a predictor of the level of a biospecimen and data collection's inherent quality. However, evidence to this effect is still required to demonstrate this hypothesis. The BBMRI/Gen2Phen Bioresource Research Impact Factor (BRIF) working group is currently collaborating to evaluate the impact of related biobanking supported research outcomes. Over time these efforts could include validating this idea and yield much information as to the actual impact of biospecimen based research, including biomarker-based studies.[11]

As awareness of the inherent scientific and market value of biospecimen and data collections has increased, stakeholders have realized that high-quality, dedicated, formalized bioresources have the potential to serve as hubs for biotechnological research and development. In an effort to aggregate and optimize resources both public (*i.e.* government network models) and private institutions, have evolved towards assembling networks of biobanks. One well-known network is the Centro Nacional de Investigaciones Oncologicas (CNIO)'s National Biobank Network which was created in September 2000 and incorporates over 25 institutions across Spain and Latin America.[12] More recently, networks have engaged in research-driven business-based collaborations in public–private consortia partnerships. There has been an increase in consortia activity cross-sector as well as crosscontinent. Examples include the National Foundation for Cancer Research's Tissue Bank Consortium in Asia (http://www.nfcr.org/index.php/research/tissue-bank-consortium-in-asia), the United States Foundation for the National Institutes of Health (FNIH) Biomarker's Consortium (http://www.biomarkersconsortium.org/), and European networks like the Biobanking and Biomolecular Research Infrastructure (BBMRI) (http://www.bbmri.nl/en-gb/home) and the String of Pearls Institute (http://www.string-of-pearls.org/). Efforts in formalizing networks and assembling consortia in general have been focused on linking bioresources and their collections demographically (country, continent, inter/intracontinental), by disease and/or research focus *via* common operating standards, unified adoption of best practice, quality-control procedures, informatics platforms (databases and specimen locators); such initiatives are still in development. Biobanking societies can also play a role in harmonization biobanking education and awareness. Societies *e.g.* the International Society of Biological and Environmental Repositories (www.isber.org), Public Population Project in Genomics (P3G) Consortia (www.p3g.org), European, Middle Eastern and African Society for Biopreservation and Biobanking (ESBB) (www.esbb.org/) have all played a role in moving the field forward.

3.2.2   Biobanking Challenges that Impede Impact of Biomarker Research Outcomes

Biobankers and biospecimen researchers in tandem with scientists who perform biomarker research have the capability to advance biomarker research and development aimed at improvement of outcomes related to incidence and prevalence of disease, diagnosis, treatment, as well as disease prevention and quality of life specific to both diseased and healthy populations. Currently, these capabilities and any related impact they may offer are impeded by significant common challenges. Some of the key challenges have been recently described.[13–15] These include, but may not be limited to: long-term sustainability of biobanking operations and infrastructure, formal execution of cost recovery, real-time implementation of best practice and quality sample management review and bioresource evaluation, expedition of "fit for purpose" and evidence-based biobanking practice, availability of tools to enhance interoperability of specimen sharing and proactive governance and utilization review.

Trans-Hit Biomarkers Inc performed a survey (internal data) to pinpoint the key issues that scientists in academia and industry face when searching for biospecimen collections.

- Biomarker scientists reported difficulties in locating sources of bio-specimens due to the lack of a dedicated comprehensive global registry, in particular biospecimens for rare disease states.
- During their searching process most scientists noted that biobanks were unable to perform rapid multicriteria search based on associated medical data. They also found that the majority of biobanks they interacted with did not have facile access to medical data. Experience demonstrated a high degree of variance of information technology and bioinformatics tools in practice (electronic sample inventory management systems) which created problems with achieving interoperability between the researcher and the biobank.
- Downstream analysis was further complicated by the fact that operating procedures for key steps in the biospecimen management process were frequently missing *e.g.* time for warm and cold ischemia for tissue.
- Requisite clinical and scientific input in the design and management of collections was inconsistent.
- For a given disease, few control groups were consistently available within the same biobank. Control groups that were available were not appropriate for assessing biomarker specificity (*e.g.* prostate cancer, benign prostatic hyperplasia, prostatitis, and other cancers).
- Additionally, the entire cycle of disease progression was not covered by collection. (For example, it is common to find collections in the cancer field for late-stage patients, but be very difficult to find specimens from cancer patient with early-stage disease.)
- Biomarker researchers noted that available specimen sets were inconsistent on overall search. (For example, some biobanks had tissues, while others had serum, DNA, *etc.*); incidence was rare to find biobanks with inventory that included a wide range of biospecimen types for the same patient.
- In addition to issues with samples, scientists repeatedly had problems with finding the complete associated clinical data sets and medical follow-up data.
- Quality management due diligence was arduous; a review of audit trails for individual biospecimen sets was not always possible.
- The most crippling issue reported was restricted access to industry partners by numerous academic biobanks.

- Lastly, the overall consensus amongst biomarker scientists was that timelines to conduct feasibility studies and the administrative process were highly unsustainable.

Understanding the challenges biobanks face helps clarify the high prevalence of reported impediments. There appears to be four key areas of dysfunction: 1) unsustainable biobanking operational models, 2) difficulties in locating and accessing biospecimen and data collections, 3) inefficient research business practices and resulting inability to engage in timely collaborations and 4) low prevalence of satisfactory, relevant biospecimen and data collections. Further insight on these issues is offered below.

- Issue #1: Unsustainable biobanking operational models

The most prevalent impediment to both quality biospecimen and biomarker research today is the disparity in biobanking business infrastructure and formal business models. Both historically and from one standpoint, biobanks have not and continue still to refrain from establishment of an econometrically data-driven business plan, outlining a short- and long-term strategy regarding the size of the biobank and return of investment. In the middle, an unconfirmed percentage of biobanks do practice cost recovery and have business practices, but business models tend to be limited in scope and/or planning. At the other end of the spectrum, some biobanks have extremely formal business and cost-recovery models with limited flexibility or that are resistant to market adaptation. Even still, many biobanks have no formal business infrastructure. The end result of lack of business planning tends to frequently result in significant lags in engagement and conduct of research, quality review and curation of specimen and data collections as well as transfer of specimens and data. Such issues can also compound risk of loss in return on investment and contribute to increased costs of scientific and clinical collaboration.

Sustainability is further retarded by insufficient strategic planning, specifically foresight regarding budgeting of funding and costs to support research projects for most biobanks, even those that generate revenue *via* cost recovery. In recent years, many governments and funding bodies have encouraged the establishment of large-population biobanks and networks. Despite continued funding, most are either underfunded or funded with unsecured revenue streams. Stakeholders are imprecise as to the significant resource investment, time and cost of operations that are required for a quality biobanking effort. Development of such a facility may require anywhere from $3 to 6 million USD depending on infrastructure. Operating costs and initial resource investment vary per bioresource depending on the size of the biobank, its business model, related activities and collections. Labor, equipment and materials tend to be the largest costs.[14,16]

- Issue #2: Difficulties in locating and accessing biospecimen and data collections

One of the key challenges to obtaining samples for biomarker research is success in locating and identifying relevant sample collections. Locating and targeting the correct set of samples and data is a pivotal requirement to support biomarker research. The primary mechanisms of acquiring specimen collections have been: 1) direct procurement from local sources (*e.g.* clinicians' colleagues within the same hospital) or 2) establishment of new relationships for individual interrogation of collections of identified sample and data sets (*via* phone, email, and web-based queries). Some repertories exist but most of these registries are national or managed by biobanks associations and access for searching is restricted. As of today, there is no exhaustive full accessible international repertory. Manual investigation of available samples is time consuming and costly and occurs prior to quality testing or downstream analysis of the sample. New trends to improve specimen searching, cultivation and access include creation of web-based technologies *i.e.* online catalogs and biolibraries, web portals and specimen locators. The NIH Specimen Resource Locator (http://pluto3.nci.nih.gov/tissue/default.htm) is one example. However specimen locator-based databases typically include access only to the resources within their biospecimen research network. Another example is the gratis Trans-Hit Biomarkers portal, B4B-Hub (Biobanks for Biomarkers: www.trans-hit.com) that allows biomarker scientists to post their biospecimen requests that are automatically sent to all biobanks members. Trans-Hit's database includes more than 1500 biobanks worldwide. In the future, greater availability of informatics tools would allow biomarker scientists rapid real-time capability to search multiple data fields and more aptly evaluate criteria for donor-specific sample sets and associated clinical data.

– Issue #3: Inefficient research business practices and resulting inability to engage in timely collaborations

Access to biobanks continues to be a prevalent barrier to research. Frequently biobanks restrict access to intramural scientists, preventing access to external scientists, particularly those from industry. Many biobanks refuse to allow access to their biospecimen collections citing institutional ethical constraints or questionable rationale. This practice limits many biomarkers to be effectively used in biomarker validation and in translation from bench to bedside. Thus, it decreases the potential for the biobank to add value to their basic research programs by unifying applied research to facilitate new biomarkers. As a result, biobank and biomarker scientists are limited in capability to increase their knowledge base requirements to perform quality, "fit for purpose" and benchmark-level biobanking and sample management. In general, "fit for purpose" is a term used to clarify that the methodology of the research design should match the intended downstream purpose of the research application. Benchmark recommendations are still being explored across the industry overall, but the concept involves the idea that performance metrics emulate the highest level of defined practice. This lack of extramural collaboration and shared learning ultimately hurts the ultimate stakeholder in biomarker research, the donors that support such research as well as prospective patients who may benefit from related discoveries. When access is available to external scientists, some biobanks implement policies with unreasonable access fees that appear to be applied solely to scientists from industry.

Another major obstacle is lag in delivery of biospecimens, even from collections already in stock and timelines to deliver biospecimen collections often due to labor required to investigate requests and delayed execution of administrative overhead. As a result, too frequently biomarker development projects are critically delayed.

In some countries, *e.g.* China, export of some types of biospecimens is not allowed. This seriously hampers validation studies and commercialization of *In Vitro* Diagnostic (IVD) tests in China, unless partnering strategies with local companies are being followed.

Overall, such constraints in biobank access, donor recruitment, and project management related timelines create extensive lags, often putting biomarker research programs at dramatic risk.

In addition to issues related to access, there are significant inefficiencies with the lags in the research administrative process. While requisite due diligence is understandable (review and approvals by the biobank's biospecimen use committee and/or institutional ethical board (IRB) approval and/or financial committee) this process tends to be overcomplicated. Global divergence in biobanking ethical, legal and regulatory frameworks induces significant delays, *e.g.* some institutions refuse to implement standardized language templates. Inefficiencies executing contractual agreements for extramural research limit industrial biomarker collaborations due to strict filing timelines that the large majority of biobanks in operation are not usually aware of.

Often, terms for collaboration tend not to be well defined or justified. The terms do not necessarily appear reasonable to the contributions garnered. The most frequent area of difficulty is the negotiation of intellectual property (IP).[17] Some biobanks insist on sharing of downstream IP in accordance with provision of biospecimens. More often biobanks desire coauthorship in future publications, which is justified mainly when there is a scientific input from the biobank. Terms for collaboration, including IP should be well discussed early on. On the other hand, when papers are published by biomarker scientists, they often fail to provide sufficient and crucial information about the source and biopsecimen.

Another prominent issue tends to be agreement on laboratory location of conduct of downstream analysis. Some biobanks will only collaborate if laboratory work can be conducted in their laboratory. This request is particularly problematic for industrial organizations to comply with due to regulatory requirements aimed at ensuring environmental quality control throughout the process chain related prefiling and precommercial activities.

Once terms are rendered and agreed upon, additional delays may occur due to lags in approval of material transfer agreement, where simple agreements may take months or up to a year. Mayol-Heath[18] and colleagues recently published an article in which they offered insight into navigating the regulatory process for biospecimen research in the United States as depicted. It is required that each biobank follow specific steps to obtain and maintain institutional regulatory board approval. This process can take an average of 1–4 months, depending on the related requirements (whether informed consent is needed), complexity of the research project, types of samples being collected and/or requested and frequency of IRB review board meetings.

- Issue #4: Low prevalence of satisfactory, relevant biospecimen and data collections

Many biobanks do not have complete or well-documented Standard Operating Procedures (SOPs). This deficiency increases difficulty in qualifying relevance and quality of biospecimen and data collections. SOPs for annotating, collecting, processing, storing biospecimen are usually different from site to site and can result in significant variation in studied biomarkers that has nothing to do with disease. Care should be taken when selecting biospecimen collections and biobanks. This is even more crucial for validation studies, when collections from different sources may be required.

Few cohort-based biobanking projects have prospectively stored an extended variety of specimens for a single subject (plasma, serum, urine, tumor, DNA, RNA, *etc.*). Most centralized tissue based biobanks are initiated in pathology departments and historically have focused on collection of fresh tissue in excess, which is usually stored in FFPE and frozen formats. Many biobanks do not collect other fluids that can be very useful for translational research and biomarker validation.

When developing and/or validating a biomarker it is important to look at biomarker profiles in control cohorts. Controls could be matched healthy subjects but may also be other types of patients who might express same profile of biomarkers or the same kind of clinical symptoms. While many biobanks may have very substantial collections for a given medical field they typically do not collect control groups.

Another key issue is that clinical and/or downstream annotation not readily available if not planned in advance – many biobanks do not have an easy and full access to clinical data. The biobanks that do have access to clinical annotation often obtain this data through retrospective medical file review, which is more difficult to validate and not necessarily "fit for purpose". As a result, data sets vary dramatically in terms of content, range of annotation, quality and completeness.

Biomarker research is often conducted utilizing aggregate samples obtained from either one or a variety of retrospective collections. Retrospective collections can certainly be used for biomarker discovery and in many cases for biomarker validation. Depending upon the final intended use, application of a biomarker, which may be submitted to different regulatory requirements (please refer to Chapter 2 of this book on regulatory aspects) retrospective samples could potentially save significant time and money compared to prospective studies. Harnessing historical collections from retrospective studies may offer the benefit of expediting such valuable research and reduce costly delays and limit the need to execute lengthy, expensive validation studies, therefore dramatically accelerating clinical validation of future diagnostic products. However, for eventual future regulatory submission to the US Food and Drug Administration (FDA), prospective collections can be mandatory.

Clinical trials often include dedicated biomarker studies and having access to archived patient specimens from prospective clinical trials could greatly contribute to the validation of biomarkers, particularly new companion or prognostic biomarkers. Pharmaceutical sponsors have dedicated high caution and budgets for well designing and conducting their clinical trials, collecting high-quality specimens and clinical data. As mentioned by Lindpainter,[19] even if only a fraction of the more than four million subjects enrolled worldwide in interventional trials were to be captured, progress would be enormous. However, regulatory and ethical issues (*i.e.* patient inform consent) considerably limit the use of biospecimen that have been collected during past clinical trials. The ability to acquire approval from sponsors of these clinical trials (which were successful or failed), to have access to these collections is also critical and uncertain.

3.2.3 Effect of Process-Chain Impediments on Impact of Biospecimen Collection Quality

The health-care industry spends billions of dollars annually on biomarker research for personalized medicine. Despite this investment, research into biomarker-disease-associated molecular changes in body tissues and fluids hasn't yet delivered on its promise. Technologies such as proteomics and genomics have contributed voluminous literature documenting thousands of claimed biomarkers;[20,21] Rare candidate biomarkers failed to replicate when tested on larger-sample cohorts and therefore have not been fully validated and/or approved for routine clinical practice. Several papers reported alarming findings concerning the reliability and reproducibility of published scientific data, which could not be replicated by a third party.[22,23] The poor reproducibility of results is alarming but may possibly be explained by differences in laboratory equipment, protocols, and personal skills and related to biomarkers, the most relevant factor is likely to be the variation in starting material that may differ between laboratories.

Besides the lack of substantial validation clinical studies, one of the primary impediments to progress the full validation for biomarkers is the lack of standardization in how specimens are collected, handled and stored. It is well known that the quality of the starting material determines the accuracy and reliability of diagnostic assay results. Therefore, it is crucial that the appropriate samples are managed in a robust manner to ensure the conclusions made based upon the derived data are reproducible and reliable. Ensuring the highest quality sample-management processes is of vital importance to ensure the robustness of the data, decision making and conclusions from studies.[24,25] Inaccurate results owing to compromised tissue quality can lead to false-positive or false-negative results with therapeutic consequences that can harm patients and affect their eventual outcome. As pointed out by Foot,[26] "if there are no samples, there is no science". Many studies have demonstrated the adverse impact of the inferior quality of samples on the generation of bias and therefore the reliability of the data derived from their analysis.[27,28]

The major impediment to progress in the validation of new biomarkers is the lack of standardization in how specimens are collected, handled and stored. It is known that at least 60% of errors occurring in routine clinical chemistry laboratories are due to mistakes made in the preanalytical phase. Most of them are attributable to mishandling procedures during collection, handling, preparing or storing the specimens. Sample integrity is of paramount importance in order to make robust conclusions from the data, from the point of collection, through handling and shipment, storage and sample management and to ensure efficient retrieval of the correct sample and ability to link with relevant clinical data. In nearly one fifth of the cases they can produce inappropriate investigations and clinical decisions.[29] If this observation is true for routine assays (already validated and commercialized), even though the laboratory itself is accredited and/or has an outstanding performance, one can imagine what happens when new biomarkers are studied in a noncontrolled quality laboratory environment. This is particularly important for biological fluids but also for tissues especially in the cancer field.[30-38] It is essential that the levels of expression of biomarkers that are used to make treatment decisions be accurately attributable to the underlying biology and not due to artifacts. Several authors have identified a variety of issues and barriers that can affect the transfer of clinical tests from research to clinical practice: differences in sample collection, handling or storage.

The effects of tissue degradation and decay have been reported for many molecules[34-36,39] Hicks and Boyce[40] recently reported how the quality of tumor tissue, determines the accuracy and reliability of companion diagnostic assay results and therefore optimal therapeutic strategies. Preanalytical variables are the "weakest link" of immunohistochemistry, as the properties of the tissues analyzed and, consequently, the results, may be affected by several factors, including the time to collection (length of time that tissues are subjected to warm ischemia between removal of the tissue at surgery and fixation), details of fixation (type of fixative agent used and length and conditions of fixation), dehydration steps, and conditions for paraffin embedding. These preanalytical parameters are beyond the control of investigators, are most often unrecorded, and constitute a major potential source of bias.[41]

Variables involved with tissue samples begin with taking the sample from the patient through medical or surgical procedures, and continue from acquisition through handling and processing, storage, distribution, analysis, and restocking unused samples.

A succession of variables can affect tissues samples:

– The type of sampling procedure: for instance, Pinhel et al.[42] comparing needle core and excision biopsy breast cancer specimens have shown the immunohistochemical reactivity for phospho-Akt and phospho-Erk1/2 was markedly reduced in the latter specimen type.

- The warm ischemia time (the time the surgeon has clamped the blood flow to the tissue and it is cut off from oxygen and nutrients before removing the sample from the patient): during this warm ischemia time the tissue remains at body temperature. Ligation of the blood supply to living tissues being excised during surgery leads to hypoxia, ischemia, metabolic stress, and the progressive degradation of macromolecules that are of potential clinical interest. During the period between artery ligation, tissue removal from the patient, and the start of fixation, the tissue remains alive and reactive, and the level of expression of gene transcripts and proteins can change significantly during this ischemic interval.[43–45]
- The cold ischemia time (the time between removal of the tissue from the body until it is fixed or frozen): immediately after removal of a tissue sample from a patient, cells in the tissue adapt to the absence of blood flow, ischemia, hypoxia, metabolic acidosis, accumulation of cellular waste, absence of electrolytes, and temperature changes.[43] Within a few minutes major changes start to occur in the protein signaling pathways of the tissue as it remains at ambient temperature. While awaiting fixation or freezing, the tissue is alive and "wounded". Post-translational-modified proteins, such as phosphoproteins are among the most labile molecular features. Espina *et al.*[43] reported fluctuations of up to more than 7-fold in some phosphoproteins when samples were incubated at room temperature for various times after excision. Studies have shown that degradation of a number of proteins and peptides starts immediately after disconnecting a tissue sample from its oxygen and nutrient supply.[46] The first changes in gene and protein expression are observed within minutes from excision of a tumor, and after a half-hour 20% of all detectable genes and proteins significantly different expression,[47] reflecting the shutdown and activation of different pathways.

Recent reports suggest that delays from tissue collection to the initiation of formalin fixation also may have an adverse impact on the analysis of hormone receptor assays.[48,49] Khoury *et al.*[49] also showed that the level of expression of estrogen receptors (ER) by immunohistochemistry starts to fall after a two-hour delay from collection to the start of fixation and similar changes are seen for progesterone receptor (PR) after only 1 h. Most surprisingly, these investigators demonstrated that HER2 fluorescence *in situ* hybridization intensity begins to show compromised interpretation after only one hour, and this reduction becomes statistically significant after two hours. The implications for these findings are that some tumors with excessive ischemic times may be classified falsely as negative for the expression of these important therapeutic targets because of protein degradation during this period. Invalid testing of breast predictive factors could have serious consequences for patients.

Time intervals from blood-vessel ligation or tissue removal to examination in the pathology laboratory may vary from minutes to hours depending upon the time of day, the type of procedure, and the proximity of the operating room while the stability of studied biomarkers might vary between sites. In order to be able to calculate and monitor cold ischemic times, the American Society of

Clinical Oncology ASCO/College of American Pathologists CAP task force has developed guidelines for ER and PR testing requiring that breast biopsy and excised breast tissue samples be placed in formalin within one hour of excision[50] This requires that the collection time for each sample be recorded in the operating room, the transport of the tissue samples to the laboratory be expedited, followed by a record of the fixation start times.

 – Other variables include the type of fixative, the fixation time, the rate of freezing, and the size of the sample. Formalin fixation and paraffin embedding (FFPE) is still the gold standard for tissue-sample preservation in the clinical setting. However, pathologists consider formalin to create "standard artifacts", but still little is known on the effects of formalin fixation on many (potential) biomarkers. Length and conditions of fixation have been branded as the "Achilles heel" of phosphoprotein assessment in clinical specimens.[42] Dehydration steps and conditions for paraffin embedding (temperature of the paraffin) may also have an influence. To complicate matters even more, different constituents of the sample will undergo decay at a different rate, phosphoproteins being among the most sensitive.

In the future, the use of reliable decay indicators of quality of biospecimen should allow simplifying and testifying as to whether a biospecimen collected and stored according to different protocols may be appropriate for biomarker research. With respect to fluids, several biomarkers have been studied but none have demonstrated so far their true usefulness. Cytokines CD40L has been proposed to be a potential marker to assess the quality of storage for blood samples.[51] For tissues, developing a list of tissue protein stability surrogate markers also remains an important goal for molecular profiling research.[52] Indeed there is an urgent need to document metrics and a cutoff at which the change is so great that it renders a specimen inadequate for diagnostic or research purposes.[52] For any of these quality-assessment assays to provide meaningful information, they must be able to guide the selection of tissue for a given assay or measurement; ideally by providing a numerical cutoff above which that assay can be used reliably.

3.3 Recommendations: Reducing Disparity of Impact and Lag in Outcomes of Biomarker Research

3.3.1 Technical and Scientific Recommendations for Biomarker Scientists

For the context of this discussion, we regard biomarker scientists as those focusing on the discovery, validation, development and implementation of biomarkers.

 – Scientific and technological advancements in genomics, transcriptomics, proteomics, metabolomics, and novel approaches in bioinformatics and systems biology have spawned new insights into the molecular processes underlying complex diseases, which are important drivers for the discovery of novel biomarker candidates. Before 2001, biomarkers appeared less than 1000 times in patents and scientific papers. In recent years more and more data have been produced about biomarkers as putative future diagnostic tests. By January 2012 PubMed referenced more than 553 000 hits, while a Google search produced 8 500 000 results for the term "biomarkers". Every week several articles are issued announcing an academic research group has identified a new protein, gene, or other types of marker to better diagnose patients with a specific medical condition. Although the vast majority of these scientific discoveries are pertinent inventions, most of these discoveries are not translated into commercial products or R&D pipelines. Results obtained with many of these biomarkers remain controversial and not conclusive. In many cases the clinical proof-of-concepts are weak, and are often based on small, poorly designed studies with limited numbers of patients and/or the use of suboptimal quality sample collections (poor quality, incomplete clinical information, combination of different sample cohorts with distinct characteristics that create bias, *etc.*). Actually, the success rate in biomarker discovery and validation versus the level of biospecimen and data collection quality has a high degree of association. Ioannidis and Panagiotou[53] described that the initial effect size of a biomarker in selected clinical populations can in the majority of cases not be confirmed in meta-analyses of subsequent studies. Surprisingly, even publications in prestigious peer-reviewed journals or from several independent groups did not ensure reproducibility. Too many scientific papers suffer from various pre- and postanalytical bias and confounders relevant to sample collection, processing and downstream testing, including assays. These biases fail to be recognized in time and often lead to invalid discoveries, *i.e.* putative biomarkers that can't withstand further testing. There is an obvious and urgent need for scientists to ensure the reliability and reproducibility of their data before publishing any new biomarker candidates. As a key factor in the value chain, biobanks can certainly perform an important role in generating much needed ancillary data to validate the robustness of the scientific findings. Unfortunately, most of the literature published by biomarker scientists contained no or little information about methods and materials and the types of biospecimens used.[21,54] As poor-quality samples can only serve to jeopardize research results, it is of the utmost importance to critically select high-quality collections before starting a biomarker project. By carefully selecting biospecimen collections and biobanks, scientists are more likely to increase their chances to adequately address scientific questions, and fulfill regulatory requirements in anticipation of bringing a biomarker to market.

– It is crucial for biomarker scientists to confirm the objectives and research hypothesis of interest upfront, and to define the future intended use for the final assay be made in order to better define the development plan and therefore inclusion criteria for patients. Consultation with biobanking colleagues will also ensure that specimen and data collections are timely, relevant and can be procured prospectively.

– It is also important to ascertain what type of collection, retrospective or prospective biobanking collections is most ideal to support biomarker research. When using biospecimen collection, *a fortiori* for eventual future regulatory submission to the FDA, great care must be taken to utilize the most appropriate samples. Both prospective and retrospective samples can be used, but each has its own set of drawbacks and regulatory concerns (Table 3.1). The path for clinical and regulatory plans must be previously defined. The use of retrospective samples is not the right solution in every case. However, depending upon the classification of product within the regulatory guidelines and intended use application, retrospective samples could potentially save significant time and money. When prospective studies are needed for the clinical validation of new biomarkers, the setup and monitoring investigator sites, respecting Good Clinical Practices guidelines and maintaining high-quality standards for biospecimen sampling, shipping and storage are crucial (Table 3.1).

– If biospecimen and data collections are needed for a biomarker project, project plans should aim to factor in time and resources required to investigate numerous biobanks before finding the right source and collaboration. Currently, too many researchers rely on whatever specimens they can obtain conveniently from local institutions. Such specimens may not be appropriate for the biomarker research project of interest. Often, biomarkers scientists may need to consider procuring collections from alternate sources. In an effort to expedite examination of

Table 3.1   Pros and cons of utilizing prospective and retrospective biobanking collections.

	Pros	*Cons*
Prospective Sample Collection	Designed to target specific clinical/therapeutical needs	Long time to collect
	Collects all required clinical data	Time-consuming legal/regulatory hurdles
	Addresses intended use	Costly
Retrospective Sample Collection	Collection readily available	May or may not have all required patient data available
	Samples targeted at the therapeutic area in general	Informed consent may be restrictive, or if no informed consent is available: must need the FDA/legal requirements for use as left-over specimens
	Less expensive	
	Addresses the intended use	

sources it is recommended to utilize existing local as well as international scientific connections. Scientists should aim to decrease searching lags *via* utilization of biobank hub portals and locator tools (*e.g.* the **B4B** hub: www.trans-hit.com/).

– In unison with biobanks, biomarker scientists should aim to define which type of donors, samples and data are needed. This involves definition and provision of inclusion and exclusion criteria, demography (*e.g.* ethnicity, gender, smoking habits, alcohol consumption, *etc.*), clinical requirements (*e.g.* disease of interest, specific patients), as well as specificities regarding type of control groups required (healthy, symptomatic or asymptomatic populations, other diseases groups of interest, *etc.*). It is important to carefully determine the number of patients and their inclusion criteria; well-designed studies are crucial to improve reliability and efficiency of research about biomarkers.[55,56]

– It is also important to define quality requirements for biospecimens and data collected relative to the biomarker of interest. Too many sources of bias in specimen impact results and interpretation of biomarker studies.[56] This includes not only consideration of clinical sampling procedures (*i.e.* application of tourniquet, resting time, posture, level of activity for blood samples) but also biobanking methodologies (mode of cryopreservation per type of sample (plasma *vs.* serum), ischemic window, length of storage, *etc.*). An example of bias is given in Table 3.2.

– Once the project requirements have been defined, due diligence steps should include formal evaluation of each biobank. While a portion of this process may be possible virtually, biobank evaluation may best be conducted *via* onsite audits. On site audits should include confirmation and review of existing SOPs (from sampling to storage and beyond) as well as accessible medical annotation per case. During the audit process, it is important to confirm the range and level of scientific expertise engaged. Efforts should be made to clarify how and in what fashion such expertise has been applied. For example, one biobank may have a satisfactory biospecimen collection in one area of expertise, but not in another one). It is essential to confirm if pathologists and/or clinicians are directly involved and the time points of direct involvement. Such details are of critical importance to ensure quality research and accurate interpretation of data. Once a biobank has been properly evaluated, feasibility planning should involve early education of ethical and regulatory requirements *e.g.* informed consent, ethical committee approval processes and related timelines, confidentiality, specimen and data-acknowledgement requirements, and any other institutional guidelines. One should understand that these requirements may vary per country, per sector, per research domain and in some cases per donor/source population, *e.g.* varying requirements for pediatric populations, postmortem samples, leftover samples, *etc.* Contractual agreements should include unified templates for material-transfer agreements. Intellectual property issues and coauthorship in case of publication should be clarified upfront. These

Table 3.2   Sources and "locations" of bias in marker research. Extracted from Ransohoff and Gourlay.[56]

Source of Bias	Location of Bias: Before or After Specimens Are Received in the Laboratory — Before	After	Example
Features of subjects, determined in selection: Age Sex Comorbid conditions Medications	X		Cancer subjects are male, whereas control subjects are mainly female. Bias: Assay results may depend on sex
Specimen collection	X		Cancer specimen come from one clinic, whereas controls come from a different clinic. Bias: Assay results may depend on conditions that differ between clinics
Specimen storage and handling	X	X	Cancer specimens are stored for 10 years because it takes longer to collect them, whereas control specimens are collected and stored over 1 year. Bias: Assay results may vary with duration of storage, or with different number of thaw–freeze cycles.
Specimen analysis		X	Cancer specimens are run on one day, whereas control specimens are run on a different day. Bias: Assay results may depend on day of analysis in a machine that "drifts" over time.

Note: The table shows examples of different sources of bias and the location of the bias before or after specimens are received in the laboratory. The list is not exhaustive: other biases may be important, and the biases listed may or may not be important in any given research study, depending on details of biology and technology (*i.e.* what is being mesured and how it might be influenced).

processes can be quite lengthy. Lags in collaboration tend to be intradependent on the biobank/institution of interest, but in some cases can be 6–18 months. Therefore, these lags should be factored in early on as part of the project planning.

– Akin to pilot testing of collections in progress, it is recommended that feasibility planning involves pilot shipping to confirm that the material-transfer process does not introduce additional variability. Not only will this process allow confirmation of quality requirements for shipping, it will also confirm reliability of timelines and preferred shipping carriers. Biomarker scientists should be aware of international best practices for shipping, which may include IATA guidelines.

- Lastly, it is recommended that whenever data obtained from materials from biobanks are published the study analysis should contain a detailed description of all the confounding parameters that may have influenced the overall results.

3.3.2   Technical and Scientific Recommendations for Biobankers

For the context of this discussion, we regard biobankers as those developing, managing and providing access to biospecimens in a biobanking environment.

Biomarker research requires timely availability of both retrospective and/or prospective collections. An unfortunate reality is that biobanking projects tend to rarely be designed with the planning considerations required to adequately support biomarker research on a broad scale. It would be highly beneficial if biobanking projects were prospectively coordinated internationally with this in mind. Such foresight may reduce the risk of monetary loss and increase return on investment.

It is recommended that biobanks evaluate the potential end products and "fit for purpose" requirements prior to biobank design and setup. Planning considerations, should aim to elucidate the basic and applied research objectives for collections. This would dramatically assist future sustainability of biomarker collaborations. Currently, the practice is quite the opposite. A large majority of biobanks appear to function as "libraries storing books that will never be read". It is estimated that biobanks are only utilizing a relatively small percentage of their biostores. Efficacy would be increased dramatically if biobanks were aware of biomarker-specific research requirements and bilateral planning considerations of biomarker scientists in the academic and industrial sectors. It is necessary that equilibrium exist to support supply and demand. Therefore, it is recommended that biobanks factor in biomarker-planning considerations early on to ensure return on investment of such collaborations and related work products.

Outlined below are primary recommendations, prerequisites and procedures that biobanks should aim to employ in order to optimize the collaborative process and related biomarker research outcomes.

3.3.2.1   Biospecimen Research and Biobank Operational Planning Efforts

- For biomarker research outcomes to have the most relevant impact the following planning efforts should be exercised regularly. Biobank planning efforts to support biomarker research should survey prospective end users to confirm project specific requirements. To remain educated and ensure relevant research projects are being selected, biobanks should also aim to keep abreast of trends in the biomarker research field. Planning should also include areas of market need; this should aim to be a key driver in prospective design of biospecimen and data collections as well as cultivation of historical archives and/or retrospective samples. Therefore, research budgets should factor in funds for labor involved in investigative R&D efforts.

- Biobanks interested in supporting biomarker research projects should aim to define an efficient pathway to aide timely biomarker collaboration. To decrease lags in research and assure quality and relevance of research outcomes, biobanking regulatory frameworks should factor in issues specific to support high-quality biomarker research. Biobanks should work to educate bioethics committees of the need to work with collaborators in industry as well as academia to ensure samples are being utilized appropriately and in a manner that offers the fullest opportunity to respect the patient's wishes. Such efforts will aim to ensure that the intent to perform research is fulfilled. Informed consent guidelines should include statements that acknowledge that third parties may have access to their information and samples, keeping this aim in mind. Regulatory advisors should be educated to understand that fulfilling market need that is commensurate with and driven by patient and/or disease incidence and prevalence-related needs.

- Biobanks can only function effectively if requisite expertise is available at the appropriate level of commitment. It is possible that the most effective biobanks are those with several dedicated managing scientists and technical teams. Experience has demonstrated that the optimal biobank collections are those tended by a pathologist or clinical chemist with high incidence of well-defined interdisciplinary activities. Career paths for Technical Directors, Biobank Managers as well as professional technical staff should also be established to ensure benchmark-level cultivation becomes a reality. When possible, it is recommended that biobanks engage biomarker experts from industry and academic on both the Scientific Advisory Board and Biospecimen Utilization Committee.

3.3.2.2 Scope of Collection

- It is recommended to keep scope of collection targeted. Biomarker collections may be less sample-size-dependent if they are well planned. Therefore, rather than aiming for broad, large, wide ranging recruitment targets in multiple disease research areas of interest, it is recommended to focus on available expertise.

3.3.2.3 Control-Sample Considerations

- Biobanks should increase the prevalence of control group biosample collection.[57] When developing and/or validating a biomarker it is important that biobanks collaborate prospectively with biomarker scientists to factor in biomarker profiles in control cohorts. Control cohorts may consist of matched healthy subjects but also other patient populations who have demonstrated expression of common profiles of biomarker or the clinical symptomology (*e.g.* pancreatic cancer versus pancreatitis). Studying other populations of patients may also prove helpful to demonstrate the specificity of a biomarker (*e.g.* pancreatic cancer *versus* colon cancer). While biobanks may plan their collections appropriate to a designated domain of interest, many fail to include necessary control groups.

3.3.2.4   Sample-Set Methodological and Annotative Considerations

- When building a biobank collection for future research it is preferable that a wide range of sample types and formats as feasible per individual/case be procured. Formats vary and can refer to procurement of primary materials, *i.e.* blood, urine and tissue with and without cryopreservative agents, but can also refer to processing formats for primary components (serum, plasma, Buffy coat) as well as secondary derivatives, *e.g.* (DNA, RNA, *etc.*). For blood, the minimum sample set suggested is a plasma and serum sample. Sample-set requirements should be carefully considered for each type of sample and its familial sample set. Source-based requirements should be assessed and factored in methods and materials, then documented in research protocols.
- Definition of biosample collections should be performed in concert with well-established best-practice-based research methodology, procedures and evidence-based protocols. To reduce preanalytical variability for a given biomarker, strict adherence to SOPs and standardized protocols is crucial. Lack of adhesion to simple guidelines and protocols regarding the preanalytical phase will compromise the quality of biospecimen collection. SOP development should mirror recommendations provided by best and evidence-based practices. Several initiatives are attempting to improve the practice of biospecimen management or biobanking. For example, the US National Cancer Institute's Cancer Human Biobank (caHUB) has historically established stringent guidelines to ensure that samples from healthy individuals and cancer patients are collected, annotated, stored and analyzed under standardized conditions and accompanied by appropriate donor medical information. NIH/NCI BBRB recommendations for minimal data sets can be found at: http://biospecimens.cancer.gov/bestpractices/Appendix1.pdf. As these guidelines were created as a broad template, it is important for biobankers to work with biomarker scientists to confirm how such data sets may need to be customized to support biomarker research specifically.
- Baseline procedures should aim to detail methodologies factoring in reporting guidelines early on and be applied during each step along the specimen lifecycle and process chain. For example, this may include instructions for biospecimen sampling. *e.g.* (identification of used collecting tube types, sample volume, *etc.*) along with annotation on sample handling (*e.g.* "place on ice" or "transport at room temperature")

as well as instructions regarding processing times including specifics (*e.g.* centrifugation specifications, aliquoting process, type of storage vials, *etc.*). For tissues, it is important to collect detailed information on ischemic times: warm ischemia time, cold ischemia time, lag between the collection of a biological sample and processing for testing in the laboratory, type of fixative, fixation protocol, fixation time, size of the tissue specimen, storage temperature and length of storage. Biobanks should ensure that their quality-management system addresses cold chain logistics along the process chain prior to time of collection through time of distribution and beyond. Any deviations should be documented to assist interpretation of downstream analysis.

– Biospecimen data collection should be accompanied by well-annotated clinical data sets. For biomarker research projects to be successful it might be necessary to collect significant clinical outcomes. Collected data should not be limited to solely pathology information. The US NIH has provided initial guidance on factors that may be beneficial to collect. This includes physiologic factors, *e.g.* gender, age, menopausal status, stage in the menstrual cycle, concomitant infection, pregnancy, fasting, exercise, smoking, consumption of alcohol or caffeine-containing beverages, time of day and year, use of a tourniquet, time of rest before sampling, posture (sitting *versus* supine), stress, medication, *etc.* Medical annotation should be complete prior to provision of samples and ideally accessible on a real-time basis when needed. The biobank must have the capability to perform expeditious multicriteria search queries to enable real-time efficient feasibility studies.

3.3.2.5   *Inventory Management, Workflow and Sample Tracking and Temperature Monitoring Practices*

– Informatics tools and software solutions for biospecimen inventory, clinical annotation, sample logistics and downstream analysis management should be coordinated to factor in workflows and annotation relative to biomarker research. A sample laboratory/biospecimen inventory management system (LIMS or BIMS) should be capable of recording and reporting on the following details on a sample-by-sample basis: biospecimen management life-cycle history, length and mode of storage, shipping and sample temperature, sample integrity and stability at time of sampling and handling intervals and long-term stability. In order to facilitate tracking of temperature and variation of individual sample stability, biobank stakeholders should aim to consider proactive implementation of temperature-sensing and -tracking techniques *e.g.* microelectromechancial systems or "MEMs"-based technologies *i.e.* those being in development by Bluechiip™.[58]

– The ideal biobank should offer an electronic sample inventory management system that eliminates paper-based tracking and reporting processes through all stages of a sample life cycle. Ideally, the system should have a reasonable level of interoperability with both internal and external collaborators, either through a common biospecimen inventory module as well as facile methods of exchanging and examining data. Best practices for biobank inventory management have been initially offered by the (NIH/NCI BBRB Best Practices (http://biospecimens.cancer. gov/bestpractices/to/bri.asp)). Biobanks should keep abreast of updates in best practices relevant to inventory management as well as guidance on annotation specific to biomarker research. Biobanks and biobanking product vendors should plan accordingly to ensure inventory management systems factor in workflows specific to biomarker research.

3.3.2.6 *Sample-Quality Management, Assurance and Auditing*

Quality practices should include established data collection procedures to promote standardized data collection. Data-collection practices should document the mode and procedures involved with data collection. Clarifications should differentiate data collection characteristics, *i.e.* electronic *versus* manual, single *versus* double capture, data cleaning and level of completeness). Logistical details should be captured, *i.e.* capability and time frame required to retrieve the latest data from the health-care system, anonymization procedure, and ability and steps needed to recontact donors for additional information and/or samples. All of this information should be presented to guide the feasibility review of biomarker projects.

Ideally, biobanks should aim to establish quality standards, *e.g.* quality decay indicators. Quality decay indicators, if available, may assess molecular integrity of the biosample and ascertain if biosample still represents the *in vivo* situation confirming that the sample is still valid for assays and reliable interpretation of results (Refer to guidelines in Section 3.3.3).

– Biobanks should aim to perform regular internal audits to assess the quality of biospecimens and data as it is relevant to biomarker research. Yearly external audits are also recommended. Quality assessments should include quality analysis of specimens as well as a review of the data collected. This may include but not be limited to details, *e.g.* adherence to tracking of sampling, sample management and storage procedures ("5W rule": to know when, who, what, why, where), the quality of related medical annotations (kind of data being retrieved, data consistency per case). Throughout this process the question of interest should confirm that both specimens and data are sufficient to support successful biomarker research outcomes.

– Biobanks interested in supporting biomarker research for long-term projects should aim to track efficiency measures. Such measures should be provided at key iterative points in project development towards guiding biomarker scientists in evaluating candidate biobanks. Outcome measures might include review of sample utilization, *e.g.* number of biospecimen used over a definedperiod: ratio used/stored, internal versus external use, types of projects performed and related outcomes (number of publications and relevant citations, number of biomarker studies resulting in successful proofs of concept, commercial end-products, *etc.*). Hofman *et al.*[59] recently proposed four major categories of indicators performance indicators to be used for the evaluation of a biobank: quality, activity, scientific production and visibility.

3.3.3 Joint Recommendations for Biomarker Scientists and Biobankers

To progress the maturation of biomarkers from bench to bedside, it is imperative that those with expertise related to biomarker discovery, validation, development and implementation work tightly together to build a successful pipeline. The biobankers, on the one hand, need to interact with a variety of biomarker scientists, both on a strategic level and on a practical level, to implement the recommendations outlined above. A few recommendations are presented below.

3.3.3.1 Content of Biospecimen Collection

The quality, relevance and content of a biobank determine its utility. Importantly, a tight interaction between the collectors and the users of bio-specimen (respectively, the biobankers and the biomarker scientists), will ensure that the content of a biobank is used to maximal extent for biomarker research and development. There could be considerable differences between biomarker scientists in their interests in clinical biospecimen. Some may just need clinical samples to confirm their preclinical findings in cellular or animal studies; others may focus on clinical impact and need samples to perform large-scale epidemiology through genetic analysis of selected genes in defined population groups. Through strategic alignment, the biomarker scientists can enhance definition of what biospecimen collections are collected, including the necessary matched control samples, thus tailoring the biobank collection to more aptly support biomarker research requirements.

3.3.3.2 Collection of Biospecimens

In various ways, the biomarker scientists mutually complement biobankers in collection and management of biospecimens. Through their functional and often global network of scientific interactions, a wide array of clinical expert centers may be approached to share clinical samples and thus enrich the collection of biospecimen from different origins and increase the intrinsic value of the biobank. The biobankers should indicate the gaps in their collections, thereby prompting the scientists to look for specific clinical samples. Interesting collection of samples can be obtained in clinical trials when the effect of pharmaceutical drugs is being tested. Particularly when serial samples are obtained from the same subject, drug effects before and after administration can be determined. Sampling from dosed healthy volunteers will potentially enable a study of mechanistic or toxic effects of a drug in a nondiseased state, or the validation of biomarkers representing those effects. Sampling in patients, however, will enable more clinically relevant studies on the mechanism of action of the drug.

Through coordinated interaction, biobankers can advise scientists how to design the clinical trials and ensure the storage of high-quality samples that can be used for predefined use in a later stage. It is important that biomarker scientists and biobankers collaborate closely on evaluation and cultivation of biomarker research utilizing biospecimen collections. In unison, biobanks and biomarker scientist should aim to work together to prospectively identify, define, elucidate and benchmark, quantify discrete outcomes and parameters for biomarker biospecimen analysis and quality review. Each stakeholder can assist the other in elucidating each other's needs and requirements, as well as success factors for increased impact of biomarker research outcomes.

3.3.3.3 Quality Control

A recurring issue with biospecimens stored in biobanks is how to ensure the quality of the sample. Through their technology platforms, biomarker scientists can support such analysis by performing a thorough analysis of control samples at regular intervals. For instance, an LC/LC-MS/MS proteomics scan of a body fluid will typically reveal 1000–5000 peptide peaks, depending on the complexity of the sample and analysis method used.[60,61] By recurrent scanning of the control samples with a fixed standard protocol at predefined times, the scientist can determine whether the abundance of some components of the proteome has changed. The subsequent identification of such components will learn whether there has been a loss of sample integrity, a protein aggregation or other events that has changed the sample over time. Recently, we have applied this method to maintain the stability of cerebrospinal fluid, demonstrating the use of this body fluid for biomarker research.[62,63]

3.3.3.4 Biomarker Research

Any biomarker research project that aims at clinical relevance and impact has to have access to the best-suited clinical samples to generate the proper results. In a powerful combination, biomarker scientists can work together with biobankers to ensure that such samples are available at the start of a joint research project. Our experience has shown that often clinicians are enthusiastic to participate in clinical biomarker projects, but despite this, may not be able to collect the required number of samples by their own. A biobank can coordinate the engagement of several of such clinicians, ensure the quality of the samples in a similar manner and thus make sure the required number for the study is met.

Both biobanks and biomarker scientists should aim to evaluate unique funding opportunities to ensure sustainability of biomarker research. Interesting opportunities arise by national and international funding of biomarker projects, such as NIH or EU-FP7 grants. Such funding often has a more long-term objective to innovate disease management in a particular area, and aims to bring together multiple scientists, clinicians and biobankers together under a functional umbrella. Creative models for collaboration can be beneficial if implemented. Moreover, increasingly such funding is through public–private partnerships, strengthening the synergy between pharmaceutical, biotech and diagnostic industry on one hand, and academic scientists and biobankers on the other hand. Research funded through such grants almost always has the objective of application in real life, and given the nature of many biomarker projects, application in clinical care is greatly stimulated through this principle. A variety of collaboration business models including the risk–benefit model should be considered on promoting sustainability.

3.3.3.5   *Crossfertilization of Best Practice*

Finally, and perhaps as important as the findings made by biomarker scientists using good samples from biobanks, there is a mutual growth in experience among the participants and stakeholders in the importance of high-quality clinical samples. From biomarker projects and sample handling, there will be a growing sense of awareness of how to collect biospecimen and to use them in biomarker research and development studies. This will result in definition of best practice that, upon publication and communication, will steer the biomedical field towards the optimal positioning of synergy between biobanks and biomarker research. Potentially, such best practice can also drive the development of reporting standards for evidence-based-practice and clinical care, *e.g.* Biospecimen Reporting for Improved Study Quality (BRISQ) and coding systems, *i.e.* SPREC or referred to as "Standard Preanalytical;[64] these variables should be jointly tracked when relevant). Recommendations outlined in this chapter may function as early best practice for biomarker–biobanking collaborations.

3.4   Utilization of Biobank Samples for Biomarker Discovery and Development

Over the past ten years there has been an accelerated growth in the biobanking field as well as in the biomarker domain. We next review aspects of recent technical developments that utilize clinical biosamples to support the discovery and development of biomarkers.

3.4.1   Biomarker Discovery

There are in general two ways in which clinical biomarkers are being discovered. First, molecular studies of cellular or animal model systems can yield candidate biomarkers that are subsequently validated in human samples of clinical relevance. Such an approach is strongly driven by the fact that preclinical model systems are more accessible and readily treatable with pharmaceutical interventions. However, in many cases there has been limited translation between preclinical and clinical behavior of the biomarker, including its level of abundance, modification, kinetics and interactions with other macromolecules[65] Alternatively, human material is used directly to identify biomarkers which circumvent the cross-species translational step. This optimal approach has led to several clinically applicable discoveries, including the increasing list of driver mutations underlying aggressive tumor development that have become the basis of personalized cancer medicine[66] Several discovery technologies have had requirements for minimal sample quality or sample amounts with the result that a study cannot be designed properly or cannot be performed at all but recent developments have improved this situation.

3.4.1.1   Sample Quality

Many biomarker studies did not include analysis of the quality of the samples and may have resulted in differential biomarkers that at least partly can be related to the sample isolation and storage process (our unpublished observations). As outlined above, the integrity of the stored samples themselves can already be improved by standardizing sample handling procedures and by including wide molecular profiling of control samples. In addition, the multiplex nature of targeted biomarker analysis methods such as proteomic/metabolomic mass spectrometry using Multiple Reaction Monitoring can now include decay indicators to determine the quality of the biosamples within the same study, which is a great advantage. Also, method developments have made formalin-fixed-paraformaldehyde-embedded (FFPE) tissue samples accessible to biomarker discovery and validation. FFPE is generally regarde as a stable preservation method, but for long the crosslinking during sample preparation has disabled the identification of novel biomarkers through molecular profiling. However, recently methods were developed to reverse the crosslinks and make fixed tissue samples amenable to hybridization-based transcriptomics profiling[67] and even to the more challenging mass-spectrometry-based proteomics.[68] In particular, in combination with isolation of histologically defined areas through laser-capture microdissection,[69] this enables enormous possibilities for retrospective molecular assessment of stored fixed samples.

3.4.1.2   Sample Amounts

Several high-end technology platforms increased their sensitivity and robustness, thereby lowering the minimally required amounts for biomarker discovery. For instance, the next-generation sequencing platforms have developed with increasing robustness and reduced cost, such that a whole genome scan is feasible for many biomarker laboratories. Depending on the research question, the entire population or subfractions of cellular DNA and RNA can be analyzed to identify variants in sequence, abundance and/or modification, including those in genomic DNA, methylated DNA, transcribed mRNA and miRNA.[70] Whereas complete genome sequencing requires substantial material, the sequencing of subpopulations can now be done with minimal amounts that is first amplified under well-controlled conditions. As such, biosamples from patients can be subjected to whole exome sequencing in a routine clinical setting to identify Mendelian genetic predisposition to disease.[71] Alternatively, novel hybridization-based microarray platforms have been developed that use only minimal amounts of material to profile DNA/RNA variants with a large variety of probes, with a similar performance as amplification-based expression profiling methods.[72,73] Recent improvements in mass spectrometry, the key technology to focus on protein, peptide and metabolite biomarkers, have increased the sensitivity of biomarker detection and consequently the minimal amounts needed for such studies.[74] For example, only 20 µl of a cerebrospinal fluid sample was sufficient to yield robust data (coefficient of variation $<7\%$) to identify potential peptide biomarkers for multiple sclerosis.[75] It can be expected that further technological developments, particularly in combination with microfluidics[76] can further decrease the sample amounts needed for biomarker analysis.

3.4.2 Biomarker Development

3.4.2.1 Biomarker Validation

To progress a biomarker from discovery to development, it is imperative to do a thorough biomarker validation in relevant clinical samples to confirm the results obtained in the initial study. Three important components strongly depend on each other in biomarker validation: the assay, the samples and the test design. The biomarker assay needs to be developed and applied following a fit-for-purpose principle, whereby limited robustness is needed in this stage of biomarker development. Clinical samples need to be selected to support the objective of the biomarker validation, need to be available in sufficient numbers and quality, and preferably be derived from different clinical sites to verify that the biomarker is specific to the disease or mechanism and not to a certain source. The biomarker test design should ensure that a sufficient number of biosamples is tested, whereby information on the technical and biological variation of the biomarker is to be obtained. Optimally, biomarker validation is to be done by several independent laboratories using shared protocols and aliquots of high-quality biosamples, ensuring that the obtained data can be compared in a meta-analysis across the participating laboratories.[77] Recently, the key biomarkers for Alzheimer's disease Aβ42, T-tau, P-Tau have been validated in such a multicenter testing, defining a best-practice example.[78]

Emerging technologies for biomarker validation tend to follow a multiplex approach, in which the biomarker of interest is compared with other biomarkers within the same assay, thus demonstrating the added value of the novel biomarker, and at the same time determine the quality of the sample through analysis of decay markers. The use of next-generation sequencing and micro-arrays to study DNA/RNA molecules intrinsically have this advantage,[70] whereas smaller-scale multiplex assays using QPCR or branched-DNA probes have become standard practice. Popular methods to analyze protein biomarkers use an immunoassay format that is bead-based,[79] microtiter plate-based[80] or planar microarray-based.[81] Of particular interest is the use of targeted mass spectrometry in which up to 100 specific analytes of proteomic or metabolomic nature are quantitated per analysis with a sensitivity that approaches that of immunoassays.[82,83] Recently, a multicenter study with eight participating laboratories showed that such targeted mass-spectrometry assays can generate highly reproducible and precise results both within and across laboratories and instrument platforms.[84]

3.4.2.2   Biomarker Test Validation

Once the biomarker is confirmed by an independent validation study, the biomarker test can then be developed for clinical application. Biomarker development will now aim to achieve increased robustness, with a focus on obtaining maximum specificity and selectivity, whilst optimizing the ease and costs in production of the final format of the test. During this process, a selected sample set will be regularly tested to verify the performance of the test. Once developed to the final format, the biomarker test will again be scrutinized using a thorough analysis of independent clinical samples, thus generating the final biomarker test performance results that will be used for marketing. Again, high-quality biosamples are imperative to be successful in this phase.

A relevant case study of clinical biomarker discovery and development that outlines the various steps mentioned above is the Mammaprint, a microarray-based biomarker test for prediction of breast-cancer recurrence (see www.mammaprint.nl). The biomarker test is based on the original discovery of a 70-gene transcript signature, identified in 98 retrospectively collected primary breast-tumor biopsies, and that distinguished patients that after five years were disease free or developed distant metastasis.[85] The selected probes enabling analysis of this gene signature were converted into a multiplex microarray test,[86] and prospectively validated in the dedicated prospective MINDACT trial.[87,88] Further validation and testing of the 70-gene Mammaprint signature indicated additional value as predictor of therapeutic efficacy in the poor outcome predicted group,[89] whereas further bioinformatics research revealed how the 70 components of the signature are related to cancer biology.[90] Finally, it was recently shown that the gene expression assay can be performed both on fresh frozen tissue as on FFPE tissue with similar results, widening the application of this approach.[91]

3.4.3   Biobanking–Biomarker Collaborations

Considering the lack of quality and standardization between biobanks and the impact on biomarker validation, one may ask if too many specimens have been collected and stored as well as if too much (public) money has been invested.

Actually, most biobanks struggle to find a suitable, sustainable business model. Moreover, looking at the success stories about biobanks but also looking at the too high numbers of biomarkers that are dying on the shelves or are not efficiently validated by lack of funds it is clear that a new business model is needed.

An interesting option is a risk–benefit sharing model where biobanks and biomarker developers will benefit.[92] It would provide biobanks with an opportunity to become a true partner with a scientific input and interest into a biomarker project, and allows the biobank to create a long-term revenue stream to establish a sustainable business. At the same time, the risk–benefit sharing model would help decrease the up-front financial requirements for the scientists developing the biomarker. Under a risk–benefit-sharing model biobanks provide access to their biospecimen collection at a substantially reduced cost, in return for a share in the revenues from a sale of the biomarker intellectual property, or royalties on the sale of the final diagnostic test. The biobank would become a partner in the biomarker research team, and be directly involved in the development program. Many biobanks have scientific specialists in their specific disease domains associated with them, who can provide valuable input into a biomarker development program, as well as access to specific patients for further testing and validation work. Such a working model will enhance collaboration between biobanks and biomarkers developers and will certainly accelerate biomarker discovery and validation.[92]

A construction to share biospecimen is particularly useful in biomarker projects by consortia funded through public–private partnerships in which multiple partners, both from industry and academia, collaborate to reach objectives that would not be achievable by the single partners alone.[93] In particular, clinical biomarker validation and development projects require multiple collaborating partners whose joint objective strongly depends on the availability of well-annotated and high-quality biosamples. Through their biosamples, biobanks can contribute well to the knowledge and output developed by the consortium, whereas they can share in potential revenue of the generated biomarker assays, creating a mutual benefit for the biobank and other partners.

Due to the large amount of public funding, it would be timely to consider coordinating a global strategy for biobanking that aligns the content and expansion of biospecimen collections with the healthcare needs in term of biomarkers. Currently, we find that too many biobanks work independently and collect the same types of biosamples, whereas it can be nearly impossible to find collections with sufficient size and content for development and validation of specific biomarkers. A global independent organization that coordinates the expansion of biobank content will avert redundancy in biobanking and decrease costs, while increasing return on investment. Equilibrium between offer and demand should exist.

On the other hand, having access to high numbers of high-quality bio-specimens collected during clinical trials with drugs that failed in development could greatly contribute to the validation of candidate biomarkers, particularly new companion or prognostic biomarkers. This source of clinical biosamples is often not accessible due to restrictions in the IRB-approved clinical protocols or the informed consent signed by the participants. Even if IRB approval and informed consent would allow such use, the sponsor who initiated these clinical trials will have to agree to make biospecimen available to third parties for biomarker research.

It is timely that all players in the biobanking field, being biobankers, biomarker scientists, research organizations, pharmaceutical and diagnostic industrial parties, funding institutions, regulatory and ethical authorities, and governments, work together in a global concerted effort to increase the amount of definitive high quality biosamples available for biomarker research and development. Ultimately, this approach could lead to the establishment of a consolidated international registry tool referencing all gold-standard biobanks aimed at expedition of biomarkers in health care.

References

1. T. Newman and J. Freitag, *Applied Clinical Trials*, July 1, 2011.
2. National Cancer Institute. Biorepositories and Biospecimen Research Branch (BBRB) Best Practices, http://biospecimens.cancer.gov/practices/2011bp.asp. 2011.
3. "2012 ISBER Best Practices for Repositories: Collection, Storage, Retrieval and Distribution of Biological Materials for Research." *Biopreservation and Biobanking*, 2012, **10**(2), 79–161.
4. The Task Force on Biological Resource Centres. "OECD Best Practice Guidelines for Biological Resource Centres." Organization for Economic Cooperation and Development (OECD), 2007, 3–113.
5. Common Minimum Standards and Protocols for Biological Resource Centers Dedicated to Cancer Research"). "World Health Organization International Agency for Research on Cancer IARC Working Group Reports, Volume 2, 2007, http://www.iarc.fr/en/publications/pdfs-online/wrk/wrk2/index.php.
6. L. M. McShane, D. G. Altman, W. Sauerbrei, S. E. Taube, M. Gion and G. M. Clark, Statistics Subcommittee of NCI-EORTC Working Group on Cancer Diagnostics (2006) Reporting recommendations for tumor marker prognostic studies (REMARK), *Breast Cancer Res. Treat.*, 2006, **100**(2), 229–235.
7. http://www.stard-statement.org.
8. http://www.strobe-statement.org.
9. H. M. Moore, A. B. Kelly, S. D. Jewell, L. M. McShane, D. P. Clark, R. Greenspan, D. F. Hayes, P. Hainaut, P. Kim, E. A. Mansfield, O. Potapova, P. Riegman, Y. Rubinstein, E. Seijo, S. Somiari, P. Watson, H. U. Weier, C. Zhuand and J. J. Vaught, Biospecimen reporting for improved study quality (BRISQ), *Cancer Cytopathol.*, 2011, **119**(2), 92–102.

10. L. Miranda "Obtaining and maintaining high-quality tissue samples: scientific and technical considerations to promote Evidence-Based Biobanking Practice (EBBP)", In: M. Wigglesworth, *Management of Chemical and Biological Samples for Screening Applications*. Wiley, 2012, 59–79.

11. A. Cambon-Thomsen and G. A. Thorisson, L. Mabile for the BRIF working group, The role of a bioresource research impact factor as an incentive to share human bioresources, *Nature Genetics*, 2011, **43**, 6.

12. Manuel M. Morente, Laura Cereceda, Francisco Luna-Crespo and Maria J. Artiga, Managing a Biobank Network, *Biopreservation and Biobanking*, 2011, **9**(2), 187–190.

13. F. Betsou, D. L. Rimm, P. H. Watson, C. Womack, A. Hubel, R. Coleman, L. Horn, S. F. Terry, N. Zeps, B. J. Clark, L. B. Miranda, R. E. Hewitt and G. D. Elliott, *Biopreservation and Biobanking*, 2010, **8**(2), 81–88.

14. J. Vaught and N. C. Lockhart, *Clin. Chim. Acta*, 2012, **413**(19–20), 1569–1575.

15. R. Dash, J. Robb, D. Booker, W. Foo, D. Witte and L. Bry, *Arch Pathol Lab Med.*, 2012, **136**, 668–678.

16. D. N. Mayol-Heath, A. S. Keck and P. Woo, *Biopreservation and Biobanking*, 2011, **9**(3), 253–257.

17. S. Pathmasiri, M. Deschenes, Y. Joly, T. Mrejen, F. Hemmings and B. M. Knoppers, *Nature Biotechnology*, 2011, **29**(4), 319–323.

18. D. N. Mayol-Heath, A. S. Keck and P. Woo, *Biopreservation and Biobanking Biorepository*, 2010, 139–145.

19. K. Lindpainter, *Nature*, 2011, **470**, 175.

20. G. Poste, *Nature*, 2011, **149**, 156.

21. D. Simeon-Dubach and A. Perren, *Nature*, 2011, **475**, 454–455.

22. F. Prinz, T. Schlange and K. Asadullah, *Nature Reviews Drug Discovery*, 2011, **10**, 712.

23. L. Osherovich, *SciBX April*, 2011, **4**(15).

24. E. P. Diamandis, *J Natl Cancer Inst*, 2010, **102**, 1462–1467.

25. S. Silberman, *Wired Magazine*, (http://www.wired.com/magazine/2010/05/ff_biobanks/all/1), June 2010.

26. E. Foot, The three C"s for biomarker development, *Scrip*, 2011.

27. R. Balasubramanian, L. Muller, K. Kugler, W. Hackl, W. L. Pleyer, M. Dehmer and A. Graber A, *Biomarkers*, 2010, **15**(8), 677–683.

28. K. G. Kugler, W. O. Hackl, L. A. Mueller, H. Fiegl, A. Graber and R. M. Pfeiffer, *Journal of Clinical Bioinformatics*, 2011, **1**, 9.

29. G. Lippi, *Clin. Chem. Lab. Med.*, 2011, **49**(7), 1113–1126.

30. M. Dresse, D. Nagel, E. M. Ganser, G. Davis, B. Dowell, R. Doss and P. Stieber, *Tumour Biol.*, 2008, **29**, 35–40.

31. A. Rai, C. Gelfand, B. Haywood, D. Warunek, J. Yi, M. Schuchard, R. Mehigh, S. Cockrill, G. Scott, H. Tammen, P. Schulz-Knappe, D. Speicher, F. Vitzthum, B. Haab, G. Siest and D. Chan, *Proteomics*, 2005, **5**(13), 3262–3277.

32. H. Tammen, *Mol. Biol.*, 2008, **428**, 35–42.

33. A.-B. Halim *Drug Discovery and Development – Present and Future* ed. M. Izet Kapetanovic Publisher: InTech, 2011, Chapter 18, 401–424.
34. K. Bensalah, K. F. Montorsi and S. Shariat, *Eur. Urol.*, 2007, **52**, 601–1609.
35. R. J. Goodwin, A. M. Lang, H. Allingham, M. Boren and A. R. Pitt, *Proteomics*, 2010, **10**, 1751–1761.
36. M. E. Hammond, W. Rees, T. Belnap, B. Rowley, S. Catmull and W. Sause, *Arch Archives of Pathology & Laboratory Medicine*, 2010, **134**(4), 606–612.
37. C. Mathay, W. Yan, R. Chuaqui, A. Skubitz, J.-P. Jeon, N. Fall, F. Betsou and M. Barnes, *Biopreservation and Biobanking*, 2012, **10**(6), 532–542.
38. S. Unhale, A. Skubitz, R. Solomon and A. Hubel, *Biopreservation and Biobanking*, 2012, **10**(6), 493–500.
39. T. Steiniche, T. B. Vainer, M. Franzmann, R. Hagemann-Madsen and M. Bak, *Ugeskr Laeger*, 2008, **170**, 1050.
40. G. Hicks and B. F. Boyce, *Biotechnic & Histochemistry*, 2012, **87**(1), 14–17.
41. C. Marchio, M. Dowsett and J. Reis–Filho, *BMC Medicine*, 2011, **9**, 41.
42. I. Pinhel, F. Macneill, M. Hills, J. Salter, S. Detre, R. A'hern, A. Nerurkar, P. Osin, I. Smith and M. Dowsett, *Breast Cancer Res.*, 2010, **12**(5), R76.
43. V. Espina, C. Mueller, K. Edmiston, M. Sciro, E. F. Petricoin and L. A. Liotta, *Proteome Clin. Appl.*, 2009, **3**, 874–882.
44. S. M. Hewitt, F. A. Lewis, Y. Cao, R. C. Conrad, M. Cronin, K. D. Danenberg, T. J. Goralski, J. P. Langmore, R. G. Raja, P. M. Williams, J. F. Palma and J. A. Warrington, *Arch. Pathol. Lab. Med.*, 2008, **132**, 1929–1935.
45. Y. Miyatake, H. Ikeda, R. Michimata, S. Koizumi, A. Ishizu, N. Nishimura and T. Yoshiki, *Exp. Mol. Pathol.*, 2004, **77**, 222–230.
46. K. Sköld, M. Svensson, M. Norrman, B. Sjögren, P. Svenningsson and P. Andrén, *Proteomics*, 2007, **7**, 4445–4456.
47. A. Spruessel, G. Steimann, M. Jung, S. Lee, T. Carr, A.-K. Fentz, J. Spangenberg1, C. Zornig, F. Juhl and K. David, *Biotechniques*, 2004, **36**, 1030–1037.
48. F. Nkoy, M. Hammond, W. Rees, T. Belnap, B. Rowley, S. Catmull and W. Sause, *Arch. Pathol. Lab. Med.*, 2010, **134**, 606–612.
49. T. Khoury, T. S. Sait, H. Hwang, G. Chandrasekhar, D. Wilding and S. Ta, *Kulkarni.Mod. Pathol*, 2009, **22**, 1457–1467.
50. *American Society of Clinical Oncology: Journal of clinical Oncology*, 2010, **28**(16), 2784–2795.
51. J. Lengellé and E. Panopoulos, *F.Betsou Cytokine*, 2008, **44**(2), 275–282.
52. J. Li, C. Kil, K. Considine, B. Smarkucki, M. Stankewich, B. Balgley and A. Vortmeyer, *Laboratory Investigation*, 2012, 1–12.
53. J. Ioannidis and O. Panagiotou, *JAMA*, 2011, **305**(21), 2200–2210.
54. D. Simeon-Dubach, A. Burt and P. Hall, *J. Pathol.*, 2012, **228**, 431–433.
55. D. Ransohoff, *Journal of Clinical Epidemiology*, 2007, **60**, 1205–1219.
56. D. Ransohoff and M. Gourlay, *Journal of Clinical Oncology*, 2010, **28**(4), 698–704.

57. J. Carpenter, H. Moore, H. Juhl, G. Thomas and L. Miranda, *Biopreservation and Biobanking*, 2011, **9**(2), 129–131.
58. L. B. Miranda, K. Wyatt, I. Johnston, M. Milljanic, J. Chaffey, "Proof of concept" pilot study: Bioprocess chain of custody and bioresource sample management temperature observations. Sample level temperature trends and stability data obtained via utilization of bluechiip® temperature tracking technology. *Biopreservation and Biobanking*, 2013, **11**(2), in press.
59. V. Hofman, M.-C. Gaziello, C. Bonnetaud, M. Ilie, V. Mauro, E. Long, E. Selva, V. Gavric-Tanga, S. Lassalle, C. Butori, C. Papin-Michaud, N. Lerda, O. Bordone, C. Coelle, J-C. Sabourin, C. Chabannon and P. Hofman, *Annales de Pathologie*, 2012, **32**, 91–101.
60. M. Mann and N. L. Kelleher, *Proc. Natl. Acad. Sci. U.S.A*, 2008, **105**(47), 18132–18138.
61. J. C. Tran, L. Zamdborg, J. E. Lee, A. D. Catherman, K. R. Durbin, J. D. Tipton, A. Vellaichamy, J. F. Kellie, M. Li, C. Wu, S. M. Sweet, B. P. Early, N. Siuti N, R. D. LeDuc, PD. Compton, P. M. Thomas and N. L. Kelleher, *Nature*, 2011, **480**(7376), 254–258.
62. M. P. Stoop, L. Coulier, T. Rosenling, S. Shi, S. A. M. Smolinska, L. Buydens, K. Ampt, C. Stingl, A. Dane, B. Muilwijk, R. L. Luitwieler, P. A. Sillevis Smitt, R. Q. Hintzen, R. Bischoff, S. S. Wijmenga, T. Hankemeier, A. J. van Gool and T. M. Luider, *Mol. Cell Proteomics*, 2010, **9**(9), 2063–2075.
63. T. Rosenling, M. P. Stoop, A. Smolinska, B. Muilwijk, L. Coulier, S. Shi, A. Dane, C. Christin, F. Suits, P. L. Horvatovich, S.S. Wijmenga, L. M. Buydens, R. Vreeken, A.J. van Gool, T. M. Luider and R. Bischoff, *Clin Chem.*, 2011, **57**(12), 1703–1711.
64. F. Betsou, S. Lehmann, G. Ashton, M. Barnes, E. Benson, D. Coppolo, Y. DeSouze, J. Eliason, B. Glazer, F. Guadagni, K. Harding, D. J. Horsfall, C. Kleeberger, U. Nanni, A. Prasad, K. Shea, A. Skubitz, S. Somiari, E. Gunter, and International Society for Biological and Environmental Repositories (ISBER) Working Group on Biospecimen Science, *Cancer Epidemiology Biomarkers and Prevention*, 2010, **19**(4), 1004–1011.
65. A. J. van Gool, B. Henry and E. D. Sprengers, *Drug Discov.Today*, 2010, **15**(3–4), 121–126.
66. J. E. Dancey, P. L. Bedard, N. Onetto and T. J. Hudson, *Cell*, 2012, **148**(3), 409–420.
67. L. Turner, J. D. Heath and N. Kurn, *Methods Mol. Biol.*, 2011, **724**, 269–280.
68. O. Azimzadeh, Z. Barjaktarovic, M. Aubele, J. Calzada-Wack, H. Sarioglu, M. J. Atkinson and S. Tapio, *J Proteome Res.*, 2010, **9**(9), 4710–4720.
69. B. L. Hood, M. M. Darfler, T. G. Guiel, B. Furusato, D. A. Lucas, B. R. Ringeisen, I. A. Sesterhenn, T. P. Conrads, T. D. Veenstra and D. B. Krizman, *Mol. Cell Proteomics*, 2005, **4**(11), 1741–1753.
70. M. L. Metzker, *Nat. Rev. Genet.*, 2010, **11**(1), 31–46.
71. C. Gilissen, A. Hoischen, H. G. Brunner and J. A. Veltman, *Eur. J Hum Genet.*, 2012, **20**(5), 490–497.

72. G. K. Geiss, R. E. Bumgarner, B. Birditt, T. Dahl, N. Dowidar, D. L. Dunaway, H. P. Fell, W. S. Ferree, R. D. George, T. Grogan, J. J. James, M. Maysuria, J. D. Mitton, P. Oliveri, J. L. Osborn, T. PengT, A. L. Ratcliffe, P. J. Webster, E. H. Davidson, L. Hood and K. Dimitrov, *Nat. Biotechnol.*, 2008, **26**(3), 317–325.

73. R. D. Canales, Y. Luo, B. Willey, C. C. Austermiller, C. Barbacioru, K. Boysen, R. V. Hunkapiller, C. R. Jensen, K. Y. Knight, Y. Lee, B. Ma, A. Maqsodi, E. H. Papallo, K. Peters, P. L. Poulter, R. R. Ruppel, L. Samaha, W. Shi, L. Yang and F. M. Zhang, *Goodsaid. Nat Biotechnol.*, 2006, **24**(9), 1115–1122.

74. P. Mallick and B. Kuster, *Nat. Biotechnol.*, 2010, **28**(7), 695–709.

75. M. P. Stoop, L. J. Dekker, M. K. Titulaer, R. J. Lamers, P. C. Burgers, P. A. Sillevis Smitt, A. J. van Gool, T. M. Luider and R. Q. Hintzen, *J. Proteome Res.*, 2009, **8**(3), 1404–1414.

76. R. Seigneuric, L. Markey, D. S. Nuyten, C. Dubernet, C. T. Evelo, E. Finot and C. Garrido, *Curr. Mol. Med.*, 2010, **10**(7), 640–652.

77. H. Mischak, G. Allmaier, R. Apweiler, T. Attwood, M. Baumann, A. Benigni, S. E. Bennett, R. Bischoff, E. Bongcam-Rudloff, G. Capasso, J. J. Coon, P. D'Haese, A. F. Dominiczak, M. Dakna, H. Dihazi, J. H. Ehrich, P. Fernandez-Llama, D. Fliser, J. Frokiaer, J. Garin, M. Girolami, W. S. Hancock, M. Haubitz, D. Hochstrasser, R. R. Holman, J. P. Ioannidis, J. Jankowski, R. A. Julian, J. B. Klein, W. Kolch, T. Luider, Z. Massy, W. B. Mattes, F. Molina, B. Monsarrat, J. Novak, K. Peter, P. Rossing, M. Sánchez-Carbayo, J. P. SchanstrA, O. J. Semmes, G. Spasovski, D. Theodorescu, V. Thongboonkerd, R. Vanholder, T. D. Veenstra, E. Weissinger, T. Yamamoto and A. Vlahou, *Sci. Transl. Med*, 2010, **2**(46), 1–6.

78. N. Mattsson, U. Andreasson, S. Persson, H. Arai, S. D. Batish, S. Bernardini, L. Bocchio-Chiavetto, M. A. Blankenstein, M. C. Carrillo, S. Chalbot, E. Coart, D. Chiasserini, N. Cutler, G. Dahlfors, S. Duller, A. M. Fagan, O. Forlenza, G. B. Frisoni, D. Galasko, D. Galimberti, H. Hampel, A. Handberg, M. T. Heneka, A. Z. Herskovits, S. K. Herukka, D. M. Holtzman, C. Humpel, B. T. Hyman, K. Iqbal, M. Jucker, S. A. Kaeser, E. Kaiser, E. Kapaki, D. Kidd, P. Klivenyi, C. S. Knudsen, M. P. Kummer, J. Lui, A. Lladó, P. Lewczuk, Q. X. Li, R. Martins, C. Masters, J. McAuliffe, M. Otto, G. P. Paraskevas, L. Parnetti, R. C. Petersen, D. Prvulovic, H. P. de Reus, R. A. Rissman, E. Scarpini, A. Stefani, H. Soininen, J. Schröder, L. M. Shaw, A. Skinningsrud, B. Skrogstad, A. Spreer, L. Talib, C. Teunissen, J. Q. Trojanowski, H. Tumani, R. M. Umek, B. Van Broeck, H. Vanderstichele, L. Vecsei, M. M. Verbeek, M. Windisch, J. Zhang, H. Zetterberg and K. Blennow, *Alzheimers Dement.*, 2011, **7**(4), 386–395.

79. R. J. Fulton, R. L. McDade, P. L. Smith, L. J. Kienker and J. R. Kettman, Jr, *Clin. Chem.*, 1997, **43**(9), 1749–1756.

80. N. Mattsson, M. Axelsson, S. Haghighi, C. Malmeström, G. Wu, R. Anckarsäter, S. Sankaranarayanan, U. Andreasson, S. Fredrikson,

A. Gundersen, L. Johnsen, T. Fladby, A. Tarkowski, E. Trysberg, A. Wallin, H. Anckarsäter, J. Lycke, O. Andersen, A. J. Simon, K. Blennow and H. Zetterberg, *Mult Scler.*, 2009, **15**(4), 448–454.

81. M. Hartmann, J. Roeraade, D. Stoll, M. F. Templin and T. O. Joos, *Annals of Bioanal. Chem.*, 2009, **393**(5), 1407–1416.

82. L. Anderson and C. L. Hunter, *Mol. Cell Proteomics*, 2006, **5**(4), 573–588.

83. S. Gallien, E. Duriez and B. Domon, *J. Mass Spectrom.*, 2011, **46**(3), 298–312.

84. T. A. Addona, S. E. Abbatiello, B. Schilling, S. J. Skates, D. R. Mani, D. M. Bunk, C. H. Spiegelman, L. J. Zimmerman, A. J. Ham, S. C. Hall, S. Allen, R. K. Blackman, C. H. Borchers, C. Buck, H. L. Cardasis, M. P. Cusack, N. G. Dodder, B. W. Gibson, J. M. Held, T. Hiltke, A. Jackson, E. B. Johansen, C. R. Kinsinger, J. Li, M. Mesri, T. A. Neubert, R. K. Niles, T. C. Pulsipher, D. Ransohoff, H. Rodriguez, P. A. Rudnick, D. Smith, D. L. Tabb, T. J. Tegeler, A. M. Variyath, L. J. Vega-Montoto, A. Wahlander, S. Waldemarson, M. Wang, J. R. Whiteaker, L. Zhao, N. L. Anderson, S. J. Fisher, D. C. Liebler, A. G. Paulovich, F. E. Regnier, P. Tempst and S. A. Carr, *Nat. Biotechnol.*, 2009, **27**(7), 633–641.

85. L. J. van't Veer, H. Dai, M. J. van de Vijver, Y. D. He, A. A. Hart, M. Mao, H. L. Peterse, H. K. van der Kooy, M. J. Marton, A. T. Witteveen, G. J. Schreiber, R. M. Kerkhoven, C. Roberts, P. S. Linsley, R. Bernards and S. H. Friend, *Nature*, 2002, **415**(6871), 530–536.

86. A. M. Glas, A. Floore, L. J. Delahaye, T. Witteveen, R. C. Pover, N. Bakx, J. S. Lahti-Domenici, T. J. Bruinsma, M. I. O. Warmoes, R. Bernards, L. F. Wessels, L. J. Van't Veer. Converting a breast cancer microarray signature into a high-throughput diagnostic test, *BMC Genomics*, 2006, **7**, 278.

87. S. Mook, L. J. Van't Veer, E. J. Rutgers, M. J. Piccart-Gebhart and F. Cardoso, *Cancer Genomics Proteomics*, 2007, **4**(3), 147–155.

88. F. Cardoso, M. Piccart-Gebhart, L. Van't Veer and E. Rutgers, *TRANSBIG Consortium. Mol. Oncol.*, 2007, **1**(3), 246–251.

89. M. E. Straver, A. M. Glas, J. Hannemann, J. Wesseling, J. E. J. van de VijJ, M. J. Vrancken Peeters, H. van Tinteren, L. J. Van't Veer and S. Rodenhuis, *Breast Cancer Res. Treat.*, 2010, **119**(3), 551–558.

90. S. Tian, P. Roepman, L. J. Van't Veer, R. Bernards, R. F. de Snoo and A. M. Glas, *Biomark Insights*, 2010, **28**(5), 129–138.

91. L. Mittempergher, J. J. de Ronde, M. Nieuwland, R. M. Kerkhoven, R. M. I. Simon, I. E. J. Rutgers, L. F. Wessels and L. J. Van't Veer, *PLoS One*, 2011, **6**(2), e17163.

92. www.trans-hit.com.

93. B. M. Knoppers and T. J. Hudson, *Hum Genet.*, 2011, **130**, 329–332.

Sample Preparation and Profiling

CHAPTER 4

Sample Preparation and Profiling: Biomarker Discovery in Body Fluids by Proteomics

N. GOVORUKHINA AND R. BISCHOFF*

University of Groningen, Centre of Pharmacy, Analytical Biochemistry, Antonius Deusinglaan 1, 9713 AV Groningen, The Netherlands
*Email: r.p.h.bischoff@rug.nl

4.1 Introduction

Discovery of biomarkers is a fast-developing field in proteomics, especially for different types of cancer.[1–7] While there is no single approach in proteomics to analyze body fluids,[8] there are challenges that are common to all body fluids such as their extremely complex composition, the wide concentration dynamic range (*e.g.* 11 orders of magnitude for human serum)[9] and the possible interference of low molecular weight compounds (*e.g.* lipids) in the analysis. Notably the presence of highly abundant proteins in serum and plasma masks proteins of lower abundance and prevents their detection and identification during nontargeted proteomics analyses. On the other hand urine is rather diluted and contains many degradation products or metabolites that are not registered in sequence databases, making protein identification by standard database searches difficult. The discrimination between variability in composition due to an ongoing disease and natural variability is another challenging aspect in the analysis of body fluids.

RSC Drug Discovery Series No. 33
Comprehensive Biomarker Discovery and Validation for Clinical Application
Edited by Péter Horvatovich and Rainer Bischoff
© The Royal Society of Chemistry 2013
Published by the Royal Society of Chemistry, www.rsc.org

Four types of body fluids will be discussed in this chapter: serum, urine, cerebrospinal fluid (CSF) and epithelial lining fluid (ELF), highlighting different aspects of body-fluid analysis. The comparative analysis of samples from different patient groups and healthy controls is often the first step in biomarker discovery. Standardized operating procedures (SOPs) are needed to produce reproducible data in order to distinguish disease-related differences from analytical or preanalytical artifacts.[10,11] Data processing and statistical validation are important parts of biomarker discovery projects. These topics will be treated in Chapters 7–9 of this book. Finally, potential biomarkers have to be subjected to biological validation across multiple sets of samples preferably from different medical centers to establish their robustness. The major goal in proteomics-based biomarker discovery is to find proteins with high sensitivity and specificity for a given condition resulting in high positive and negative predictive values that are suitable for the target group or even for population-wide screening.[12]

In this chapter we present practical examples of proteomics of serum samples from cervical cancer patients, of urine from patients with proteinuria, of ELF from Chronic Obstructive Pulmonary Disease (COPD) patients and of CSF from a preclinical animal model of Multiple Sclerosis (MScl). Our descriptions focus on sample handling, sample preparation, protein and peptide separation and mass spectrometry with some discussion of issues related to data processing.

4.2 Samples

4.2.1 Serum

4.2.1.1 Sample Preparation

The preparation of serum by coagulation of whole blood is a biochemical process that is difficult to control, since cellular metabolism and the activity of enzymes continue for many hours after the collection of the blood sample.[13,14] Clotting time affects the serum proteome and especially the low molecular weight fraction (peptidome).[3,15,16] Previous studies showed that clotting time needs to be controlled when studying the low molecular weight fraction of the proteome and notably peptides that are related to the clotting process.[3,10,17]

Our own study showed that the larger part of the measured proteome (about the top 100 proteins) did not change significantly when allowing blood to coagulate for at least 2 h at room temperature in a tube containing a clotting activator.[18] The effect of clotting time on LC-MS profiles of serum, obtained from healthy volunteers, was studied after digestion with trypsin, followed by univariate and multivariate statistics. The supernatant of acid-precipitated serum samples, containing low molecular weight proteins and peptides (peptidome)[3,19,20] was analyzed as well. Serum from a single healthy volunteer was prepared by letting fresh blood coagulate at room temperature for 1, 2, 4, 6 or 8 h followed by centrifugation at room temperature for 10 min at 3000 rpm.

Comparative quantitative analyses were performed by label-free as well as stable-isotope labeling (iTRAQ™)[21] – nano-LC-MS. A set of peptide peaks was observed in the supernatant of TCA-precipitated serum samples. The major peak at m/z 1465.8, which was identified as an N-terminally truncated form of Fibrinopeptide A (FPA) (DSGEGDFLAEGGGVR, doubly charged ion at m/z 733.8) by MS/MS, was shown to be influenced by clotting time.[18] The majority of the other peptides were derived from FPA while sequences of Fibrinopeptide B (FPB) were identified as well as peptides influenced by clotting time. Clotting time is thus an important factor during sample preparation notably when focusing on the low molecular weight peptidome. The results showed further an absence of major changes in protein levels in serum samples with clotting times of 1 to 8 h. We recommend to use serum samples that have clotted for at least 2 h. When shorter clotting times are used, they need to be standardized and controlled.

A number of other parameters may also have an effect on the abundance of the measured peptides and proteins, thus affecting the outcome of the statistical analysis. Important factors such as the type of the blood collection tube, the ratio of trypsin/protein during digestion, the number of freeze–thaw cycles, depletion of high-abundance proteins, stopping the trypsin digestion with acid or the residence time in the autosampler at 4 °C may all have an effect on the final outcome and confound disease-related biomarkers with experimental artifacts. We approached this problem in a Factorial Design to prioritize various preanalytical factors with respect to their influence on the final data. The main result of our study (unpublished data) is that the hemolysis level, which is a parameter that is not controlled during sampling of serum, is the most important factor affecting the proteome meaning that serum with a high hemolysis level has to be discarded from the analysis. We are currently assessing ways to define a threshold for the level of hemolysis based on the abundance of hemoglobin-related peptides in the digested serum samples, since it is not possible to change the level of hemolysis in serum samples in retrospect.

4.2.1.2   Depletion of High-Abundance Proteins

Twenty two proteins represent 99% of the total amount of protein in serum.[22] These proteins, and peptides derived from them, often mask the detection of less-abundant proteins. In order to improve concentration sensitivity, high-abundance proteins can be removed by immunoaffinity chromatography based on commercial products from a number of manufacturers.[23–25] For example, we used a Multiple Affinity Removal column (Agilent, 4.6×50 mm, Part # 5185-5984 Palo Alto, California, USA). The flow-through fraction, serum that has been depleted of the 6 most-abundant proteins [albumin, IgG, alpha-1-antitrypsin, IgA, transferrin and haptoglobin], contained only 5% of the original protein amount. It should be taken into account that depletion of high-abundant proteins may lead to the loss of important information. For example, it was previously shown that proteolytic fragments of abundant serum proteins may have diagnostic and prognostic value.[26] On the other hand, some of the low-abundance proteins of interest, like cytokines, may be codepleted.[27–31]

4.2.1.3   Protein/Peptide Prefractionation and Enrichment

The complexity of serum samples with or without depletion can also be reduced in other ways. The "classical" approach of 2D gel electrophoresis followed by LC-MS/MS after in-gel digestion with trypsin is still of interest, notably when it comes to relating protein species to disease. As this approach is rather work intensive and suffers from a limited concentration dynamic range, it is advisable to use a pooling strategy for biomarker candidate discovery and to work with depleted serum.

Chromatographic prefractionation at the protein level opens further possibilities to enhance concentration sensitivity and to detect protein isoforms. High-temperature reversed-phase chromatography (Macroporous Reversed-Phase mRP-C18 column, Agilent, 4.6 × 50 mm, Part # 5188-5231 Palo Alto, California,

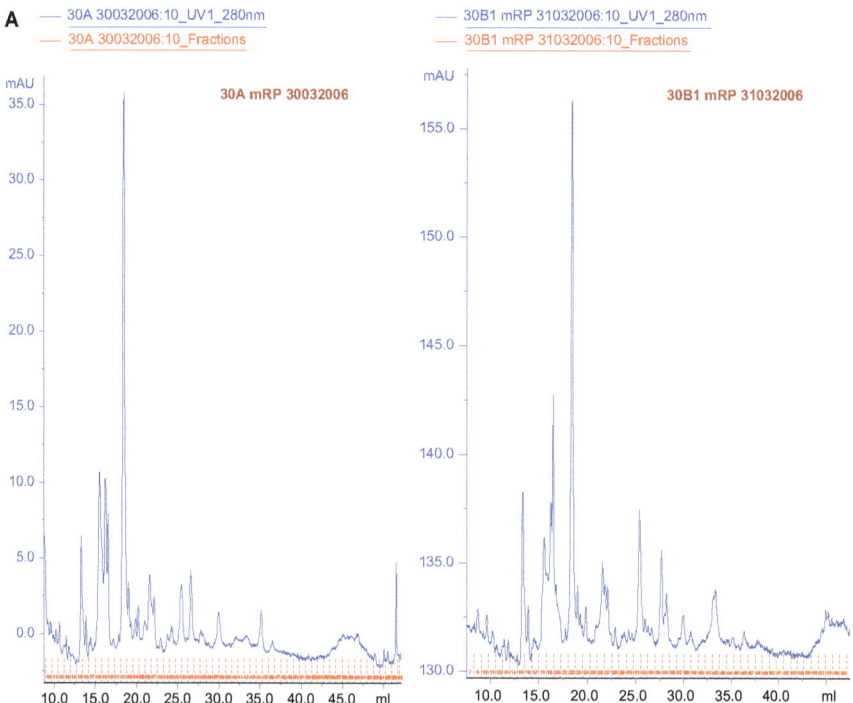

Figure 4.1   (A) Prefractionation of a depleted (Multiple Affinity Removal column [Agilent, 4.650 mm, Part # 5185-5984 Palo Alto, California, USA], removing the 6 most-abundant proteins [albumin, IgG, alpha-1-antitrypsin, IgA, transferrin and haptoglobin]) serum sample of a cervical cancer patient before (left) and after (right) medical treatment by reversed-phase chromatography at 80 1C on an mRP column.

USA) under denaturing conditions at 80 °C results in fractions that can be further analyzed by gel electrophoresis, Western blotting or LC-MS/MS after proteolytic digestion.[32,33] The recommended column load is ~300 µg (about 300 µl) of depleted serum containing 0.48 g urea and 13 µL of glacial acetic acid (see Figures 4.1B, Coomassie staining and 4.2B, silver staining, as examples and see www.agilent.com/chem/bioreagents for details). Fractionation at 80 °C and a flow rate of 0.75 mL/min (detection at 280 nm) showed excellent reproducibility[32] over at least 200 runs (unpublished data). An example of protein fractionation by high-temperature reversed-phase chromatography in serum from cervical cancer patients before and after medical treatment and subsequent analysis by 1D SDS-PAGE shows that certain proteins are clearly increased in patient serum after treatment (see Figures 4.1 and 4.2) while others show the opposite behavior (not shown).

The combination of high-abundance protein depletion with high-temperature reversed-phase separation of proteins, SDS-PAGE, in-gel digestion[34] and nano-LC-MS/MS allows sufficient sensitivity to identify even low-abundance proteins such as the E1 protein of human papilloma virus (Figure 4.3, left lower panel shows the Mascot identification result). As this is a very powerful but time-consuming way of analyzing the serum proteome, we are currently investigating whether a sample pooling strategy is feasible. Our

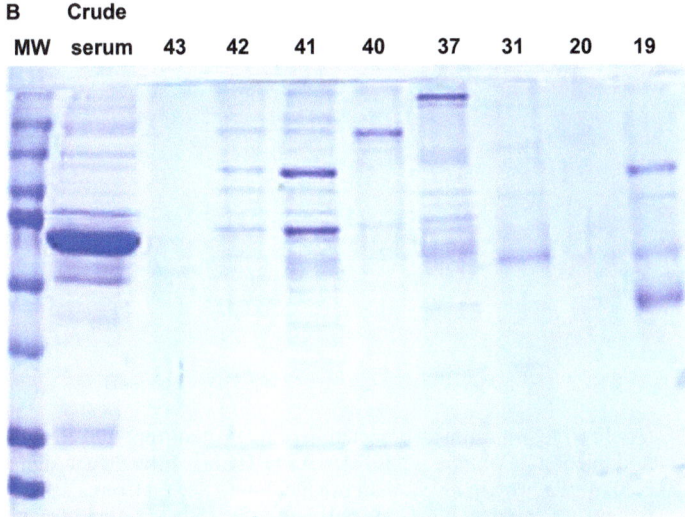

Figure 4.1   (B) Example of 1D gel electrophoresis of mRP column fractions 19–20, 31, 37 and 40–43 demonstrates good fractionation and reduction of the complexity (see line 2 for crude serum). SDS PAGE was performed in a Mini-Protein III cell (Bio-Rad, www.biorad.com) using a 12% gel with 0.1% SDS according to the manufacturer's instructions. The samples were boiled with standard loading buffer containing 0.02M DTT for 1 min. Staining of gels was performed with Coomassie Brilliant Blue R concentrate (Sigma, www.sigmaaldrich.com) as prescribed by the manufacturer.

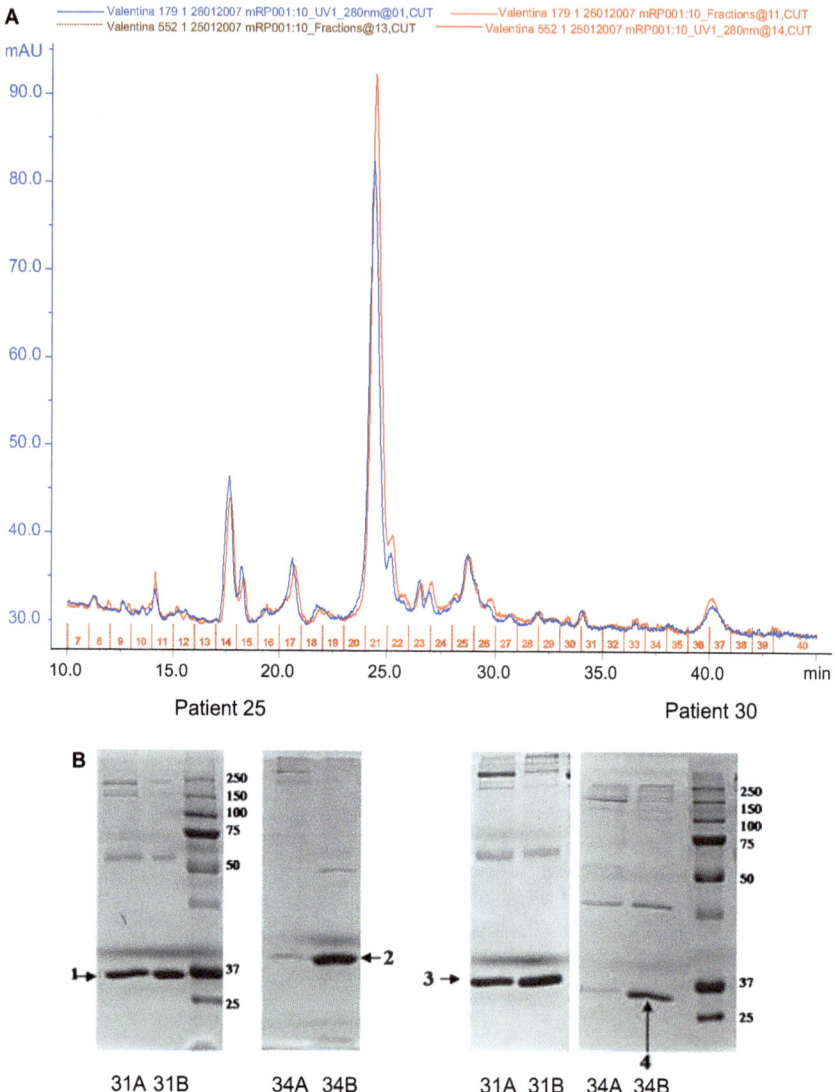

Figure 4.2   (A) Prefractionation of depleted serum (ProteoPrep® 20 Plasma Immunodepletion Columns, SIGMA-ALDRICH, http://www.sigmaaldrich.
com) at the protein level on an mRP column. Serum from a cervical cancer
patient was analyzed before (red) and after (blue) medical treatment. (B)
1D gel electrophoresis of mRP column fractions (cervical cancer patients
25 and 30):
31A: patient before medical treatment, fraction #31 (mRP column)
31B: patient after medical treatment, fraction #31 (mRP column)
34A: patient before medical treatment, fraction #34 (mRP column)
34A: patient after medical treatment, fraction #34 (mRP column).
Arrows (1–4) indicate differently expressed proteins.

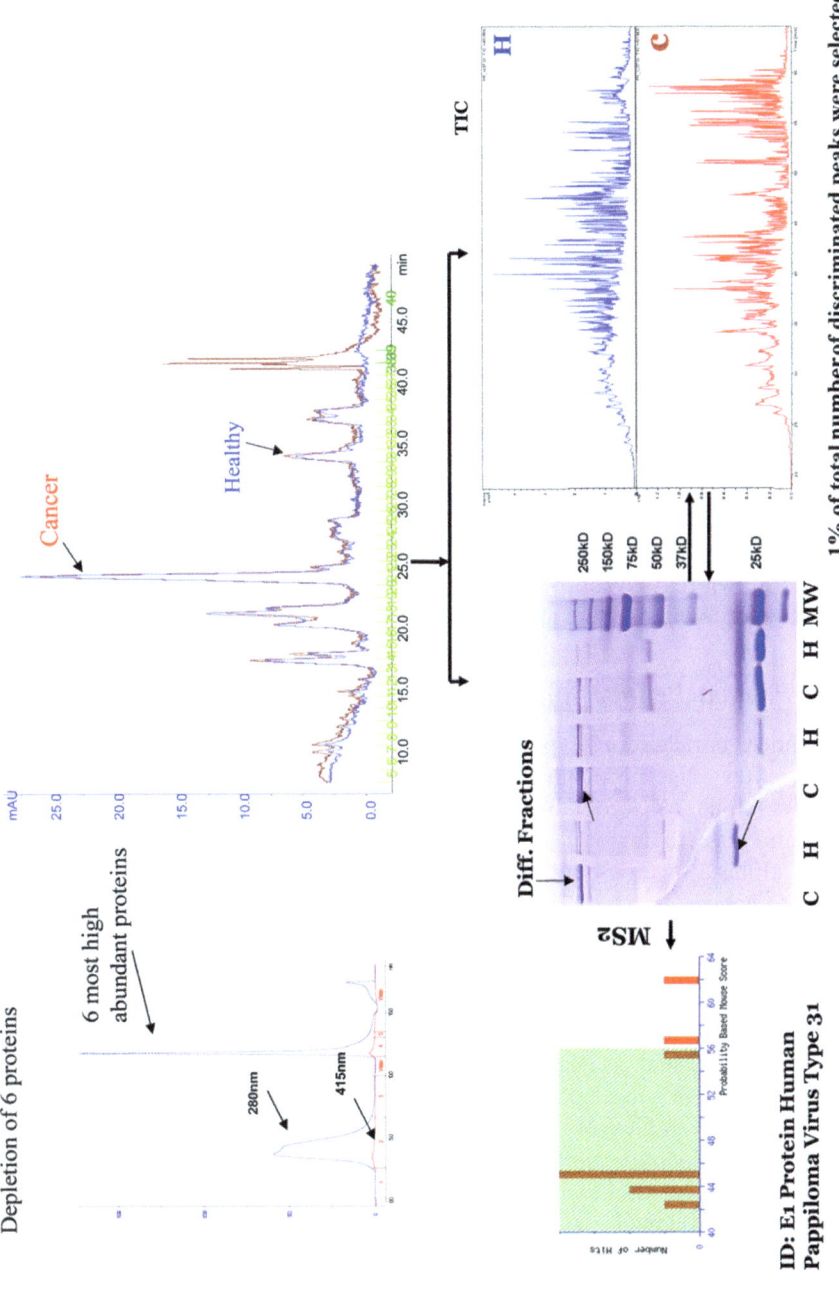

Figure 4.3 Example of a protein prefractionation workflow for the analysis of pooled serum samples. Red chromatograms indicated with a C show data from patients with cervical cancer, while blue chromatograms indicated with an H show data from cancer patients that were considered cured after treatment.

initial data indicate that this is a viable approach but verification analyses of biomarker candidates in individual serum samples are still ongoing.

Prefractionation at the level of peptides by Strong Cation Exchange (SCE) chromatography (PolySULFOETHYL, 4.6 × 200 mm column, column volume: 3.32 mL (PolyLC, Columbia, Maryland, USA)) after digestion is widely used and known as so-called "shotgun proteomics". It is particularly popular in combination with chemical stable-isotope labeling of which the isobaric tagging reaction for accurate quantification (iTRAQ) is a widely applied approach.[18] SCE allows excess iTRAQ reagent to be removed and to prefractionate the extremely complex peptide mixture prior to reversed-phase nano-LC-MS/MS analysis. Peptides are fractionated using a salt (*e.g.* KCl) gradient at acidic pH prior to reversed-phase LC-MS/MS.

Another approach to simplify the serum or plasma proteome and to improve the concentration sensitivity is based on lectin-affinity chromatography to enrich glycoproteins.[35–42] Using different lectin columns prior to deglycosylation, trypsin digestion and LC-MS/MS analysis allows addressing different sugar motives, which results not only in better concentration sensitivity but also in enhanced selectivity. Since glycoproteins constitute by far the largest part of the serum or plasma proteome (except for albumin, which is nonglycosylated and makes up 50% of the total amount of protein in blood serum/plasma), lectin-affinity enrichment is also an effective way of depleting albumin.[43] It has been furthermore shown that glycoproteins may be sensitive markers[35–37] for various diseases including ovarian,[38] colorectal,[40] pancreatic[39] and breast cancers.[38,41,42]

4.2.1.4   *Protein/Peptide Enrichment*

Immunoaffinity enrichment of proteins or peptides with specific antibodies before or after digestion allows to enhance concentration sensitivity significantly. Alpha-2 macroglobulin is a serum protease inhibitor that binds cytokines and growth factors.[44] Immunoaffinity enrichment and proteomic analysis of alpha-2 macroglobulin complexes from the serum of six patients with prostate cancer *versus* six controls without malignancy were reported by Burgess *et al.*[44] Endogenous immunoglobulins were removed prior to immunoaffinity enrichment by Protein A affinity chromatography and the "precleaned" samples were subjected to immunoprecipitation with an anti-alpha-2 macroglobulin antibody (Ab7337, Abcam). SDS-PAGE analysis showed that coenriched proteins, such as heat-shock protein 90 alpha (HSP90α), were only found in the serum of cancer patients.

The inclusion of stable-isotope-labeled internal standards in combination with capture by antipeptide antibodies (SISCAPA) prior LC-MS/MS allows to reach sub-ng/mL sensitivities even in protein-rich biological matrices.[45–49] SISCAPA was developed to capture individual tryptic peptides from a complete human plasma digest with the help of magnetic beads. SISCAPA followed by targeted LC-MS/MS in the selected reaction monitoring (SRM) mode showed an enrichment of alpha 1-antichymotrypsin and lipopolysaccharide-binding protein by 1800 and 18 000 times, respectively.[47]

Nonspecific binding to the SISCAPA antibody or the magnetic beads was not reported. Peptides with stable-isotope labels (^{13}C, ^{15}N) were used with a mass shift of +8 or +10 Da, respectively, and the three most-abundant transitions were monitored for each peptide. One of the advantages of SISCAPA is the possibility of peptide enrichment from a large volume of plasma (1 mL) with limits of detection and quantification in the pg/mL range. SISCAPA can further be automated and multiplexed (*e.g.* 9 targets in one assay).[46,48] It was recently reported that more than half of the antipeptide antibodies (220 out of 403) are applicable for the detection of peptides at concentrations of less than 0.5 fmol/μL in human plasma corresponding to protein concentrations of less than 25 ng/mL.[47] More than 1000 reported biomarker candidates for breast cancer obtained in 13 independent sets of data including tissue data (based on a mouse model) were studied to determine how reliable they are in predicting tumor development. 88 new SRM assays were developed and evaluated in 80 plasma samples confirming the presence of the tumor in 36 out of 80 plasma samples of tumor-bearing animals.[47] The authors assume that a similar approach is applicable for the analysis of large numbers of human serum samples. High-throughput technologies for multiple SISCAPA-based SRM analyses open considerable opportunities to measure hundreds of candidate protein biomarkers for different diseases in a reasonable time and with sub-nM sensitivity.[47] It is, however, clear that further methodological improvements are necessary, since most current cancer biomarkers have threshold values at or below the ng/mL range.

Another technological development to perform enrichment of proteins and peptides is based on the "Nanotrap Biomarker Discovery Platform" (www.BiomarkerDiscovery.com). This approach allows capturing low-abundant and excluding highly abundant proteins. The core–shell hydrogel particles are composed of different affinity monomers and reactive dyes that allow fractionation and enrichment of low-abundance proteins. Different classes of particles contain affinity baits inside the core and on the shells with a variety of pore structures. Those particles are designed for the enrichment of low-abundant proteins based on size and affinity. NIPAm (N-isopropylacrylamide) nanoparticles with a Cibacron BlueF3GA bait were successfully used to enrich human growth hormone (hGH) in urine from an initial concentration level of 1 pg/mL.[50] The hydrogel microparticles were also applied to enrich a bacterial antigen related to Lyme disease in urine samples.[51] This approach further allowed identification of fragments from low-abundant proteins that were not detectable in serum samples from patients with early-stage ovarian cancer without enrichment.[52]

Details concerning peptide analysis by LC-MS will be treated in Chapter 5 and details about LC-MS data analysis will be dealt with in Chapters 7–9. We therefore refrain from discussing these important topics at this point.

Equal amounts of protein have to be used for comparative analyses. For this purpose we determine protein concentrations in body fluids by high-temperature, macroporous reversed-phase chromatography with concomitant desalting and denaturing of the proteins.[95]

4.2.2   Urine

Urine is a complex body fluid of variable composition, as it contains the metabolic end products of biological processes that are ongoing in the body. The protein composition of urine is hard to predict from nucleic acid sequence data due to many modifications including proteolysis. Changes in the composition of urine are primarily related to diseases that are associated with the urogenital tract and especially the kidneys.[53–57] There have been recent attempts to obtain an overview over the various studies that focused on the urine proteome and peptidome in order to facilitate comparisons and to avoid unnecessary repetition.[58]

Biomarker studies in urine are complicated by its complexity and by the large interindividual variability, since the composition of urine does not underlie strict homeostatic control such as blood. However, since large amounts of urine can be easily obtained, it is a body fluid that is widely used in the diagnosis of for example metabolic disorders.[59]

Next to water-soluble proteins and peptides, urine also contains vesicles, the so-called urinary exosomes, which harbor membrane-associated proteins.[60–62] Exosomes can be enriched by differential centrifugation and may provide access to a different class of biomarkers than the soluble part. This is currently under investigation, notably with respect to cancer of the urogenital tract.[55,57]

The comparative analysis of urine has been extensively used for the discovery of disease biomarkers.[63–68] While a number of interesting leads were found, further validation is needed. Notably large interindividual variation, standardized urine collection and compound stability are issues that require careful consideration before validated biomarkers can enter clinical practice.

Different separation methods have been used in combination with mass spectrometry to study the urinary proteome, namely one- or two-dimensional gel electrophoresis (1D/2D SDS-PAGE), reversed-phase liquid chromatography (RPLC), and capillary electrophoresis (CE).[69–73] An array of methods is used for sample preparation of the urinary proteome including protein precipitation, ultracentrifugation,[72] solid-phase extraction[59] or enrichment of smaller proteins and peptides by restricted access chromatography.[74,75]

Data processing includes the alignment of chromatograms in the time domain,[76–80] the discrimination of peaks from background noise and spikes,[81] the normalization of intensities and the generation of a common peak matrix from all analyzed samples.[59,82–84] These aspects will be presented in detail in Chapters 7–9 of this book.

4.2.3   Epithelial Lining Fluid (ELF)

4.2.3.1   Sample Preparation

Epithelial lining fluid (ELF) constitutes the interface between the lung surface and the airspace. It is thus a fluid that likely reflects changes in the environment through exchange of compounds between inhaled air and ELF as well as the response of notably the epithelial cell layer to compounds in the air. ELF is therefore ideally suited to study lung diseases such as Chronic Obstructive Pulmonary Disease (COPD) that is caused by the inhalation of cigarette smoke or due to air pollution.

ELF is sampled during bronchoscopy using a specially designed double-walled optical bronchoscope harboring a microsampling probe with an adsorptive tip.[85] Protein analyses in tissue biopsies, biofluids such as serum, urine, bronchoalveolar lavage fluid (BALF) or sputum have been performed in view of discovering and validating biomarkers for pulmonary diseases such as COPD or asthma. ELF is attractive for proteomics studies due to the higher concentration of proteins compared to BALF and the additional option of sampling from well-defined, multiple locations in the lung. Usually 0.2–0.4 mL of ELF can be sampled per probe, which means that the volume per person is around 1 mL. Extraction of proteins and other aspects of sample preparation have been described in detail.[85] Blood contamination is one of the potential caveats of ELF sampling due to scratching the surface of the lung with the probe. Visibly contaminated samples should be discarded from further analysis. We are currently working on establishing a more objective way of assessing blood contamination using quantitative proteomics data.

4.2.3.2   Quantitative Proteomics (iTRAQ Labeling)

Stable-isotope labeling through chemical derivatization of peptides is widely used to assess whether there is a statistically significant difference in protein levels between patients and controls. Isobaric tagging with labels that contain reporter ions that are liberated during collision-induced dissociation (CID) (*e.g.* iTRAQ) is well suited for the multiplexed, comparative analysis of proteomic samples. We applied the 8-plex iTRAQ approach to compare protein profiles in ELF from healthy individuals and COPD patients. ELF samples containing 50 µg protein each were used for iTRAQ labeling. Samples were dissolved in 0.5 M triethylammonium bicarbonate (TEAB) followed by denaturation in 2% SDS, reduction with 50 mM tris-(2-carboxyethyl)phosphine (TCEP) and alkylation with 200 mM methyl methanethiosulfonate (MMTS). Each sample was digested overnight at 37 °C with trypsin (sequencing grade modified trypsin, # V5111, www.promega.com) at an enzyme to protein ratio of 1:6. Subsequently, each sample was labeled (see www.appliedbiosystems.com for details). Four COPD (labeled with tags 113, 114, 115, 116, respectively) and four healthy control samples (labeled with tags 117, 118, 119 and 121, respectively) were compared. The individually labeled protein digests were combined and the peptide mixture was subjected to SCE chromatography (PolyLC, 2.1 × 200 mm column, Columbia, Maryland, USA). The SCE fractions were subjected to nano-LC-MS/MS analysis (see Figure 4.4 for an example). Prefractionation by SCE allows removing excess iTRAQ reagent and simplifying reversed-phase nano-LC-MS/MS analysis.

To obtain reliable results on both the quantification and the identification level, it is critical to exclude all peptides without iTRAQ modification and to

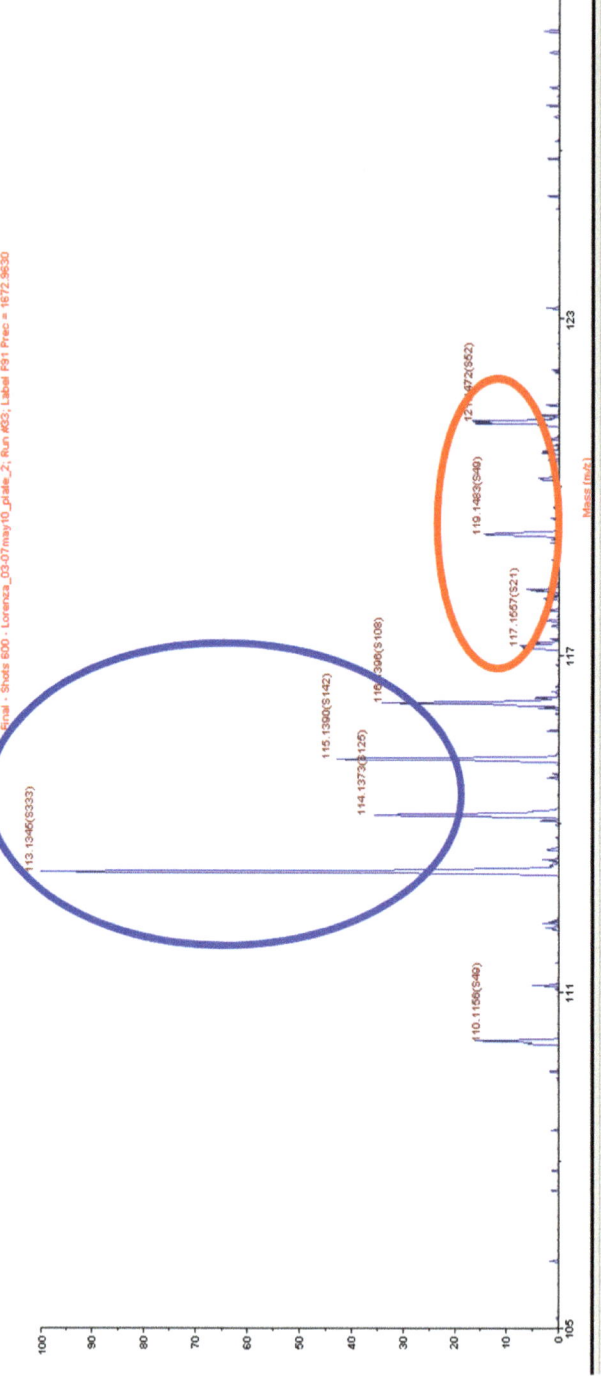

Figure 4.4 Comparative protein quantification in ELF by iTRAQ labeling. Example of a protein that is increased in four COPD patients (113–116 reporter ions, blue ellipse) versus four healthy controls (117–119 and 121 reporter ions, red ellipse). The peak at 120 is an immonium ion and not related to the iTRAQ labeling.

include only peptides that have been identified with a high confidence level
(>95% in our case). Calculation of peak areas and peak-area ratios for all tags
(normalized to iTRAQ-113 in this case) is followed by calculation of *p*-values of
the log-transformed ratios between COPD and control samples for each
peptide excluding peptides with *p*-values > 0.05. A comparison of reporter ions
from a protein that is increased in COPD patients is shown in Figure 4.4.
Further validation of potential candidate biomarkers by Western blot and
immunohistochemistry or ELISA is currently ongoing. The iTRAQ approach
is a powerful but costly and time-consuming way of analyzing the proteome. It
may thus be advisable to employ a pooling strategy followed by the targeted
validation of biomarker candidates in individual samples. A risk of isobaric
tagging is furthermore that imprecise precursor ion selection may "contaminate"
the reporter ion readout, leading to erroneous quantitative results.[86,87]

4.2.4   Cerebrospinal Fluid (CSF)

4.2.4.1   Stability and Sample Handling

Successful biomarker discovery in body fluids depends on a good understanding and control of the many preanalytical factors that may affect protein,
peptide or metabolite stability. Our work on CSF serves as an example to
outline strategies to standardize sample storage and handling based on
proteomics and metabolomics analyses.[88–90] The challenge is to tackle this
problem not only for a single target analyte but for a multitude of proteins,
peptides and metabolites, each with its own potential stability issues.

Sampling CSF from experimental animal models or patients is an invasive
procedure that requires considerable skill. In a clinical setting it is often not
possible to snap-freeze the collected CSF in liquid nitrogen immediately after
collection. Furthermore, there are cells in CSF that must be removed by
centrifugation prior to freezing. In order to study the effect of cell removal and
the ensuing delay time between sample collection and freezing, we performed a
controlled proteomic study in pigs.[91] Porcine CSF was sampled from the
cerebromedullary cistern of the subarachnoid space in the cervical region and
snap-frozen in liquid nitrogen directly after collection or after 30 min
or 120 min, respectively. Proteomic analysis after trypsin digestion by
microfluidics-based nano-LC-MS showed that certain peptides were susceptible
to changes due to delay time (Figure 4.5). These peptides were derived from
Prostaglandin-D synthase, a protein that is particularly sensitive to delay time
and should not be considered a suitable biomarker unless this time is tightly
controlled. This study was followed by collecting human CSF from patients
without a clear diagnosis of a neurological disorder.[92] To reduce the risk of
stability problems, cells were removed immediately after sample collection and
CSF was snap-frozen in liquid nitrogen after centrifugation at time points 0, 30
and 120 min. Stability was indeed much improved and clustering occurred
primarily according to the individual patient rather than to delay time for both
proteomics and metabolomics data (see Figure 4.6 for an example of CSF that

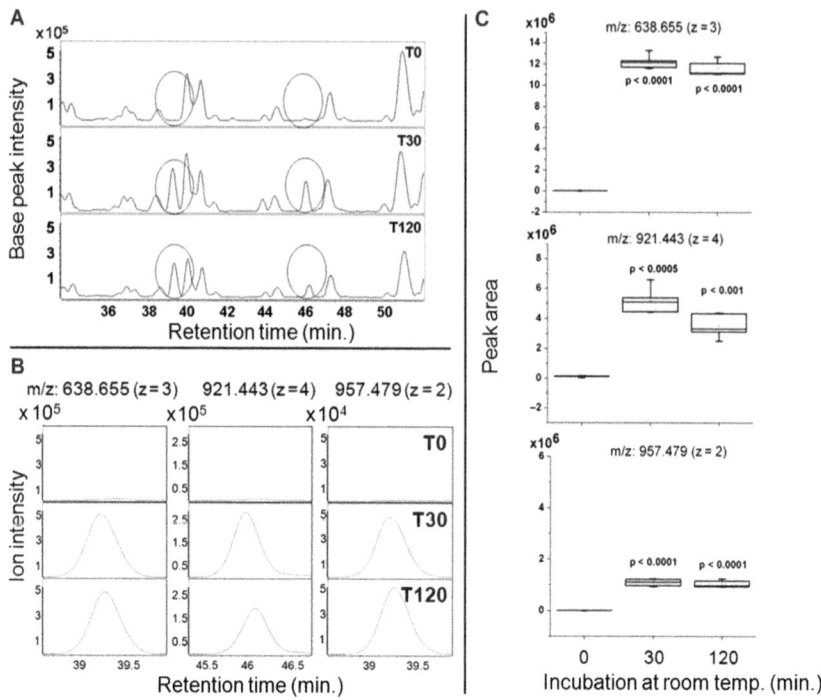

Figure 4.5   (A) Base peak chromatograms of the chipLC-MS analysis of depleted, trypsin-digested, porcine CSF with delay times of 0, 30 or 120 min at room temperature (T0, T30, T120) between CSF collection and freezing in liquid nitrogen. The encircled regions show peptide peaks at ~39 min and ~46 min retention time that change with increasing delay time. (B) Extracted ion chromatograms of three peaks that change with increasing delay time (see panel A). Peaks at $m/z = 638.655$ (Mw: 1912.943) and 957.479 (Mw: 1912.943) have the same retention time and represent different charge states a peptide that was assigned to Prostaglandin-D synthase by MS/MS. The peak at $m/z = 921.443$ (Mw: 3681.743) was also assigned to Prostaglandin-D synthase by MS/MS. (C) Univariate statistical analysis of the peak areas of peptides that change significantly with respect to delay time. Analysis is based on two-tailed Students t-tests of 5 repetitions of the LC-MS analysis of depleted, trypsin-digested, porcine CSF. Data are represented as box and whisker plots with p-values (T0 *vs.* T30 and T0 *vs.* T120, respectively). Peak areas between T30 and T120 were not significantly different (p-value > 0.05) (reprinted with permission from ref. 88 copyright American Chemical Society).

was analyzed for metabolites by NMR spectroscopy). Concerning protein analyses, we confirmed the profiling data for individual proteins using quantitative LC-MS/MS in the selected reaction monitoring (SRM) mode with stable-isotope-labeled internal standard peptides. Cystatin C, a protein that was earlier shown to be rather instable, varied by less than 10% relative standard deviation. While these results cannot give assurance that no other

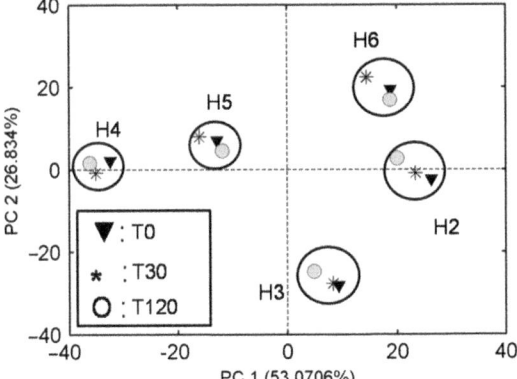

Figure 4.6 Analysis of human CSF by NMR spectroscopy after a delay time of 0, 30 or 120 min. Multivariate statistical analysis by PCA of metabolomics data derived from five patients (H2–H6) based on 51 detected metabolites (reprinted with permission from ref.,[89] copyright American Association for Clinical Chemistry).

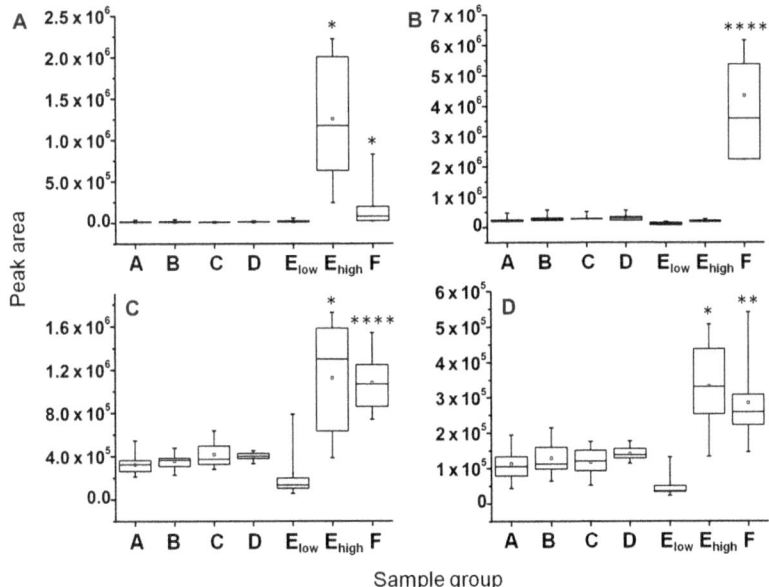

Figure 4.7 Box and Whisker plots of the sums of the peak areas of all detected peptides from 4 discriminatory proteins. Group E is divided in two groups (E_{high} and E_{low}) based on the area under the curve of the Total Ion Current (TIC). (A) Serum paraoxonase/arylestarase 1, (B) Lysozyme C1, (C) Complement C3, (D) Afamin (reprinted with permission from ref.,[93] copyright American Chemical Society).

proteins or metabolites change with delay time, they provide guidelines on how to handle CSF for proteome and metabolome studies.

4.2.4.2   *Applications*

The experimental autoimmune encephalomyelitis (EAE) model in rats reflects certain aspects of the pathophysiology of Multiple Sclerosis (MScL). A proteomic study in CSF from rats that were injected with myelin basic protein (MBP) together with complete Freund's adjuvant showed clear changes in protein patterns, some of which followed the course of EAE, while others preceded the development of overt neurological symptoms (Figure 4.7).[93] This shows that CSF proteomics is a useful approach to find biomarker candidates some of which may allow following therapeutic interventions, classify patients or assess the efficacy of novel experimental drugs.

4.3   Conclusion

The proteomic analysis of body fluids is likely the most challenging application of this approach due to the extremely wide dynamic concentration range of proteins, the overwhelming presence of a few "high-abundance" proteins and the potential for many unchartered post-translational modifications that result in possibly half a million protein species. Biomarker discovery research in body fluids has to deal with these fundamental difficulties as well as with a range of challenges due to the often wide biological variation of proteins between apparently healthy individuals and the need for extensive sample preparation and prefractionation to detect a relatively small part of the actual proteome. Body-fluid proteomics generates an enormous amount of data for each analysis and presents a formidable challenge to data processing and statistics to select discriminating features out of tens of thousands of signals, while analyzing at best a few hundred samples.

   This list of possible caveats might discourage anyone to even think of doing biomarker discovery in body fluids. However, successful examples exist where biomarker candidates have been discovered and validated in independent sets of samples, although these examples are rare. Following strict guidelines with respect to patient and control selection, sampling and sample handling, analytical method validation and data processing are critical to increasing the chances for success in combination with thorough validation of the statistical analysis to avoid false-positives due to overfitting the data.[94] It is critical that initial findings that were obtained in a limited set of samples are validated in sufficiently large sample sets from different sources (*e.g.* different hospitals) to assess whether the discovered biomarker candidates are robust and generally applicable. To this end, it is necessary to switch from the often work-intensive "discovery methodology" to simpler methods that can be automated such as targeted LC-MS/MS in the SRM mode or immunoassays. There is still a considerable need for improved methods in both the biomarker discovery and validation areas.

In this chapter we focused on a few representative examples of body fluids and notably on the often neglected area of evaluating the influence of preanalytical factors such as sample collection, sample storage and sample handling on the proteome and in one case also the metabolome. In fact, many of the issues that are covered in this chapter are also of relevance to metabolomics although the methodology for analysis may differ.

Acknowledgements

Work in the Department of Analytical Biochemistry at the University of Groningen is supported by grants from the Dutch Cancer Society (KWF), the Netherlands Proteomics Center (NPC), the Netherlands Bioinformatics Center (NBIC), the Top Institute Pharma (TIP), the Dutch Technology Foundation (STW) and the Dutch Research Organization (NWO).

References

1. Y. Yasui, M. Pepe, M. L. Thompson, B. L. Adam, G. L. Wright Jr, Y. Qu, J. D. Potter, M. Winget, M. Thornquist and Z. Feng, *Biostatistics*, 2003, **4**, 449–463 (DOI:10.1093/biostatistics/4.3.449).
2. M. H. Fortier, E. Bonneil, P. Goodley and P. Thibault, *Anal. Chem.*, 2005, **77**, 1631–1640 (DOI:10.1021/ac048506d).
3. H. Tammen, I. Schulte, R. Hess, C. Menzel, M. Kellmann, T. Mohring and P. Schulz-Knappe, *Proteomics*, 2005, **5**, 3414–3422 (DOI:10.1002/pmic.200401219).
4. J. M. Jacobs, J. N. Adkins, W. J. Qian, T. Liu, Y. Shen, D. G. Camp 2nd and R. D. Smith, *J. Proteome Res.*, 2005, **4**, 1073–1085 (DOI:10.1021/pr0500657).
5. M. Gronborg, T. Z. Kristiansen, A. Iwahori, R. Chang, R. Reddy, N. Sato, H. Molina, O. N. Jensen, R. H. Hruban, M. G. Goggins, A. Maitra and A. Pandey, *Mol. Cell. Proteomics*, 2006, **5**, 157–171 (DOI:10.1074/mcp.M500178-MCP200).
6. R. R. Packard and P. Libby, *Clin. Chem.*, 2008, **54**, 24–38 (DOI:10.1373/clinchem.2007.097360).
7. N. Wentzensen and M. von Knebel Doeberitz, *Dis. Markers*, 2007, **23**, 315–330.
8. N. Govorukhina and R. Bischoff, in, ed. Visith Thongboonkerd, Humana Press, Totowa, New Jersey, USA, 2007, p. 31.
9. N. L. Anderson and N. G. Anderson, *Mol. Cell. Proteomics*, 2002, **1**, 845–867.
10. J. Villanueva, J. Philip, C. A. Chaparro, Y. Li, R. Toledo-Crow, L. DeNoyer, M. Fleisher, R. J. Robbins and P. Tempst, *J. Proteome Res.*, 2005, **4**, 1060–1072 (DOI:10.1021/pr050034b).
11. M. Schiffman, N. Wentzensen, S. Wacholder, W. Kinney, J. C. Gage and P. E. Castle, *J. Natl. Cancer Inst.*, 2011, **103**, 368–383 (DOI:10.1093/jnci/djq562).

12. M. Arbyn, G. Ronco, J. Cuzick, N. Wentzensen and P. E. Castle, *Int. J. Cancer*, 2009, **125**, 2489–2496 (DOI:10.1002/ijc.24774).

13. L. Ekblad, B. Baldetorp, M. Ferno, H. Olsson and C. Bratt, *J. Proteome Res.*, 2007, **6**, 1609–1614 (DOI:10.1021/pr060633y).

14. L. P. Aristoteli, M. P. Molloy and M. S. Baker, *J. Proteome Res.*, 2007, **6**, 571–581 (DOI:10.1021/pr0602996).

15. M. Raida, P. Schulz-Knappe, G. Heine and W. G. Forssmann, *J. Am. Soc. Mass Spectrom.*, 1999, **10**, 45–54 (DOI:10.1016/S1044-0305(98)00117-2).

16. R. R. Drake, E. E. Schwegler, G. Malik, J. Diaz, T. Block, A. Mehta and O. J. Semmes, *Mol. Cell. Proteomics*, 2006, **5**, 1957–1967 (DOI:10.1074/mcp.M600176-MCP200).

17. H. Tammen, I. Schulte, R. Hess, C. Menzel, M. Kellmann and P. Schulz-Knappe, *Comb. Chem. High Throughput Screen*, 2005, **8**, 725–733.

18. N. I. Govorukhina, M. de Vries, T. H. Reijmers, P. Horvatovich, A. G. van der Zee and R. Bischoff, *J. Chromatogr. B. Analyt. Technol. Biomed. Life. Sci.*, 2009, **877**, 1281–1291 (DOI:10.1016/j.jchromb.2008.10.029).

19. P. Schulz-Knappe, M. Schrader and H. D. Zucht, *Comb. Chem. High Throughput Screen*, 2005, **8**, 697–704.

20. E. F. Petricoin, C. Belluco, R. P. Araujo and L. A. Liotta, *Nature Rev. Cancer*, 2006, **6**, 961–967 (DOI:10.1038/nrc2011).

21. P. L. Ross, Y. N. Huang, J. N. Marchese, B. Williamson, K. Parker, S. Hattan, N. Khainovski, S. Pillai, S. Dey, S. Daniels, S. Purkayastha, P. Juhasz, S. Martin, M. Bartlet-Jones, F. He, A. Jacobson and D. J. Pappin, *Mol. Cell. Proteomics*, 2004, **3**, 1154–1169 (DOI:10.1074/mcp.M400129-MCP200).

22. T. D. Veenstra, T. P. Conrads, B. L. Hood, A. M. Avellino, R. G. Ellenbogen and R. S. Morrison, *Mol. Cell. Proteomics*, 2005, **4**, 409–418 (DOI:10.1074/mcp.M500006-MCP200).

23. N. I. Govorukhina, T. H. Reijmers, S. O. Nyangoma, A. G. van der Zee, R. C. Jansen and R. Bischoff, *J. Chromatogr. A*, 2006, **1120**, 142–150 (DOI:10.1016/j.chroma.2006.02.088).

24. L. J. Dekker, J. Bosman, P. C. Burgers, A. van Rijswijk, R. Freije, T. Luider and R. Bischoff, *J. Chromatogr. B. Analyt. Technol. Biomed. Life. Sci.*, 2007, **847**, 65–69 (DOI:10.1016/j.jchromb.2006.09.038).

25. K. Bjorhall, T. Miliotis and P. Davidsson, *Proteomics*, 2005, **5**, 307–317 (DOI:10.1002/pmic.200400900).

26. J. Villanueva, D. R. Shaffer, J. Philip, C. A. Chaparro, H. Erdjument-Bromage, A. B. Olshen, M. Fleisher, H. Lilja, E. Brogi, J. Boyd, M. Sanchez-Carbayo, E. C. Holland, C. Cordon-Cardo, H. I. Scher and P. Tempst, *J. Clin. Invest.*, 2006, **116**, 271–284 (DOI:10.1172/JCI26022).

27. M. Zhou, D. A. Lucas, K. C. Chan, H. J. Issaq, E. F. Petricoin 3rd, L. A. Liotta, T. D. Veenstra and T. P. Conrads, *Electrophoresis*, 2004, **25**, 1289–1298 (DOI:10.1002/elps.200405866).

28. R. S. Tirumalai, K. C. Chan, D. A. Prieto, H. J. Issaq, T. P. Conrads and T. D. Veenstra, *Mol. Cell. Proteomics*, 2003, **2**, 1096–1103 (DOI:10.1074/mcp.M300031-MCP200).
29. J. Granger, J. Siddiqui, S. Copeland and D. Remick, *Proteomics*, 2005, **5**, 4713–4718 (DOI:10.1002/pmic.200401331).
30. M. F. Lopez, A. Mikulskis, S. Kuzdzal, D. A. Bennett, J. Kelly, E. Golenko, J. DiCesare, E. Denoyer, W. F. Patton, R. Ediger, L. Sapp, T. Ziegert, C. Lynch, S. Kramer, G. R. Whiteley, M. R. Wall, D. P. Mannion, G. Della Cioppa, J. S. Rakitan and G. M. Wolfe, *Clin. Chem.*, 2005, **51**, 1946–1954 (DOI:10.1373/clinchem.2005.053090).
31. M. S. Lowenthal, A. I. Mehta, K. Frogale, R. W. Bandle, R. P. Araujo, B. L. Hood, T. D. Veenstra, T. P. Conrads, P. Goldsmith, D. Fishman, E. F. Petricoin 3rd and L. A. Liotta, *Clin. Chem.*, 2005, **51**, 1933–1945 (DOI:10.1373/clinchem.2005.052944).
32. J. Martosella, N. Zolotarjova, H. Liu, G. Nicol and B. E. Boyes, *J. Proteome Res.*, 2005, **4**, 1522–1537 (DOI:10.1021/pr050088l).
33. N. Govorukhina, P. Horvatovich and R. Bischoff, *Methods Mol. Biol.*, 2008, **484**, 67–77 (DOI:10.1007/978-1-59745-398-1_5).
34. A. Shevchenko, M. Wilm, O. Vorm and M. Mann, *Anal. Chem.*, 1996, **68**, 850–858.
35. D. Loo, A. Jones and M. M. Hill, *J. Proteome Res.*, 2010, **9**, 5496–5500 (DOI:10.1021/pr100472z).
36. P. M. Drake, W. Cho, B. Li, A. Prakobphol, E. Johansen, N. L. Anderson, F. E. Regnier, B. W. Gibson and S. J. Fisher, *Clin. Chem.*, 2010, **56**, 223–236 (DOI:10.1373/clinchem.2009.136333).
37. P. M. Drake, B. Schilling, R. K. Niles, M. Braten, E. Johansen, H. Liu, M. Lerch, D. J. Sorensen, B. Li, S. Allen, S. C. Hall, H. E. Witkowska, F. E. Regnier, B. W. Gibson and S. J. Fisher, *Anal. Biochem.*, 2011, **408**, 71–85 (DOI:10.1016/j.ab.2010.08.010).
38. K. L. Abbott, J. M. Lim, L. Wells, B. B. Benigno, J. F. McDonald and M. Pierce, *Proteomics*, 2010, **10**, 470–481 (DOI:10.1002/pmic.200900537).
39. S. Barrabes, L. Pages-Pons, C. M. Radcliffe, G. Tabares, E. Fort, L. Royle, D. J. Harvey, M. Moenner, R. A. Dwek, P. M. Rudd, R. De Llorens and R. Peracaula, *Glycobiology*, 2007, **17**, 388–400 (DOI:10.1093/glycob/cwm002).
40. S. Kellokumpu, R. Sormunen and I. Kellokumpu, *FEBS Lett.*, 2002, **516**, 217–224.
41. Z. Zeng, M. Hincapie, B. B. Haab, S. Hanash, S. J. Pitteri, S. Kluck, J. M. Hogan, J. Kennedy and W. S. Hancock, *J. Chromatogr. A*, 2010, **1217**, 3307–3315 (DOI:10.1016/j.chroma.2009.09.029).
42. Z. Zeng, M. Hincapie, S. J. Pitteri, S. Hanash, J. Schalkwijk, J. M. Hogan, H. Wang and W. S. Hancock, *Anal. Chem.*, 2011, **83**, 4845–4854 (DOI:10.1021/ac2002802).
43. Z. Zhang, R. C. Bast Jr, Y. Yu, J. Li, L. J. Sokoll, A. J. Rai, J. M. Rosenzweig, B. Cameron, Y. Y. Wang, X. Y. Meng, A. Berchuck, C. Van Haaften-Day, N. F. Hacker, H. W. de Bruijn, A. G. van der Zee,

I. J. Jacobs, E. T. Fung and D. W. Chan, *Cancer Res.*, 2004, **64**, 5882–5890 (DOI:10.1158/0008-5472.CAN-04-0746).

44. E. F. Burgess, A. J. Ham, D. L. Tabb, D. Billheimer, B. J. Roth, S. S. Chang, M. S. Cookson, T. J. Hinton, K. L. Cheek, S. Hill and J. A. Pietenpol, *Proteomics Clin. Appl.*, 2008, **2**, 1223 (DOI:10.1002/prca.200780073).

45. N. L. Anderson, N. G. Anderson, L. R. Haines, D. B. Hardie, R. W. Olafson and T. W. Pearson, *J. Proteome Res.*, 2004, **3**, 235–244.

46. N. L. Anderson, A. Jackson, D. Smith, D. Hardie, C. Borchers and T. W. Pearson, *Mol. Cell. Proteomics*, 2009, **8**, 995–1005 (DOI:10.1074/mcp.M800446-MCP200).

47. J. R. Whiteaker, C. Lin, J. Kennedy, L. Hou, M. Trute, I. Sokal, P. Yan, R. M. Schoenherr, L. Zhao, U. J. Voytovich, K. S. Kelly-Spratt, A. Krasnoselsky, P. R. Gafken, J. M. Hogan, L. A. Jones, P. Wang, L. Amon, L. A. Chodosh, P. S. Nelson, M. W. McIntosh, C. J. Kemp and A. G. Paulovich, *Nature Biotechnol.*, 2011, **29**, 625–634 (DOI:10.1038/nbt.1900; 10.1038/nbt.1900).

48. J. R. Whiteaker, L. Zhao, S. E. Abbatiello, M. Burgess, E. Kuhn, C. Lin, M. E. Pope, M. Razavi, N. L. Anderson, T. W. Pearson, S. A. Carr and A. G. Paulovich, *Mol. Cell. Proteomics*, 2011, **10**, M110.005645 (DOI:10.1074/mcp.M110.005645).

49. J. R. Whiteaker, L. Zhao, L. Anderson and A. G. Paulovich, *Mol. Cell. Proteomics*, 2010, **9**, 184–196 (DOI:10.1074/mcp.M900254-MCP200).

50. C. Fredolini, D. Tamburro, G. Gambara, B. S. Lepene, V. Espina, E. F. Petricoin 3rd, L. A. Liotta and A. Luchini, *Drug Test. Anal.*, 2009, **1**, 447–454 (DOI:10.1002/dta.96).

51. T. A. Douglas, D. Tamburro, C. Fredolini, B. H. Espina, B. S. Lepene, L. Ilag, V. Espina, E. F. Petricoin 3rd, L. A. Liotta and A. Luchini, *Biomaterials*, 2011, **32**, 1157–1166 (DOI:10.1016/j.biomaterials. 2010.10.004).

52. M. F. Lopez, A. Mikulskis, S. Kuzdzal, E. Golenko, E. F. Petricoin 3rd, L. A. Liotta, W. F. Patton, G. R. Whiteley, K. Rosenblatt, P. Gurnani, A. Nandi, S. Neill, S. Cullen, M. O'Gorman, D. Sarracino, C. Lynch, A. Johnson, W. Mckenzie and D. Fishman, *Clin. Chem.*, 2007, **53**, 1067–1074 (DOI:10.1373/clinchem.2006.080721).

53. W. Mullen, C. Delles, H. Mischak and EuroKUP COST action, *Curr. Opin. Nephrol. Hypertens.*, 2011, **20**, 654–661 (DOI:10.1097/MNH. 0b013e32834b7ffa).

54. A. Albalat, H. Mischak and W. Mullen, *Expert Rev. Proteomics*, 2011, **8**, 615–629 (DOI:10.1586/epr.11.46).

55. H. Mischak, C. Delles, J. Klein and J. P. Schanstra, *Adv. Chronic Kidney Dis.*, 2010, **17**, 493–506 (DOI:10.1053/j.ackd.2010.09.004).

56. H. Mischak, V. Thongboonkerd, J. P. Schanstra and A. Vlahou, *Proteomics Clin. Appl.*, 2011, **5**, 211–213 (DOI:10.1002/prca.201190031; 10.1002/prca.201190031).

57. J. J. Coon, P. Zurbig, M. Dakna, A. F. Dominiczak, S. Decramer, D. Fliser, M. Frommberger, I. Golovko, D. M. Good, S. Herget-Rosenthal, J. Jankowski, B. A. Julian, M. Kellmann, W. Kolch, Z. Massy, J. Novak, K. Rossing, J. P. Schanstra, E. Schiffer, D. Theodorescu, R. Vanholder, E. M. Weissinger, H. Mischak and P. Schmitt-Kopplin, *Proteomics Clin. Appl.*, 2008, **2**, 964 (DOI:10.1002/prca.200800024).

58. A. Marimuthu, R. N. O'Meally, R. Chaerkady, Y. Subbannayya, V. Nanjappa, P. Kumar, D. S. Kelkar, S. M. Pinto, R. Sharma, S. Renuse, R. Goel, R. Christopher, B. Delanghe, R. N. Cole, H. C. Harsha and A. Pandey, *J. Proteome Res.*, 2011, **10**, 2734–2743 (DOI:10.1021/pr2003038).

59. R. F. Kemperman, P. L. Horvatovich, B. Hoekman, T. H. Reijmers, F. A. Muskiet and R. Bischoff, *J. Proteome Res.*, 2007, **6**, 194–206 (DOI:10.1021/pr060362r).

60. T. Pisitkun, R. Johnstone and M. A. Knepper, *Mol. Cell. Proteomics*, 2006, **5**, 1760–1771 (DOI:10.1074/mcp.R600004-MCP200).

61. T. Pisitkun, R. F. Shen and M. A. Knepper, *Proc. Natl. Acad. Sci. U. S. A.*, 2004, **101**, 13368–13373 (DOI:10.1073/pnas.0403453101).

62. H. Zhou, T. Pisitkun, A. Aponte, P. S. Yuen, J. D. Hoffert, H. Yasuda, X. Hu, L. Chawla, R. F. Shen, M. A. Knepper and R. A. Star, *Kidney Int.*, 2006, **70**, 1847–1857 (DOI:10.1038/sj.ki.5001874).

63. D. Fliser, J. Novak, V. Thongboonkerd, A. Argiles, V. Jankowski, M. A. Girolami, J. Jankowski and H. Mischak, *J. Am. Soc. Nephrol.*, 2007, **18**, 1057–1071 (DOI:10.1681/ASN.2006090956).

64. H. Roelofsen, G. Alvarez-Llamas, M. Schepers, K. Landman and R. J. Vonk, *Proteome Sci.*, 2007, **5**, 2 (DOI:10.1186/1477-5956-5-2).

65. O. V. Nemirovskiy, D. R. Dufield, T. Sunyer, P. Aggarwal, D. J. Welsch and W. R. Mathews, *Anal. Biochem.*, 2007, **361**, 93–101 (DOI:10.1016/j.ab.2006.10.034).

66. V. B. Lokeshwar and M. G. Selzer, *Urol. Oncol.*, 2006, **24**, 528–537 (DOI:10.1016/j.urolonc.2006.07.003).

67. H. H. Ngai, W. H. Sit, P. P. Jiang, R. J. Xu, J. M. Wan and V. Thongboonkerd, *J. Proteome Res.*, 2006, **5**, 3038–3047 (DOI:10.1021/pr060122b).

68. D. Theodorescu, S. Wittke, M. M. Ross, M. Walden, M. Conaway, I. Just, H. Mischak and H. F. Frierson, *Lancet Oncol.*, 2006, **7**, 230–240 (DOI:10.1016/S1470-2045(06)70584-8).

69. J. Adachi, C. Kumar, Y. Zhang, J. V. Olsen and M. Mann, *Genome Biol.*, 2006, **7**, R80 (DOI:10.1186/gb-2006-7-9-R80).

70. D. Fliser, S. Wittke and H. Mischak, *Electrophoresis*, 2005, **26**, 2708–2716 (DOI:10.1002/elps.200500187).

71. R. Pieper, C. L. Gatlin, A. M. McGrath, A. J. Makusky, M. Mondal, M. Seonarain, E. Field, C. R. Schatz, M. A. Estock, N. Ahmed, N. G. Anderson and S. Steiner, *Proteomics*, 2004, **4**, 1159–1174 (DOI:10.1002/pmic.200300661).

72. V. Thongboonkerd, K. R. McLeish, J. M. Arthur and J. B. Klein, *Kidney Int.*, 2002, **62**, 1461–1469 (DOI:10.1111/j.1523-1755.2002.kid565.x).

73. H. Idborg-Bjorkman, P. O. Edlund, O. M. Kvalheim, I. Schuppe-Koistinen and S. P. Jacobsson, *Anal. Chem.*, 2003, **75**, 4784–4792.

74. K. Wagner, T. Miliotis, G. Marko-Varga, R. Bischoff and K. K. Unger, *Anal. Chem.*, 2002, **74**, 809–820.

75. E. Machtejevas, H. John, K. Wagner, L. Standker, G. Marko-Varga, W. G. Forssmann, R. Bischoff and K. K. Unger, *J. Chromatogr. B. Analyt. Technol. Biomed. Life. Sci.*, 2004, **803**, 121–130 (DOI:10.1016/j.jchromb. 2003.07.015).

76. P. H. Eilers, *Anal. Chem.*, 2004, **76**, 404–411 (DOI:10.1021/ac034800e).

77. A. M. van Nederkassel, M. Daszykowski, P. H. Eilers and Y. V. Heyden, *J. Chromatogr. A*, 2006, **1118**, 199–210 (DOI:10.1016/j.chroma. 2006.03.114).

78. S. S. Chen and R. Aebersold, *J. Chromatogr. B. Analyt. Technol. Biomed. Life. Sci.*, 2005, **829**, 107–114 (DOI:10.1016/j.jchromb.2005.09.039).

79. K. M. Pierce, L. F. Wood, B. W. Wright and R. E. Synovec, *Anal. Chem.*, 2005, **77**, 7735–7743 (DOI:10.1021/ac0511142).

80. D. Bylund, R. Danielsson, G. Malmquist and K. E. Markides, *J. Chromatogr. A*, 2002, **961**, 237–244.

81. P. H. Eilers, *Anal. Chem.*, 2003, **75**, 3631–3636.

82. P. Horvatovich, N. I. Govorukhina, T. H. Reijmers, A. G. van der Zee, F. Suits and R. Bischoff, *Electrophoresis*, 2007, **28**, 4493–4505 (DOI:10.1002/elps.200600719).

83. P. L. Horvatovich, R. F. Kemperman and R. Bischoff, , 2009, **10**, 977.

84. P. Horvatovich, B. Hoekman, N. Govorukhina and R. Bischoff, *J. Sep. Sci.*, 2010, **33**, 1421–1437 (DOI:10.1002/jssc.201000050).

85. L. Franciosi, N. Govorukhina, N. Ten Hacken, D. Postma and R. Bischoff, *Methods Mol. Biol.*, 2011, **790**, 17–28 (DOI:10.1007/ 978-1-61779-319-6_2).

86. L. Ting, R. Rad, S. P. Gygi and W. Haas, *Nature Methods*, 2011, **8**, 937–940 (DOI:10.1038/nmeth.1714; 10.1038/nmeth.1714).

87. C. D. Wenger, M. V. Lee, A. S. Hebert, G. C. McAlister, D. H. Phanstiel, M. S. Westphall and J. J. Coon, *Nature Methods*, 2011, **8**, 933–935 (DOI:10.1038/nmeth.1716; 10.1038/nmeth.1716).

88. T. Rosenling, C. L. Slim, C. Christin, L. Coulier, S. Shi, M. P. Stoop, J. Bosman, F. Suits, P. L. Horvatovich, N. Stockhofe-Zurwieden, R. Vreeken, T. Hankemeier, A. J. van Gool, T. M. Luider and R. Bischoff, *J. Proteome Res.*, 2009, **8**, 5511–5522 (DOI:10.1021/pr9005876).

89. T. Rosenling, M. P. Stoop, A. Smolinska, B. Muilwijk, L. Coulier, S. Shi, A. Dane, C. Christin, F. Suits, P. L. Horvatovich, S. S. Wijmenga, L. M. Buydens, R. Vreeken, T. Hankemeier, A. J. van Gool, T. M. Luider and R. Bischoff, *Clin. Chem.*, 2011, **57**, 1703–1711 (DOI:10.1373/ clinchem.2011.167601).

90. M. P. Stoop, L. Coulier, T. Rosenling, S. Shi, A. M. Smolinska, L. Buydens, K. Ampt, C. Stingl, A. Dane, B. Muilwijk, R. L. Luitwieler, P. A. Sillevis Smitt, R. Q. Hintzen, R. Bischoff, S. S. Wijmenga, T. Hankemeier, A. J. van Gool and T. M. Luider, *Mol. Cell. Proteomics*, 2010, **9**, 2063–2075 (DOI:10.1074/mcp.M900877-MCP200).
91. T. Rosenling, C. L. Slim, C. Christin, L. Coulier, S. Shi, M. P. Stoop, J. Bosman, F. Suits, P. L. Horvatovich, N. Stockhofe-Zurwieden, R. Vreeken, T. Hankemeier, A. J. van Gool, T. M. Luider and R. Bischoff, *J. Proteome Res.*, 2009, **8**, 5511–5522 (DOI:10.1021/pr9005876).
92. T. Rosenling, M. P. Stoop, A. Smolinska, B. Muilwijk, L. Coulier, S. Shi, A. Dane, C. Christin, F. Suits, P. L. Horvatovich, S. S. Wijmenga, L. M. Buydens, R. Vreeken, T. Hankemeier, A. J. van Gool, T. M. Luider and R. Bischoff, *Clin. Chem.*, 2011, **57**, 1703–1711 (DOI:10.1373/clinchem.2011.167601).
93. T. Rosenling, M. P. Stoop, H. Attali, E. van Aken, C. Suidgeest, C. Christin, F. Suits, P. L. Horvatovich, R. Q. Hintzen, T. Tuinstra, R. Bischoff and T. M. Luider, *J. Proteome Res.*, 2012 (DOI:in press).
94. H. Mischak, G. Allmaier, R. Apweiler, T. Attwood, M. Baumann, A. Benigni, S. E. Bennett, R. Bischoff, E. Bongcam-Rudloff, G. Capasso, J. J. Coon, P. D'Haese, A. F. Dominiczak, M. Dakna, H. Dihazi, J. H. Ehrich, P. Fernandez-Llama, D. Fliser, J. Frokiaer, J. Garin, M. Girolami, W. S. Hancock, M. Haubitz, D. Hochstrasser, R. R. Holman, J. P. Ioannidis, J. Jankowski, B. A. Julian, J. B. Klein, W. Kolch, T. Luider, Z. Massy, W. B. Mattes, F. Molina, B. Monsarrat, J. Novak, K. Peter, P. Rossing, M. Sanchez-Carbayo, J. P. Schanstra, O. J. Semmes, G. Spasovski, D. Theodorescu, V. Thongboonkerd, R. Vanholder, T. D. Veenstra, E. Weissinger, T. Yamamoto and A. Vlahou, *Sci. Transl. Med.*, 2010, **2**, 46ps42 (DOI:10.1126/scitranslmed.3001249).
95. A. Boichenko, N. Govorukhina, A. G. van der Zee and R. Bischoff, *Anal. Bioanal. Chem.*, 2013, **405**, 3195–3203.

CHAPTER 5

Sample Preparation and Profiling: Mass-Spectrometry-Based Profiling Strategies

YEOUN JIN KIM AND BRUNO DOMON*

Luxembourg Clinical Proteomics Center, CRP-Santé, 1A-B rue Thomas Edison, L-1445 Strassen, Luxembourg, Email: yeounjin.kim@crp-sante.lu
*Email: bruno.domon@crp-sante.lu

5.1 Introduction

Biomarkers are indicators of normal biological processes, pathogenic processes, or pharmacologic responses to a therapeutic intervention as defined by the National Institutes of Health (NIH).[1] Biomarkers should be objectively measurable in clinical subjects. In clinical practice, molecular biomarkers enable clinicians to detect and identify a disease at an early stage, to stratify patients for a suitable treatment, and to follow disease progression/regression.

Although the drug development process is well established, based on the United States' Federal Food and Drug Administration (FDA) regulations, the process for biomarker development still requires clarification, with terminologies to be defined and milestones to be established. In general, the development of protein biomarkers is divided into three or four phases before the clinical implementation. A three-step definition, namely discovery, verification, and validation, is used in this chapter (Figure 5.1). In this three-stage classification scheme, discovery is the first phase where candidate biomarkers associated with certain diseases are identified. Verification is achieved if the

RSC Drug Discovery Series No. 33
Comprehensive Biomarker Discovery and Validation for Clinical Application
Edited by Péter Horvatovich and Rainer Bischoff
© The Royal Society of Chemistry 2013
Published by the Royal Society of Chemistry, www.rsc.org

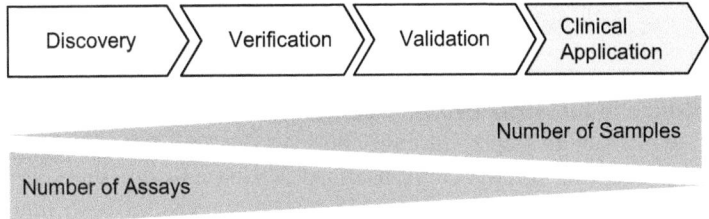

Figure 5.1 A general scheme of the protein biomarker development workflow. The number of clinical samples required in each step increases along the process advancement while the number of assays decreases as the candidate biomarkers are triaged.

differential abundance of the biomarker is confirmed in samples used for diagnosis (*e.g.* human plasma), and validated if the biomarker specificities are assessed quantitatively with statistical significance. In the validation step, large-scale studies based on robust assays have to be carried out as well.[2]

The design of experiments in the discovery phase depends on the clinical purpose; diagnosis, prediction, or prognosis. When the aim of discovery is to find a diagnostic marker indicating the presence of certain diseases, the discovery phase can be described as identifying novel protein candidates that exhibit a disease-specific expression level, preferably detectable in bodily fluids such as blood or urine. With respect to predictive markers that indicate a likely therapeutic outcome in individual patients, their identification relies on finding proteins whose concentration profiles or structural variations correlate to specific clinical outcomes, such as the extent of disease-free survival after drug treatment. Prognostic markers estimate the overall patient outcome regardless of therapy,[3] and thus the discovery of prognostic markers focuses on identifying proteins whose concentration levels or structural variations correlate well with overall survival.

FDA regulations reinforce the need for companion diagnostics to properly identify patients most likely to respond to treatment.[4] Therefore, biomarker discovery studies along with those addressing therapeutic response will become an important tool for pharmaceutical companies seeking final approval of new drugs or new applications for older ones.

New proteomics technologies have accelerated the discovery of biomarker candidates by monitoring the entire proteome in parallel and by employing systematic and integrative strategies. Mass spectrometry (MS) serves as the cornerstone of proteomics-based discovery as it allows both qualitative and quantitative analyses of proteins on a large scale, which is difficult to achieve with any other technology. In addition to measuring a characteristic molecular mass, peptide fragmentation in the gas phase enables the identification of the primary sequence of peptide with the help of a protein search algorithm using genomic and proteomic databases. In conjunction with proper separation methods, such as liquid chromatography or electrophoresis, systematic peptide identification expedites the profiling of protein components in biological samples. Numerous studies employed MS-based proteome profiling for the discovery of useful targets. The robustness of mass-spectrometric methods facilitates large-scale analyses for the verification and validation stages in which large sample numbers are required to generate statistically meaningful results.

Several workflows have been developed for biomarker discovery leveraging recent advances in mass-spectrometry designs. Biomarker studies usually start with a protein-profiling step to discover biomarker candidates in a predefined group of samples. For practical reasons, most MS-based studies are performed at the peptide level. The endpoint of MS analyses provides both qualitative (identity) and quantitative (abundance) information with respect to peptides obtained from the original proteins. In this chapter, liquid chromatography-mass spectrometry (LC-MS)-based profiling strategies for the discovery of protein biomarkers are discussed. In addition to the experimental strategies widely employed in the field, newer methods, such as targeted approaches in conjunction with data mining, are included. Although targeted approaches are more often applied to a post-discovery phase, its application as a hypothesis-driven platform presents distinct benefits in discovery.

5.2   LC-MS-Based Proteomics Applied to Biomarker Discovery

Mass spectrometry plays an essential role in the discovery of new biomarkers by profiling the protein content of clinical samples. The biomedical research community anticipates that useful biomarkers will be detected directly by LC-MS, since several proteins detectable with mass spectrometry, such as CEA and CA129,[5] are already used in clinical settings. The first *in vitro* multivariate index assay of biomarkers discovered through an MS-based platform received FDA clearance for use in diagnosing ovarian cancer in 2009.[6] However, the disappointing fact that proteomics did not reveal many other significant biomarkers despite nearly two decades of research necessitates a change of paradigm with more careful study designs. It is therefore of importance to comprehend the various analytical strategies and the use of new technologies in order to design discovery experiments that enable more effective profiling of proteins present in clinical samples.

Extensive profiling, or the comprehensive cataloging of a proteome, is a common practice in discovery and an essential step to avoid iterative redundant experiments. In the late 1990s and early 2000s mass spectrometry emerged as a powerful tool for biomarker discovery. In those early days, one of the major interests of MS-based proteomics was the ability of identifying and analyzing large lists of proteins from biological subjects. Several strategies for systematically identifying proteins in biological samples were developed. The quantification of peptides often is achieved during the separation and identification processes.

5.2.1   Sample Preparation: Prefractionation and Enrichment

The biological samples often used in discovery experiments are human plasma, serum, urine, and tissues from diseased organs. The proteomes of these samples are extremely complex, with the abundant variations including post-translational modifications, alternative splicing, and point mutations. Furthermore, the dynamic concentration range of the proteins present in the samples is wide. In the case of human plasma, the range of concentrations between the most-abundant—and the lowest-abundant proteins (*e.g.* cytokines)—is more than ten orders of magnitude. One of the best-known biomarkers currently used in a clinical setting is prostate specific antigen (PSA), which can help to diagnose prostate cancer. The serum concentration of PSA is less than 4 ng/mL while that of human serum albumin is more than 20 mg/mL. The analysis of low-abundant proteins can be easily impeded by high-abundant species during mass-spectrometry-based proteomics studies.

The comprehensive profiling of proteins/peptides for biomarker discovery studies is a challenging task, particularly when the samples are very complex. Mass-spectrometry-based analyses exhibit a limit of detection not inherent to the intrinsic instrument sensitivity but rather due to the background of analytes attributing to the sample complexity. Despite the advances in mass-spectrometry technology, direct analysis of proteins present at less than 1 µg/ml in plasma without any analytical pretreatment remains challenging.[7,8] In LC-MS analyses, the more-abundant peptides, derived from abundant proteins, are more readily detected as compared to the lower-abundant peptides. It addition, if the number of coeluting peptides exceeds the capacity of the performing MS/MS events where peptides are selected for identification according to their intensity rank, the low-concentration peptides are simply not selected, even if their signal would be sufficient for MS/MS analysis. This bias towards abundant peptides leads to a problematic undersampling of low-concentration peptides. For extensive profiling approaches, the interrelated issues of high complexity and wide dynamic range of the analytes need to be addressed. Three strategies for reduction of sample complexity before LC-MS/MS analysis are listed in Table 5.1: selective enrichment, selective depletion, and extensive fractionation.

Rather than analyzing a complete proteome, a fraction of the proteome (subproteome) can be selectively isolated based on its structural characteristics. Commonly chosen subproteomes are proteins with specific and functional modifications such as glycosylation and phosphorylation. Several methods have been developed for glycoprotein/glycopeptides enrichment. Lectin affinity is used to enrich glycoproteins containing specific glycan types.[9] To isolate glycoproteins regardless of the presence of specific carbohydrate moieties, chemical methods such as periodate oxidation of carbohydrates followed by hydrazide chemistry on a solid support is widely used (Figure 5.2).[10,11] Periodate oxidation under mild condition can enrich specifically sialylated glycoproteins by selectively oxidizing the terminal sialic acids in carbohydrate moieties.[12] Although less specific, hydrophilic interaction liquid chromatography (HILIC) is readily available to enrich glycopeptides.[13] The reduction of sample complexity by glycoprotein/glycopeptides enrichment is highly effective, especially for plasma samples, because human serum albumin, the most-abundant plasma protein is not glycosylated.

On the other hand, up to 75% of the plasma proteins (by number) are N-glycosylated.[11] From the *in silico* tryptic digest of plasma proteins, less than

Table 5.1 Enrichment and fractionation strategies to reduce sample complexity.

Strategy	Targeted Analytes	Strategy	
Selective enrichment	Glycoprotein Glycopeptide	Glycan capture	Lectin affinity, high specificity. Hydrazide conjugation, high coverage. Hydrophilic interaction, simple implementation.
	Cys-peptide	Thiol-reactive	Biotinylation, requiring avidin-based purification. Disulfide bond exchange (Thiopropyl sepharose).
	Phosphoprotein Phosphopeptide	Metal affinity	IMAC, nonspecificity toward negative charged peptides. TiO_2, nonspecificity toward negative charged peptides.
		Immunoaffinity	pY peptide Abs, phospho protein Abs.
Selective depletion	Most-abundant proteins		Negative selection by immunoaffinity of 2–20 abundant proteins in blood.
Extensive fractionation	Proteome	Online SCX	Peptide fractionation by MudPIT.
		Offline SCX	Protein/peptide fractionation by SCX for further LC-MS/MS.
		In-gel IEF	Protein separation by pI, requiring in-gel digestion.
		Off-gel IEF	Protein/peptide fractionation based on pI in solution for further analysis.
		High-pH RPLC	Peptide fractionation by hydrophobicity using basic (pH 10) mobile phase.

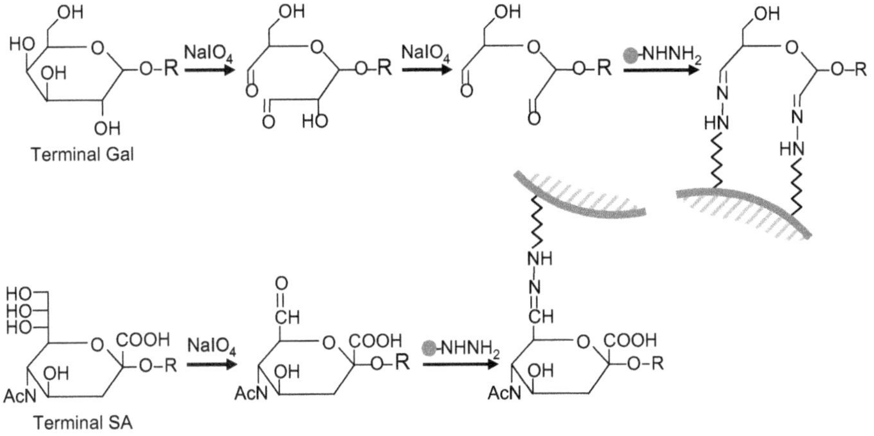

Figure 5.2 Glycocapture based on periodate oxidation and hydrazide conjugation. Upper part: Cis-diol groups of terminal galactose (Gal) in carbohydrate moieties of glycoproteins are oxidized by periodate to generate aldehydes.[10] Newly formed aldehydes react with hydrazide groups immobilized on a solid support to covalently link to the solid support *via* hydrazone bonds. This allows isolation of glycoproteins from a complex protein mixture. Lower part: If carbohydrate moieties are capped by sialic acids (SA), aldehydes are formed at the C-7 position of the sialic acid residue. Since sialic acids can be oxidized under milder condition, sialylated glycoproteins can be selectively enriched through a milder periodate oxidation.[12]

5% of the tryptic peptides contain N-glycosylation sites. In theory, 75% of plasma proteins can be analyzed by measuring only 5% of the tryptic peptides. For more specific applications focusing on functional activities of proteins, such as cellular signal transduction, phosphoproteins/phosphopeptides are enriched *via* metal affinity using immobilized metal affinity chromatography (IMAC), titanium-dioxide-based columns (TiO$_2$), immunoaffinity toward phosphoproteins/phosphopeptides, or a combination of these methods.[14,15]

Isolation of peptides that contain specific amino acids such as cysteine (cys-peptides) and histidine, or collecting only N-terminal peptides can yield subproteomes that reasonably represent the whole proteome. Theoretically, 96% of human proteins contain at least one cysteine in their sequence, yet only 22.5% of tryptic digest contain cysteine.[16] Cys-peptide enrichment is an attractive strategy for subproteome profiling; specific and efficient enrichment methods are available due to the cysteine's highly reactive thiol group. Thiol-reactive reagents linked to biotin are used to specifically label cysteines. Biotin-labeled cys-peptides are then isolated with an immobilized avidin column using avidin-biotin interactions. Another method is to employ thiopropyl Sepharose to couple cysteines though disulfide bond formation.[16,17] Reduction of complexity by collecting N-terminal peptides of each protein is in principle an effective strategy since all proteins yield only one tryptic peptide. A method of isolating N-terminal peptides by hydrophobic labeling of newly generated free amines following proteolysis has been developed.[18]

Alternatively, negative selection of proteins by depleting abundant yet nonrelevant components reduces sample complexity. Immunoaffinity-based depletion of abundant proteins in human blood is routinely applied in proteomics-based biomarker studies. There are immediate benefits to depletion of such proteins during downstream analysis. By depleting only the two most-abundant proteins, albumin and immunoglobulin G (IgG), up to 80% of protein amount (by mass) is removed from plasma, and 99% removal is achieved after depleting the 22 most-abundant proteins.[19] Antibody columns designed to deplete 2–20 abundant plasma proteins are commercially available.[20]

The third category of sample preparation described in Table 5.1 relies on extensive fractionation. This approach divides the complete proteome into several fractions. To reproducibly fractionate and evenly distribute the analytes, physiochemical properties of the proteins/peptides such as net electric charge states, hydrophobicity, or molecular size are employed. Strong cation exchange chromatography (SCX), isoelectric focusing (IEF), and basic reverse-phase (RP) LC are commonly used methods in discovery experiments.

SCX-based separation can be incorporated in the sample preparation orthogonal to an RP-LC separations coupled to a mass spectrometer, resulting in a 2-dimentional (2D) separation.[21] Figure 5.3 illustrates the direct impact of

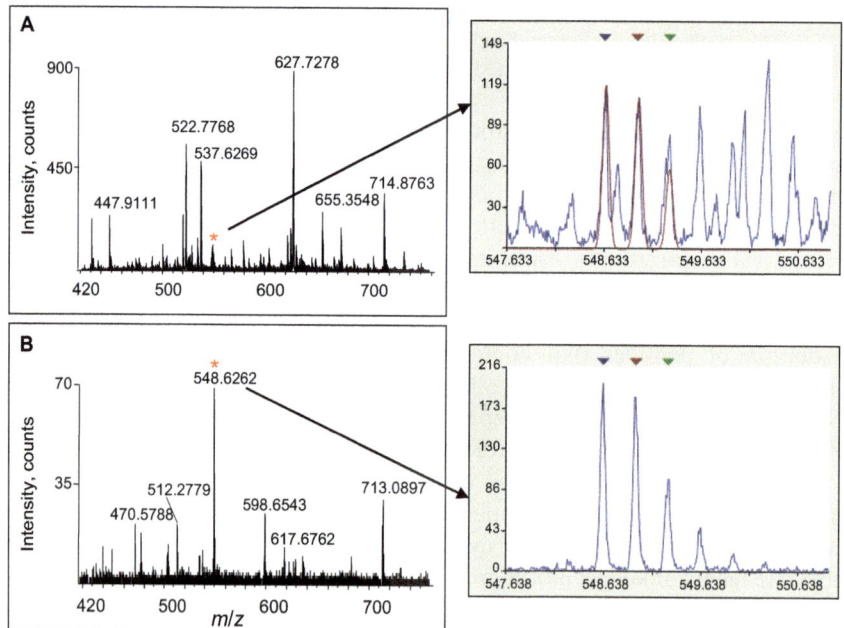

Figure 5.3   MS spectra of the peptide FICIYPAYLNNKK in whole-cell lysate before and after SCX fractionation (adapted from ref. 22). (A) The peptide peak is indicated with an asterisk. Sequencing the peptide peak in this complex environment was not successful. (B) MS spectra of the peptide FICIY-PAYLNNKK in the SCX fraction. Each peptide peak was magnified in the inserted boxes. The monoisotopic (^{12}C), ^{13}C and ^{13}C$_2$ peaks were indicated with a blue, red, and green triangle, respectively.

SCX-based fractionation on an MS-based profiling platform. The example shown in Figure 5.3A is a mass spectrum of the tryptic peptides from a whole-cell lysate, eluted during a certain time window. The peak indicated with an asterisk in the spectrum corresponds to the peptide FICIYPAYLNNKK. The detailed view of the m/z range around this ion is shown in the panel to the right. During the LC-MS/MS analysis using DDA mode, this peptide was not selected as a precursor for collision-induced dissociation (CID) due to the presence of multiple coeluting ions with much higher intensities.[22] A second attempt to analyze this peptide by forcing the m/z range of the ion to be selected using an inclusion list was not successful either, mainly because of interfering ions close to the precursor m/z. Figure 5.3B is a mass spectrum of the same peptide after prefractionation by SCX. The spectrum shows that this peptide was detected with the highest intensity in the shown mass range. In the LC-MS/MS analysis with a setting identical to that used in Figure 5.3A, the peptide in the SCX fraction was successfully identified with high confidence because the overlapping ions had been separated and collected into the other SCX fractions.

One notable development of 2D-LC is the multidimensional protein identification technology (MudPIT) where the capillary columns of SCX and RP are packed consecutively. This online 2D-LC-MS/MS dramatically increases the number of peptides to be analyzed in a single experiment, and facilitates the detection of low-abundant peptides by an expanded separation.[23,24]

Although prefractionation by SCX has been widely used for deep profiling of a proteome, isoelectric focusing (IEF) is increasingly used, mainly due to the introduction of the off-gel system.[25] The off-gel IEF allows in-solution preparation of proteins and peptide fractions, separated according to their pI values, which are orthogonal to hydrophobicity. IEF-based separations provide better reproducibility and resolution than SCX-based fractionations, with similar recovery.[26,27] Sample preparation using off-gel prefractionation can be adapted to a large-scale format, which suits the biomarker discovery.

Lastly, the high-pH RP-LC can be used for prefractionation and is partially orthogonal to the conventional low-pH RP-LC.[28,29] This technology uses the same type of columns as conventional RP-LC but separates peptides with a basic mobile phase (pH 10). Under basic conditions, peptides carry significantly different charge states as compared to low-pH RP-LC due to the ionization of the carboxylic group, phenolic hydroxyl groups and the deprotonation of ammonium groups. Consequently, the elution order of peptides is changed.[28] Online or offline 2D-RP-LC using high-pH followed by low-pH separation provide an alternative to extensive fractionation. The advantages of using high-pH RP-LC, as compared to SCX, are a better resolving power and the use of volatile salts facilitating downstream MS analysis. It is a simpler procedure that leads to both increased separation and recovery.[29]

5.2.2   Peptide Separation by Liquid Chromatography

Extensive fractionation and/or selective enrichment greatly improve both identification and quantification of low-abundant peptides in LC-MS analysis due to a much lower background. At the same time, improvements in liquid chromatography and mass spectrometry increase the throughput of the separation and confidence in peptide identifications. High analytical performance of LC systems entails high peak capacity, high resolution, and lower consumption of samples for MS analysis. Nanoflow LC systems with capillary columns of 75–300 μm inner diameter provide greater sensitivity in MS detection than the conventional systems requiring higher flow rate. Although the normal length of LC columns used in protein profiling is 10–15 cm, the use of longer columns and longer elution gradients increases the number of peptides identified in one LC-MS by increasing the separation power.[30]

The reproducibility of an LC system warrants the reproducible elution times of peptides obtained from the LC experiment. Elution time is an important parameter to characterize peptides according to their hydrophobicity. In addition to maintaining the reproducibility of the platform by reinforcing standardized procedures, the drift of elution times can be monitored and corrected by using reference peptides spiked into the sample. Reference peptides are the set of peptides (natural or artificial sequence) whose hydrophobicity covers the entire range of elution times used during the analyses. Figure 5.4 shows an example of reference peptides used in proteomic profiling. The elution times and intensities of the reference peptides spiked in each sample can be monitored prior to the actual analysis in order to assess the LC-MS performance (Figures 5.4B). More importantly, based on the elution times of

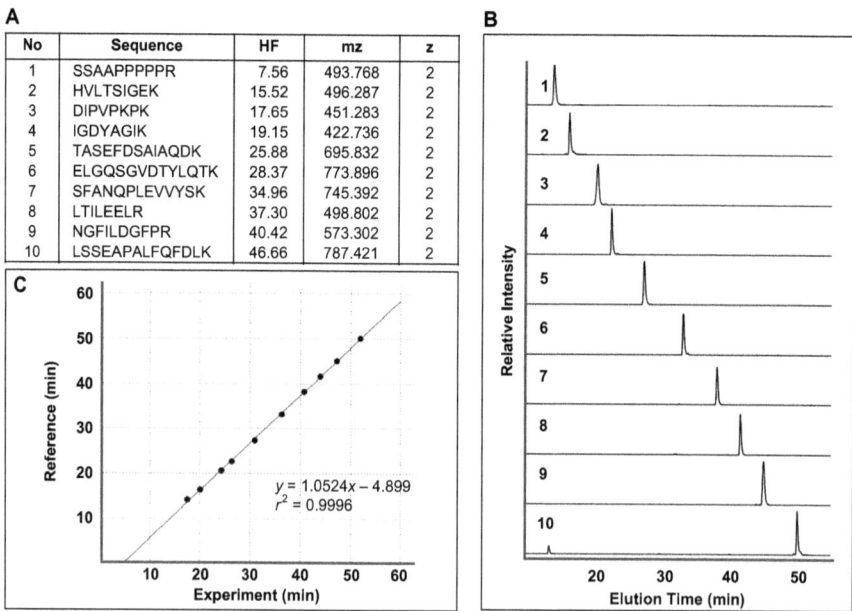

A

No	Sequence	HF	mz	z
1	SSAAPPPPPR	7.56	493.768	2
2	HVLTSIGEK	15.52	496.287	2
3	DIPVPKPK	17.65	451.283	2
4	IGDYAGIK	19.15	422.736	2
5	TASEFDSAIAQDK	25.88	695.832	2
6	ELGQSGVDTYLQTK	28.37	773.896	2
7	SFANQPLEVVYSK	34.96	745.392	2
8	LTILEELR	37.30	498.802	2
9	NGFILDGFPR	40.42	573.302	2
10	LSSEAPALFQFDLK	46.66	787.421	2

C

$y = 1.0524x - 4.899$
$r^2 = 0.9996$

Figure 5.4   Use of reference peptides in proteome profiling. (A) A set of reference peptides with hydrophobicity factor. (B) Extracted ion chromatograms (XICs) of each peptide showing corresponding elution times. (C) Correlation of elution times of reference peptides spiked in two biological samples. The correlation can be used to normalize the elution time variation between the LC-MS runs.

the reference peptides in an experiment, the elution times of analytes (peptides) can be corrected, which enables a normalization of the elution times across the multiple experiments (Figure 5.4C).

5.3   MS-Based Discovery Platforms: Unsupervised Profiling

Peptide mass finger printing (PMF) was introduced in the early 1990s to identify an individual protein using MS and a protein database.[31] Later, fragment ion peaks of a single peptide generated by CID were used for identification by matching the masses of fragments to the theoretical values derived from protein databases. The MS-based method enabled systematic protein identification while only consuming small amounts of samples. MS-based peptide and protein identification became routine thereafter once the complete human genome sequence was publicly available. Rapid advances in instrumentation, together with the improvement of search algorithms, expedited the development of large-scale proteomics workflows that represent an integral part of biomarker discovery today.

5.3.1   Data-Dependent Acquisition (DDA)

MS-based proteomic profiling is often performed by a "shotgun" method that includes proteolytic digestion of the protein mixture, separation of the resulting peptide mixtures by HPLC, and analysis of the eluting peptides by MS/MS.[32] The identification of proteins is inferred from the peptides observed. The mass accuracy and the resolving power of the mass analyzers are critical parameters. High mass accuracy decreases the false-positive assignment rate of the search algorithm, both for the precursor ions and the fragment ions. The higher resolving power of the mass analyzer discriminates interfering species, and thus increases the overall quality of LC-MS/MS results.

The automated data-dependent acquisition (DDA, also called information-dependent acquisition, IDA) is a widely used method in shotgun-based profiling. A series of precursors for MS/MS analyses are selected based on preset parameters, such as ion intensity and charge state, acquired in survey scans. In this mode, once the MS/MS spectrum or accumulated spectra of the peptide provides sufficient signals to enable the peptide search process, the analyzed ions are excluded from the selection of precursors within a certain time window. This allows analysis of peaks with lower ranking of the intensity.

DDA was first introduced for analysis of small molecules,[33] and became routine in the proteomics field for peptide profiling.[34] The logical selection of precursors in DDA mode enables a systematic acquisition and efficient data analysis in shotgun profiling. The method, however, has some drawbacks. During the chromatographic elution of the peptides, the best MS/MS quality is obtained at the apex of the precursor signal in the elution profile. However, the position where a precursor ion is actually selected for MS/MS in DDA mode does not necessarily coincide with this optimum location. Furthermore, low-intensity peaks are often missed during analyses even after excluding the high-abundant peaks already analyzed.

In addition to the identification, quantification of the proteins is often performed during the profiling since knowledge of the protein concentration is crucial in selecting the candidate markers to be evaluated further. The quantitative method often used in DDA-based profiling is called "spectral counting", which relies on the number of MS/MS spectra assigned to each peptide in the LC-MS/MS experiment.[35] The higher the concentration of a peptide, the larger the number of associated spectra (spectral counts). While spectral counting of peptides estimates a relative abundance of protein,[36] more rigorous quantification based on ion counts provides more reliable information.

5.3.2 Data-Independent Acquisition (DIA)

Instead of the sequential MS/MS analysis for selected precursors, parallel MS/MS analysis methods selecting several precursor ions concomitantly, have emerged. The workflow in this mode skips the specific precursor selection step by fragmenting the entire mass range or using wide (10–25 Da) precursor windows stepwise, providing convoluted MS/MS spectra containing ion fragments derived from multiple precursors.[37–40] Fragments and their matching precursor ions are identified after acquisition, by decoding their coelution profiles (chromatogram reconstruction) in MS/MS spectra over the chromatographic separation. The wideband precursor fragmentation mode is also known as data independent acquisition (DIA). This method significantly increases the analysis throughput. More sophisticated software to deconvolute fragment ions from all of the precursors is essential for this technology. Figure 5.5 illustrates the workflows of DDA *vs.* DIA.

5.4 MS-Based Discovery Platforms: Supervised Profiling

Even with high-performing instrumentation and advanced software tools facilitating sophisticated data analysis, the unsupervised profiling for biomarker discovery faces several limitations. One of them is the redundancy of identification both within the dataset of one LC-MS/MS and between datasets. If the goal of the LC-MS/MS experiment is simply to obtain a snapshot of a proteome, a set of peptide/protein profiles based on a shotgun result may suffice. However, in order to build a comprehensive protein catalog used as a basis for discovery studies, repeated shotgun experiment using multiple samples returns a large number of redundant peptides identified without adding much new information. Furthermore, this format tends to lose the low-abundant proteins that are important for such studies. Despite its robustness, nonsupervised profiling may not be sufficient for an indepth study. To address this limitation, a number of discovery workflows have been developed. Three

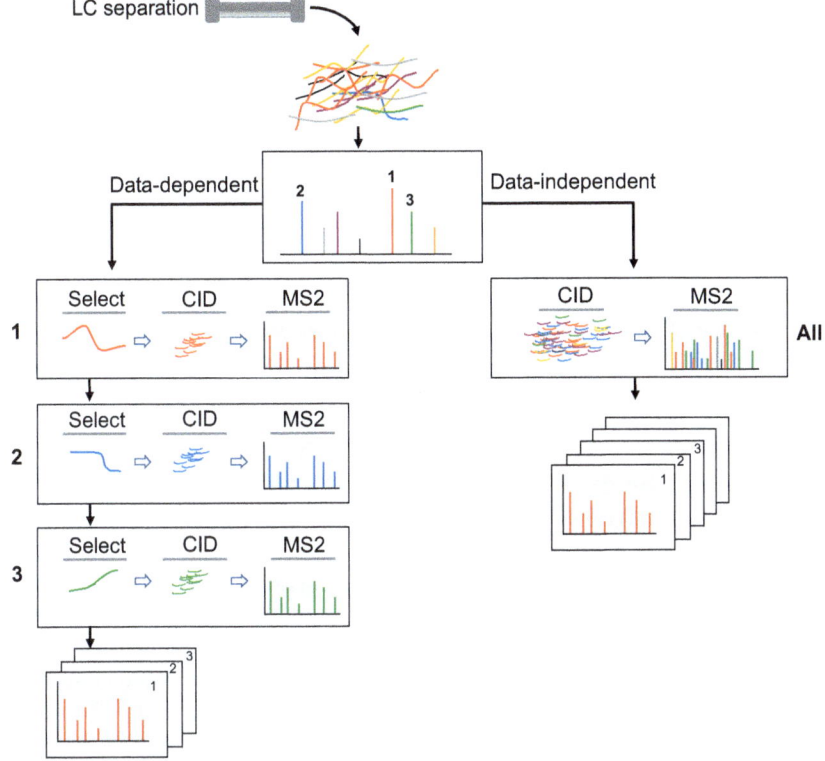

Figure 5.5   Data-dependent acquisition (DDA) and data-independent acquisition (DIA) for unsupervised discovery platforms. For both methods, after a full-scan MS1 spectrum (survey scan) is acquired, subsequent CID-based fragmentations are conducted. In DDA mode, precursors detected in the survey scan are selected sequentially to be subjected to CID. Peptide sequences are searched based on the individual precursor ion and its corresponding fragments. In DIA mode, all ions in the mass window are subjected to CID to generate a composite spectrum of fragment ions. Deconvolution of the fragment ions to find a corresponding precursor ion is performed by identifying matching chromatographic profiles of fragment ions.

strategies may be considered: (1) Peptide identification facilitated by accurate mass and elution times; (2) Quantification performed prior to identification to select differentially expressed peptides as an inclusion list to be sequenced; and (3) Targeted screening based on existing information and biological hypothesis.

5.4.1   Surrogate Identification by Accurate Mass and Elution Time

The mass-to-charge ratio (m/z) and the elution time of peptides observed in LC-MS experiments represent the molecular mass and hydrophobicity of the analytes. The combination of these characteristic parameters creates a unique value for each peptide, which can be used to identify the peptide sequence without CID-based MS/MS. The idea of using m/z values and elution times as a surrogate identification has emerged in the accurate mass and time (AMT) tag approach using the high-accuracy mass measurements of Fourier transform-ion cyclotron resonance mass spectrometry (FTICR-MS) and nano-LC systems.[33] Since then, the AMT strategy has been applied to numerous proteomic analyses.[41,42]

Two tasks are essential in order to implement an AMT strategy. First, a comprehensive list of peptides with their LC-MS attributes [m/z, z, elution time] present in a given biological system needs to be constructed as a peptide database. The entries are created typically by shotgun-based LC-MS/MS, using either DDA or DIA modes. The precursor m/z values are supplemented with the theoretical values after peptide sequences are identified, and the detected elution times are normalized using reference peptides as shown in Figure 5.3. Extensive prefractionations are often performed in order to include peptides of low abundance or those subjected to interference in complex mixtures as discussed in Section 5.2.1. The reproducibility of LC system and the mass accuracy of the analyzer are key components for this strategy.

Secondly, a data-processing algorithm is necessary for matching the LC-MS attributes to those stored in the peptide database. Once the peptide DB is built, and the matching algorithm validated, LC-MS analysis is carried out without extensive fractionation or MS/MS, as illustrated in Figure 5.6. This workflow can expedite large-scale proteomics analyses in discovery by eliminating redundant MS/MS data and by leveraging the high-quality identification data acquired in extensively fractionated samples. In the case of the peptide shown in Figure 5.3A, where DDA fails to sequence the ion detected in a whole-cell lysate, this ion can be identified solely by its m/z and elution time by matching them to those in the peptide database. In the database, the successful MS/MS spectrum already acquired with the fractionated sample (Figure 5.3B) should be ready for this analysis.

5.4.2   Selective Identification Using Inclusion List

An important milestone in the biomarker discovery phase is the identification of the aberrant proteins associated with diseases. Instead of profiling all proteins by indepth sequencing followed by quantification, one may rather focus on proteins exhibiting differential expression, and conduct a selective MS/MS analysis accordingly. This strategy involves a prequantification step using the mass-spectrometric signal intensities decoupled from the identification step. After the quantitative analysis, the ions of interest are selected and added to an inclusion list containing m/z values and predicted elution times of the peptides to be subjected to MS/MS (Figure 5.7).[43,44] The LC-MS/MS analyses are conducted in a directed manner, in which only the ions specified in the acquisition methods are elected for sequencing. Since the precursor selection and CID are focused only on particular ions, MS/MS is more efficient and redundancy is avoided. Consequently, the indepth analysis is possible because it targets the ions that otherwise may be missed during a DDA

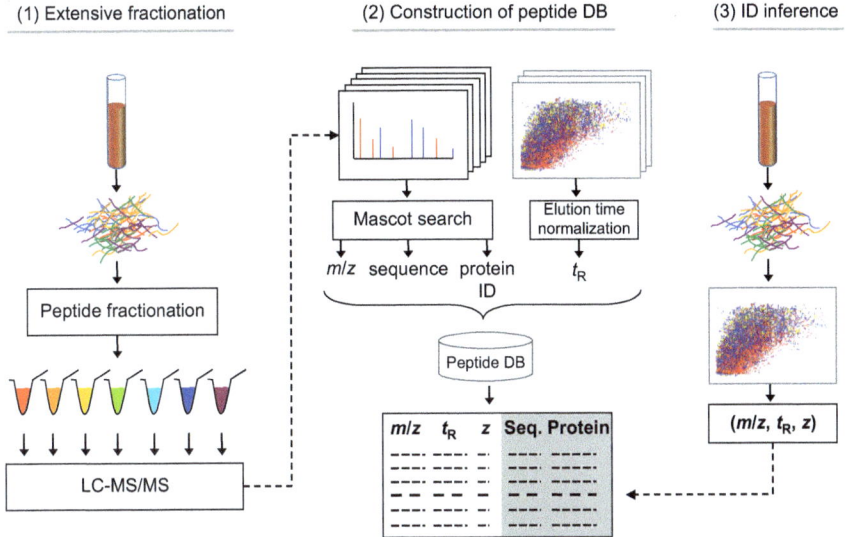

Figure 5.6 General workflow of using accurate mass and elution time tags as surrogates for peptide identification. Peptide database containing peptide sequences and the corresponding LC-MS attributes [m/z, normalized elution time (t_R), z] is created using indepth LC-MS analyses of a proteome involving extensive fractionations and MS/MS-based sequencing. Subsequently, peptide ions detected in LC-MS experiments are analyzed based on their [m/z, t_R, z] values by matching with the information stored in the database.

experiment. However, for studies of cellular pathways or molecular interaction networks, where specific changes may affect proteins not present in the inclusion list, it could foreclose the opportunity of discovering new targets.

Quantitative information can be obtained by various strategies based on labeling methods combined with isotope dilution, or with label-free methods. The term "label" in quantitative proteomics has been used in different contexts, which includes chemical derivatization and/or incorporation of stable isotopes. In this chapter, we consistently refer to isotope incorporation. Figure 5.7A shows the prequantification strategy in the label-free method. The mass-spectrometric signal intensities of peptide ions in different samples are obtained by generating extracted ion chromatograms (XICs) for each ion. Signal intensities are then measured and used to perform a relative quantification across the samples. A set of peptide ions is selected to address the biological interest (*e.g.* overexpressed proteins in a particular system). Figure 5.7B is the same workflow applied in the stable-isotope labeling-based quantification. The peptides in sample 1 are modified with a reagent composed of $^{12}C/^{14}N$ (light isotopes); those in sample 2 are modified with the reagent, composed of stable isotopes, $^{13}C/^{15}N$ (heavy isotopes).

The labeled peptides are pooled to generate a combined sample, and analyzed together in a single experiment. Relative quantification is achieved by

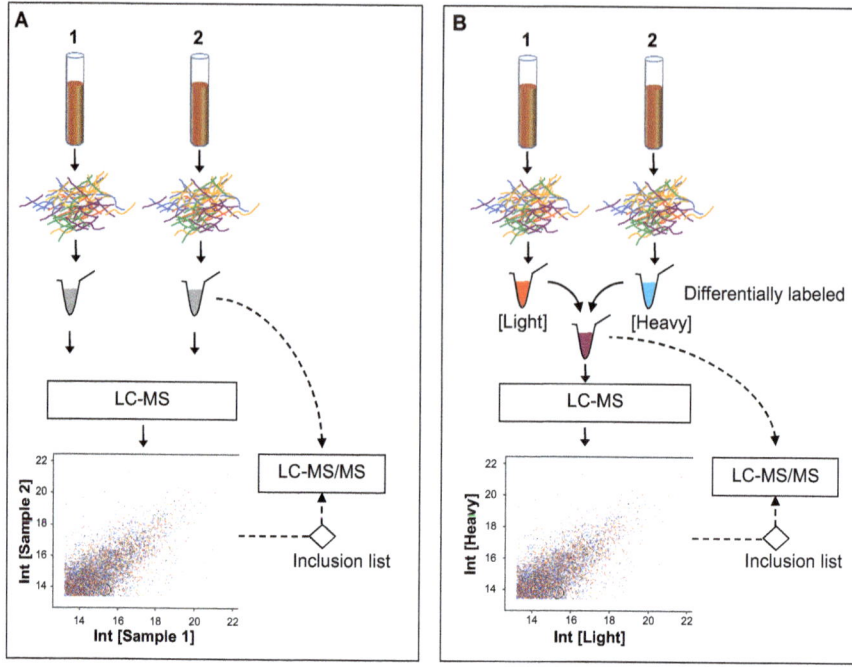

Figure 5.7   Typical workflows of the selective identification using an inclusion list. Quantitative analysis is performed prior to identification to create an inclusion list containing relevant ions to be sequenced. This list is used to subsequently direct the LC-MS/MS analysis, performed on the same sample. (A) Label-free-based quantification. The signal intensities of the corresponding peptide ions from two samples (sample 1 and 2) are compared. (B) Quantification using isotopically labeled peptides. The tryptic peptides are derivatized with light ($^{12}C/^{14}N$) and the heavy ($^{13}C/^{15}N$) reagents in sample 1 and 2, respectively, prior to pooling. The quantification is obtained by comparing the signal intensities of the coeluting light and heavy peptides in a single LC-MS run.

measuring the intensity ratio of light and heavy peptide ions detected in the LC-MS experiment as illustrated in Figure 5.8. Although the light/heavy pair of peptide ions are used for quantification, usually only one of the ions is added to the inclusion list. In the case of an overexpressed protein in heavy form, it is reasonable to choose the heavy peptides for MS/MS analysis as they have more intense signals than the light peptides. Several stable-isotope labeling methods have been developed including isotope-coded affinity tag (ICAT) targeting cysteines,[45] proteolytic ^{18}O labels incorporated to all tryptic digests during the proteolysis,[46] and metabolic labeling at the protein level with the method called stable-isotope labeling by amino acids in cell culture (SILAC).[47]

In both approaches, label-free and stable-isotope labeling, the identification step is decoupled from the quantification process. The shorter duty cycle increases the number of chromatographic data points acquired, and thus

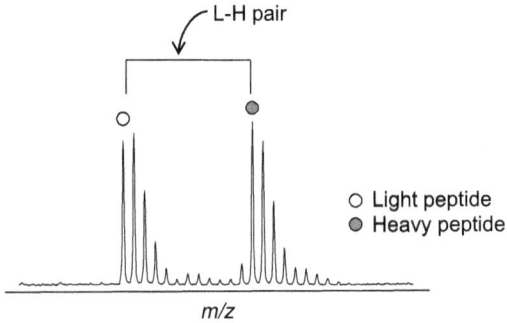

Figure 5.8   An example of a partial MS spectrum of a light–heavy peptide pair present in a pooled sample. The peak intensity ratio is used for relative quantification.

increases the precision of the analyses. Because of the inclusion list involved in this method to direct the identification step, it has been called a directed acquisition mode or directed proteomics.[48,49]

5.4.3 Hypothesis-Driven Discovery Proteomics

The *de novo* discovery experiments with extensive peptide sequencing commonly used in proteomics-based biomarker studies, have generated a wealth of information over the last decade that can now be used to design specific analytical methods. The concept of targeted proteomics driven by hypothesis has emerged, and is focused on the analysis of a subset of proteins representing candidate biomarkers.

The general workflow in targeted proteomics is depicted in Figure 5.9. The first step of the workflow is defining the target proteins. The target list comprises proteins derived from various discovery studies, public genomics and transcriptomics data sets, proteins associated with disease pathogenesis, and bioinformatics networking analysis. The emergence of this strategy has prompted the creation of numerous public and commercial repositories providing useful -omics data and tools to be used in proteomic analysis. Among them, the Human Protein Atlas (www.proteinatlas.org) is a protein information portal focusing on the antibody-based tissue expression profiles of target proteins.[50] Currently (version 10.0) this database covers > 14 000 protein expression profiles in multiple cancer tissues.[51] PRoteomics IDEntifications (PRIDE) (http://www.ebi.ac.uk/pride/) is a public data repository more focused on the mass-spectrometry-based experiments. PRIDE contains proteins, peptides, and corresponding MS/MS spectra, which can be used in targeted proteomics workflow. Various disease-based protein databases are useful in providing disease-specific information to prepare the target lists.[52] Functional information of targets in cellular networks with gene ontology and pathway analysis using public tools (www.babelomics.bioinfo.cipf.es) can be

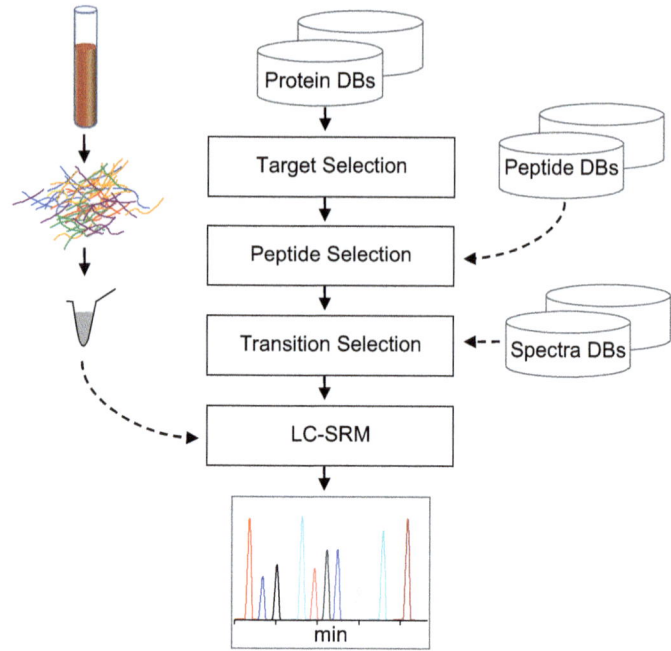

Figure 5.9   Workflow of targeted proteomics strategy. Proteins that have been identified as potential biomarkers in discovery studies are selected for a target list. PTPs of proteins in the target list are chosen based on uniqueness of sequence and the degree of detectability. For the SRM analysis, transitions of each PTP need to be defined. Peptide and spectra databases either built inhouse or from the public domain, are used to select the best transitions for each peptide. Expected elution times are taken into account during the LC-MS/MS.

used. This information, together with the actual detectability of peptides in specific specimen types, plasma (www.plasmaproteomedatabase.org) and urine (www.proteome.biochem.mpg.de/urine), are available to be used for target selection. Polanski and Anderson[53] published a list of potential cancer biomarkers derived from the literature. The information of >1200 genes/ proteins in this report has facilitated targeted proteomics approaches for cancer biomarkers. Commercial biomarker databases such as GVK bio online biomarker database (GOBIOM) (www.gobiomdb.com/gobiom) or Thomson Reuters Integrity database (www.thomsonreuters.com/products_services/ science/science_products/a-z/integrity) are also available, providing vast amounts of information on protein/gene and related compounds.

After targets have been defined, the second step is to generate a set of peptides detectable during LC-MS/MS analysis. The primary sequences of target proteins in protein databases such as UniProt (Universal Protein Resource, www.uniprot.org) are analyzed to prepare the peptide sets by *in silico* digestion. Since these selected peptides represent the target proteins, the amino acid sequence of each peptide should relate uniquely to the corresponding protein. A unique peptide whose physical properties (hydrophobicity, length, and charge states) are suitable for the LC-MS analysis is called a proteotypic peptides (PTP). To select the best PTPs representing target proteins efficiently, several public databases of peptides have been built including PeptideAtlas (www.peptideatlas.org), PRIDE (www.ebi.ac.uk/pride/), and the Global Proteome Machine (GPM, www.thegpm.org/). In these databases the experimental information with respect to the peptides detected on mass-spectrometry platforms are stored. For example, PRIDE database (version 2.8.18) contains more than six million unique peptides already analyzed with mass spectrometry, PeptideAtlas provides scored peptide entries in various human subject including plasma, urine, and brain. This information is useful for selecting peptides likely to be measured in a LC-MS/MS experiment.

Once PTPs of targets are defined, the LC-MS methods of selected reaction monitoring (SRM) are generated. The SRM is performed in a triple-quadrupole (QQQ) instrument where the m/z selection of precursor ions at the first quadrupole (Q1) is synchronized with a fragment ion of this precursor at Q3, after fragmentation at Q2. The set of m/z values corresponding to the precursor ion and one of its fragment ions is defined as a transition. The two-stage mass filtering in SRM ensures high sensitivity, selectivity and a broad dynamic range. Public databases as well as inhouse data are used in selecting SRM transitions. SRMAtlas (www.srmatlas.org) provides SRM transitions from a variety of experimental data sources covering > 60% of human proteome and from theoretical analyses covering > 99% human proteome. The overall throughput of LC-SRM analysis can be increased by employing elution-time prediction to schedule the measurement only at the time when the targeted ions are present.[54]

Although triple-quadrupole mass spectrometers are the reference instruments for performing SRM experiments, quadrupole-linear ion-trap mass spectrometers have been also used. More recently, a fragment ion-based quantification method emerged using quadrupole-time of flight (Q-TOF) and quadrupole-orbitrap instruments leveraging their capabilities of measuring ions in high resolution and with high accuracy.[42,55] The method called Parallel Reaction Monitoring (PRM) was developed on a quadrupole-orbitrap instrument, in which quantification is performed with the set of fragment ions measured in orbitrap.[56,57] Since the signals of all fragment ions are already recorded in the PRM experiment, there is no need for *a priori* transition selection, a requisite for SRM experiment.

In the context of targeted proteomics, the Accurate Inclusion Mass Screening (AIMS) approach is also used. AIMS generates the inclusion list based on hypothesis, and subjects it to the LC-MS/MS analysis. If the targeting precursor ions exist in the sample, MS/MS analyses are triggered for identification.[58] It is often used as a triaging tool between discovery and verification steps in biomarker studies.[59] Table 5.2 summarizes the various strategies in profiling efforts described in Sections 5.3 and 5.4.

Table 5.2 Summary of profiling strategies.

Strategy	Category	Description	Mode	Major platforms	Amendable to large scale
Unsupervised profiling	Shotgun by data-dependent acquisition	Select precursor ions for sequential peptide sequencing	DDA	Q-TOF, LIT-Orbitrap	+
	Shotgun by data-independent acquisition	Parallel sequencing of coeluting peptides without precursor selection	DIA	Q-TOF, Q-Orbitrap	++
Supervised profiling	Surrogate identification by AMT	Build a database for identified peptides and leverage the information for further ID	AMT	Q-TOF, LIT-Orbitrap	++++
	Prequantification	Perform quantification based on the mass intensity and identify only relevant ions using inclusion list	Directed	Q-TOF	+++
	Targeted proteomics	Define target proteins and compute measurable peptides to follow	AIMS	Q-TOF, LIT-Orbitrap	++
			SRM	QQQ, Q-LIT	+++
			PRM	Q-TOF, Q-Orbitrap	++++

5.5   Example: Directed Proteomics and Enrichment Strategies Applied to a Colon Cancer Stem-Cell Biomarker Study

One example of LC-MS-based strategies found in biomarker discovery is shown in Figure 5.10. The goal of this study is to identify cell-surface biomarkers specific to cancer stem cells whose presence in tumor cells may cause drug resistance and tumor recurrence after the treatment of colorectal cancer. In this study, proteins in the primary tumor cells derived from colorectal cancer patients and those in the tumor spheroid cells (representing cancer stem cells) from the same patients were compared to identify cell surface proteins specifically expressed in cancer stem cells.[60] The sample preparation in this study employed glycoprotein-enrichment technology to capture the glycoproteins on the plasma membrane. Sample complexity was further

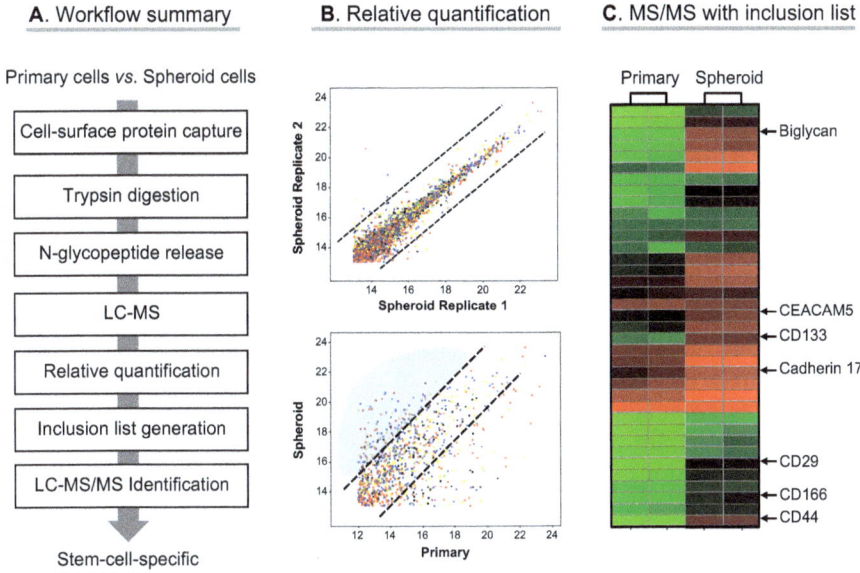

Figure 5.10   An example of a biomarker discovery study that applied strategies described in this chapter (adapted from ref. 60). (A) The overall workflow of the study consisting of glycoprotein enrichment and directed proteomics. (B) Mass-spectrometric intensity plots of glycopeptides. Peptide signal intensities in process replicates of spheroid cells (top) show a tight correlation, indicating a reproducible sample process. The intensity profile changed dramatically from spheroid to primary cell populations (bottom), indicating that glycoprotein expression profiles are different in the two populations. Among those differentially expressed proteins, peptides present in spheroid sample > 4-fold (in blue shade) were selected as an inclusion list, and subjected to MS/MS. (C) A heat map of the peptides identified based on the inclusion list. Seven peptide ions are assigned to the corresponding proteins in the sample.

reduced by isolating tryptic peptides that contain N-glycosylation sites. These strategies are described in Section 5.2.1. For LC-MS (/MS) analysis, the directed proteomics was applied involving selective identification using an inclusion list as discussed in Section 5.4.2.

Figure 5.10A shows the workflow for this process combining sample preparation and LC-MS analysis. Figure 5.10B is an example of the prequantification performed to select peptides of interest. In label-free-based LC-MS, the reproducibility of the intensity measurement should be confirmed prior to the differential analysis. The upper scatter plot shows the signal intensities of each ion measured in two LC-MS experiments from the same tumor spheroid cells as process duplicates. This step confirms the reproducibility of the process and defines the sample variations associated with the analytical process. The bottom scatter plot in Figure 5.10B shows the intensities of each ion measured in two different samples. Differential analysis was performed to find peptide ions detected with higher intensities in cancer stem cells compared to the primary tumor cells. The area shaded blue where the peptide intensities are more than four-fold different were selected for MS/MS as an inclusion list (see also Figure 5.7A in Section 5.4.2). The MS/MS data acquired based on the inclusion list identified peptides and proteins in the list (Figure 5.10C). These cell-surface proteins are specifically expressed in tumor spheroid cells, and were further evaluated as biomarkers for colon cancer stem cells.[60]

5.6 Perspectives

In this chapter, major analytical strategies commonly used in LC-MS-based biomarker discovery are discussed, focusing on strategic sample preparation and efficient protein profiling. In order to design efficient workflows and to select optimal technologies, it is important to understand the ultimate deliveries of the studies (*e.g.* the disease-specific proteins detectable in plasma) at the end of the profiling. At the same time, understanding the purpose of applications of the biomarkers is essential to design the study and to coordinate the clinical sample collection and their annotation with clinical data.

There are clear trends in developing LC-MS-based profiling technologies. First, more-targeted approaches are employed during the discovery phase. This is synergistic with the increasing availability of public databases and bioinformatics tools used to triage the candidate proteins as mentioned in Section 5.4.3. In addition, the ongoing efforts to annotate all human proteins (~ 20 300) with respect to gene location, cellular distribution and quantification, will ultimately provide a useful resource for targeted approaches. In contrast to conventional proteomic workflow that largely depends on protein-search algorithms, targeted methods can focus on proteins with structural variations such as modifications, mutations, and deletions that are potential targets of identifying disease onsets and therapeutic responses.[61]

A second trend is the emphasis on standardization to ensure the portability of the procedures between the laboratories. Definitions of standard protocols on sample collection and preparation, LC-MS analysis, and data processing are essential. This affects more on the later stages of the biomarker studies, verification and evaluation, where a larger set of samples are collected and processed in multiple sites.[62] Standardization is also important in the discovery stage, as it allows transparent and consistent analytical procedures for laboratories working on the same protein candidates.

Lastly, it is anticipated that mass-spectrometry-based technologies are going to gain traction in the future not only for the verification and evaluation phases of biomarker discovery but to actually perform the routine molecular tests. In this context, the integration of the discovery and verification platforms with standardized protocols will streamline the general implementation of the LC-MS-based assays in clinical setting as well as in the translational research for personalized medicine.

Acknowledgement

This work is supported by the Ministry for Higher Education (MESR) of Luxembourg and the Fonds National de la Recherche (FNR).

References

1. Biomarkers Definitions Working, G. Biomarkers and surrogate endpoints: preferred definitions and conceptual framework. *Clinical Pharmacology and Therapeutics* 2001, **69**, 89–95.
2. N. Rifai, M. A. Gillette and S. A. Carr, Protein biomarker discovery and validation: the long and uncertain path to clinical utility, *Nature Biotechnol.*, 2006, **24**, 971–983.
3. C. N. Oldenhuis, S. F. Oosting, J. A. Gietema and E. G. de Vries, Prognostic versus predictive value of biomarkers in oncology, *European Journal of Cancer*, 2008, **44**, 946–953.
4. C. Schmidt, Larger companies dominate cancer companion diagnostic approvals, *Nature Biotechnol.*, 2011, **29**, 955–957.
5. S. M. Hanash, S. J. Pitteri and V. M. Faca, Mining the plasma proteome for cancer biomarkers, *Nature*, 2008, **452**, 571–579.
6. Z. Zhang and D. W. Chan, The road from discovery to clinical diagnostics: lessons learned from the first FDA-cleared in vitro diagnostic multivariate index assay of proteomic biomarkers, *Cancer Epidemiol. Biomarkers Prev.*, 2010, **19**, 2995–2999.
7. H. Keshishian, T. Addona, M. Burgess, E. Kuhn and S. A. Carr, Quantitative, multiplexed assays for low abundance proteins in plasma by targeted mass spectrometry and stable isotope dilution, *Mol. Cell Proteomics*, 2007, **6**, 2212–2229.
8. D. R. Barnidge, M. K. Goodmanson, G. G. Klee and D. C. Muddiman, Absolute quantification of the model biomarker prostate-specific antigen in serum by LC-Ms/MS using protein cleavage and isotope dilution mass spectrometry, *J. Proteome Res.*, 2004, **3**, 644–652.

9. K. L. Abbott and J. M. Pierce, Lectin-based glycoproteomic techniques for the enrichment and identification of potential biomarkers, *Methods in Enzymology*, 2010, **480**, 461–476.
10. J. M. Bobbitt, Periodate oxidation of carbohydrates, *Advances in Carbohydrate Chemistry*, 1956, **48**, 1–41.
11. Y. J. Kim, Z. Zaidi-Ainouch, S. Gallien and B. Domon, Mass spectrometry-based detection and quantification of plasma glycoproteins using selective reaction monitoring, *Nature Protocols*, 2012, **7**, 859–871.
12. Y. Zeng, T. N. Ramya, A. Dirksen, P. E. Dawson and J. C. Paulson, High-efficiency labeling of sialylated glycoproteins on living cells, *Nature Methods*, 2009, **6**, 207–209.
13. P. Hagglund, J. Bunkenborg, F. Elortza, O. N. Jensen and P. Roepstorff, A new strategy for identification of N-glycosylated proteins and unambiguous assignment of their glycosylation sites using HILIC enrichment and partial deglycosylation, *J. Proteome Res.*, 2004, **3**, 556–566.
14. M. Mann, *et al.*, Analysis of protein phosphorylation using mass spectrometry: deciphering the phosphoproteome, *Trends in Biotechnology*, 2002, **20**, 261–268.
15. B. Blagoev, S. E. Ong, I. Kratchmarova and M. Mann, Temporal analysis of phosphotyrosine-dependent signaling networks by quantitative proteomics, *Nature Biotechnol.*, 2004, **22**, 1139–1145.
16. T. Liu, *et al.*, Improved proteome coverage by using high efficiency cysteinyl peptide enrichment: the human mammary epithelial cell proteome, *Proteomics*, 2005, **5**, 1263–1273.
17. T. Liu, W. J. Qian, D. G. Camp, 2nd and R. D. Smith, The use of a quantitative cysteinyl-peptide enrichment technology for high-throughput quantitative proteomics, *Methods in Molecular Biology*, 2007, **359**, 107–124.
18. K. Gevaert, *et al.*, Exploring proteomes and analyzing protein processing by mass spectrometric identification of sorted N-terminal peptides, *Nature Biotechnol.*, 2003, **21**, 566–569.
19. L. A. Echan, H. Y. Tang, N. Ali-Khan, K. Lee and D. W. Speicher, Depletion of multiple high-abundance proteins improves protein profiling capacities of human serum and plasma, *Proteomics*, 2005, **5**, 3292–3303.
20. L. Steinstrasser, *et al.*, Immunodepletion of high-abundant proteins from acute and chronic wound fluids to elucidate low-abundant regulators in wound healing, *BMC Research Notes*, 2010, **3**, 335.
21. G. J. Opiteck, K. C. Lewis, J. W. Jorgenson and R. J. Anderegg, Comprehensive on-line LC/LC/MS of proteins, *Analytical Chemistry*, 1997, **69**, 1518–1524.
22. Y. J. Kim, *et al.*, Reference map for liquid chromatography-mass spectrometry-based quantitative proteomics, *Analytical Biochemistry*, 2009, **393**, 155–162.
23. A. J. Link, *et al.*, Direct analysis of protein complexes using mass spectrometry, *Nature Biotechnol.*, 1999, **17**, 676–682.

24. C. Delahunty and J. R. Yates, Protein identification using 2D-LC-MS/MS, *Methods*, 2005, **35**, 248–255.
25. M. Heller, *et al.*, Two-stage Off-Gel isoelectric focusing: protein followed by peptide fractionation and application to proteome analysis of human plasma, *Electrophoresis*, 2005, **26**, 1174–1188.
26. L. N. Waller, K. Shores and D. R. Knapp, Shotgun proteomic analysis of cerebrospinal fluid using off-gel electrophoresis as the first-dimension separation, *J. Proteome Res.*, 2008, **7**, 4577–4584.
27. Z. Cao, H. Y. Tang, H. Wang, Q. Liu and D. W. Speicher, Systematic Comparison of Fractionation Methods for In-depth Analysis of Plasma Proteomes, *J Proteome Res*, 2012.
28. N. Delmotte, M. Lasaosa, A. Tholey, E. Heinzle and C. G. Huber, Two-dimensional reversed-phase x ion-pair reversed-phase HPLC: an alternative approach to high-resolution peptide separation for shotgun proteome analysis, *J. Proteome Res.*, 2007, **6**, 4363–4373.
29. Y. Wang, *et al.*, Reversed-phase chromatography with multiple fraction concatenation strategy for proteome profiling of human MCF10A cells, *Proteomics*, 2011, **11**, 2019–2026.
30. T. Kocher, R. Swart and K. Mechtler, Ultra-high-pressure RPLC hyphenated to an LTQ-Orbitrap Velos reveals a linear relation between peak capacity and number of identified peptides, *Analytical Chemistry*, 2011, **83**, 2699–2704.
31. V. C. Wasinger, *et al.*, Progress with gene-product mapping of the Mollicutes: Mycoplasma genitalium, *Electrophoresis*, 1995, **16**, 1090–1094.
32. M. P. Washburn, D. Wolters and J. R. Yates, Large-scale analysis of the yeast proteome by multidimensional protein identification technology, *Nature Biotechnol.*, 2001, **19**, 242–247.
33. P. R. Tiller, Z. El Fallah, V. Wilson, J. Huysman and D. Patel, Qualitative Assessment of Leachables Using Data-dependent Liquid Chromatography/Mass Spectrometry and Liquid Chromatography/ Tandem Mass Spectroemtry, *Rapid Commun. Mass Spec.*, 1997, **11**, 1570–1574.
34. M. Mann, R. C. Hendrickson and A. Pandey, Analysis of proteins and proteomes by mass spectrometry, *Annual Review of Biochemistry*, 2001, **70**, 437–473.
35. H. Liu, R. G. Sadygov and J. R. Yates, A model for random sampling and estimation of relative protein abundance in shotgun proteomics, *Analytical Chemistry*, 2004, **76**, 4193–4201.
36. W. Zhu, J. W. Smith and C. M. Huang, Mass spectrometry-based label-free quantitative proteomics, *Journal of Biomedicine & Biotechnology*, 2010, **2010**, 840518.
37. J. C. Silva, M. V. Gorenstein, G. Z. Li, J. P. Vissers and S. J. Geromanos, Absolute quantification of proteins by LCMSE: a virtue of parallel MS acquisition, *Mol. Cell Proteomics*, 2006, **5**, 144–156.

38. J. D. Venable, M. Q. Dong, J. Wohlschlegel, A. Dillin and J. R. Yates, Automated approach for quantitative analysis of complex peptide mixtures from tandem mass spectra, *Nature Methods*, 2004, **1**, 39–45.
39. T. Geiger, J. Cox and M. Mann, Proteomics on an Orbitrap benchtop mass spectrometer using all-ion fragmentation, *Mol. Cell Proteomics*, 2010, **9**, 2252–2261.
40. L. C. Gillet, *et al.*, Targeted data extraction of the MS/MS spectra generated by data-independent acquisition: a new concept for consistent and accurate proteome analysis, *Mol. Cell Proteomics*, 2012, **11**, O111 016717.
41. R. D. Smith, *et al.*, An accurate mass tag strategy for quantitative and high-throughput proteome measurements, *Proteomics*, 2002, **2**, 513–523.
42. J. S. Zimmer, M. E. Monroe, W. J. Qian and R. D. Smith, Advances in proteomics data analysis and display using an accurate mass and time tag approach, *Mass Spectrom. Rev.*, 2006, **25**, 450–482.
43. B. Domon and R. Aebersold, Mass spectrometry and protein analysis, *Science*, 2006, **312**, 212–217.
44. T. He, Y. J. Kim, J. L. Heidbrink, P. A. Moore and S. M. Ruben, Drug target identification and quantitative proteomics, *Expert Opin. Drug Discov.*, 2006, **1**, 477–489.
45. S. P. Gygi, *et al.*, Quantitative analysis of complex protein mixtures using isotope-coded affinity tags, *Nature Biotechnol.*, 1999, **17**, 994–999.
46. X. Yao, A. Freas, J. Ramirez, P. A. Demirev and C. Fenselau, Proteolytic 18O labeling for comparative proteomics: model studies with two serotypes of adenovirus, *Analytical Chemistry*, 2001, **73**, 2836–2842.
47. S. E. Ong, *et al.*, Stable isotope labeling by amino acids in cell culture, SILAC, as a simple and accurate approach to expression proteomics, *Mol. Cell Proteomics*, 2002, **1**, 376–386.
48. B. Domon and S. Broder, Implications of new proteomics strategies for biology and medicine, *J. Proteome Res.*, 2004, **3**, 253–260.
49. A. Schmidt, *et al.*, An integrated, directed mass spectrometric approach for in-depth characterization of complex peptide mixtures, *Mol. Cell Proteomics*, 2008, **7**, 2138–2150.
50. M. Uhlen, *et al.*, A human protein atlas for normal and cancer tissues based on antibody proteomics, *Mol. Cell Proteomics*, 2005, **4**, 1920–1932.
51. M. Uhlen, *et al.*, Towards a knowledge-based Human Protein Atlas, *Nature Biotechnol*, 2010, **28**, 1248–1250.
52. L. Wang, *et al.*, HLungDB: an integrated database of human lung cancer research, *Nucleic Acids Research*, 2010, **38**, D665–669.
53. M. Polanski and N. A. Anderson, A list of candidate cancer biomarkers for targeted proteomics, *Biomarker Insights*, 2006, **1**, 1–48.
54. S. Gallien, *et al.*, Highly multiplexed targeted proteomics using precise control of peptide retention time, *Proteomics*, 2012, **12**, 1122–1133.
55. J. H. Baek, H. Kim, B. Shin and M. H. Yu, Multiple products monitoring as a robust approach for peptide quantification, *J. Proteome Res.*, 2009, **8**, 3625–3632.

56. S. Gallien, *et al.*, Targeted proteomic quantification on quadrupole-orbitrap mass spectrometer, *Mol. Cell Proteomics*, 2012, **11**, 1709–1723.

57. A. C. Peterson, J. D. Russell, D. J. Bailey, M. S. Westphall and J. J. Coon, Parallel reaction monitoring for high resolution and high mass accuracy quantitative, targeted proteomics, *Mol. Cell Proteomics*, 2012.

58. J. D. Jaffe, *et al.*, Accurate inclusion mass screening: a bridge from unbiased discovery to targeted assay development for biomarker verification, *Mol. Cell Proteomics*, 2008, **7**, 1952–1962.

59. J. R. Whiteaker, *et al.*, A targeted proteomics-based pipeline for verification of biomarkers in plasma, *Nature Biotechnol.*, 2011, **29**, 625–634.

60. D. D. Fang, *et al.*, Expansion of CD133(+) colon cancer cultures retaining stem cell properties to enable cancer stem cell target discovery, *British Journal of Cancer*, 2010, **102**, 1265–1275.

61. M. Stastna and J. E. Van Eyk, Secreted proteins as a fundamental source for biomarker discovery, *Proteomics*, 2012, **12**, 722–735.

62. T. A. Addona, *et al.*, Multi-site assessment of the precision and reproducibility of multiple reaction monitoring-based measurements of proteins in plasma, *Nature Biotechnol*, 2009, **27**, 633–641.

CHAPTER 6

Sample Preparation and Profiling: Probing the Kinome for Biomarkers and Therapeutic Targets: Peptide Arrays for Global Phosphorylation-Mediated Signal Transduction

JASON KINDRACHUK[a] AND SCOTT NAPPER*[b,c]

[a] Emerging Viral Pathogens Section, National Institute of Allergy and Infectious Diseases, National Institutes of Health, Bethesda, Maryland, USA, 20892; [b] Department of Biochemistry, University of Saskatchewan, Saskatoon, Saskatchewan, S7N 5E5 Canada; [c] Vaccine and Infectious Disease Organization, University of Saskatchewan, Saskatchewan, S7N 5E3 Canada
*Email: napper@usask.ca

6.1 Introduction

Post-translation modification through phosphorylation is the predominant mechanism for control of protein function that in turn regulates cellular responses and phenotypes. Protein phosphorylation events are catalyzed by a class of enzymes known as the kinases. Kinases occupy a central role in the regulation of virtually every cellular process including, but not limited to,

RSC Drug Discovery Series No. 33
Comprehensive Biomarker Discovery and Validation for Clinical Application
Edited by Péter Horvatovich and Rainer Bischoff
© The Royal Society of Chemistry 2013
Published by the Royal Society of Chemistry, www.rsc.org

metabolism, transcription, cell-cycle progression, cytoskeletal rearrangement, immune defense, apoptosis and differentiation.[1] Kinases are also intimately associated with disease, both as causative agents (many diseases are related to kinase defects) as well as therapeutic targets (kinases are now the second most frequently targeted class of enzymes).[2] In addition to these well-recognized contributions to human health there is growing appreciation that characterization of global cellular kinase activity, the kinome, is an effective approach to understand complex biological responses. Further, it is understood that "kinomic fingerprints" or "kinotypes" have considerable potential to serve as biomarkers of various physiological and pathophysiological states. In this capacity kinotyping can be used to predict the propensity for, onset of and/or outcome of various disease states.

The growing interest in kinases has driven efforts to develop research tools that enable characterization of phosphorylation-mediated signal transduction activity. Peptide arrays have demonstrated remarkable potential for characterizing of the kinome in a robust, cost-effective fashion.[3] In particular, recent advances enable researchers to adopt increasingly global perspectives in a wider range of species and to more effectively extract biological information from kinomic data. In this chapter the opportunities for utilizing peptides arrays for kinome analysis are discussed including specific examples in which peptide arrays have: contributed to the identification of disease-associated processes/biomarkers, provided therapeutic targets, and offered novel insight into complex biological processes.

6.2   Understanding Complex Biology through Phosphorylation-Mediated Signal Transduction

The rapid proliferation of various "omics" technologies has fostered considerable debate as to the most appropriate biological level at which to define cellular responses. Within any investigation there are typically a number of considerations (scientific, practical and economical) that predicate the selection of a particular approach.

Transcriptional analysis, based largely on affordability, availability and experimental maturity, remains one of the most widely applied techniques for global analysis of cellular responses. However, there are concerns that sometimes descriptions of gene expression patterns, no matter how comprehensive, do not accurately describe nor predict cellular phenotypes. This primarily reflects the inability of transcriptional analysis to consider a multitude of post-transcriptional regulatory events that separate a gene transcript from a phenotype. This includes, but is not limited to, gene silencing, mRNA stability, unique translational efficiencies, protein turnover, segregation of enzymes from their substrates and post-translational modification of proteins to regulate their function.

In contrast, kinases function at a biological level subsequent to these regulatory elements and are therefore functionally closer to the phenotype. Indeed, protein phosphorylation is often the defining event for initiating cellular responses and phenotypes. That kinases represent the foundation of cellular signal transduction, with seminal roles in regulating virtually every cellular behavior, lends support for defining cellular responses through characterization of changes in kinase activity. Characterization of phosphorylation-mediated signal transduction has the potential to offer important, and predictive, insight into phenotypes.

Defining cellular responses at the level of kinase activity is also of priority as disruption of these activities is associated with a number of pathophysiological states including cancer, inflammation, neurological disorders and diabetes.[4] Indeed over 250 protein kinase genes map to disease loci.[5] The causal roles of kinases in many diseases, in addition to their central regulatory role in many cellular pathways and processes, make them highly desirable targets for therapeutic intervention.[6] As such, in addition to serving as informative indicators of biological responses, protein kinases also offer the opportunity to not only understand, but also influence, biology.

There is considerable interest and potential to utilize kinase inhibitors as therapeutic agents. Protein kinases have been described as highly "druggable" due to their suitability for drug-design efforts.[7] Specifically, the conserved catalytic cleft of the protein kinases has proven an effective target for drug development.[2] Kinase inhibitors are emerging as a specific, and practical, mechanism of therapeutic intervention. Kinases and their inhibitors represent a top priority of the pharmaceutical industry as the most frequently targeted gene class for cancer therapies and second only to G protein-coupled receptors across all therapeutic areas.[2,6-8] From these efforts there are a large number of kinase inhibitors that have either received Food and Drug Administration (FDA) approval or are in clinical trials for disease treatments including leukemia and gastrointestinal stromal tumors (Gleevec (imatinib)),[8-10] diabetic retinopathy (ruboxistaurin),[11] atopic dermatitis (safingol)[12] and cerebral ischemia (fasudil)[13] (Table 6.1). It is certain that there will be a large number of new kinase inhibitors that will be developed, licensed and applied to a spectrum of human-health conditions.

Recently, there has also been an appreciation that kinase inhibitors with FDA approval may have alternate therapeutic applications including anti-infective therapies. Reeves *et al.*[14] have demonstrated that Gleevec protected mice from Vaccinia infection and reduced viral dissemination by five orders of magnitude. As the associated costs of moving a new drug to market are estimated at over a billion dollars the use of approved kinase inhibitors in novel applications is enticing.

6.3   Kinome *vs.* Phosphoproteome

The experimental approaches available for investigation of phosphorylation-based signaling can be classified as either phosphoproteome or kinome analysis.[15] The distinction between the approaches is that kinome analysis quantifies the activities of particular kinases, whereas phosphoproteome investigations identify and characterize patterns of protein phosphorylation.

Table 6.1 Kinase inhibitors with clinical approval.

Drug (Trade Name)	Structure	FDA Approval	Targets	Indications
Dasatinib (Sprycel)		2007	Bcr-Abl, Src family, Tec family	Imatinib-resistant chronic myelogenous leukemia
Erlotinib (Tarceva)		2004	EGFR, ErbB2 (Her2)	Pancreatic cancer, nonsmall cell lung cancer
Gefitinib (Iressa)		2004	EGFR, ErbB2 (Her2), Her4	Nonsmall cell lung cancer
Imatinib (Gleevec)		2001	Bcr-Abl, c-Abl, PDGFR, c-Kit, DDR1	Chronic myelogenous leukemia; gastrointestinal stromal tumors; hypereosinphilic syndrome

Table 6.1 (*Continued*)

Drug (Trade Name)	Structure	FDA Approval	Targets	Indications
Lapatinib (Tykerb)		2007	ErbB2 (Her2)	Breast cancer
Nilotinib (Tasigna)		2007	Bcr-Abl, c-Abl, PDGFR, c-Kit, DDR1	Imatinib-resistant chronic myelogenous leukemia
Sorafenib (Nexavar)		2006	c-Kit, PDGFR, VEGFR, b-Raf	Renal cancer

Sirolimus (Rapamune)	1999	mTOR	Immunosuppressant for preventing organ transplant rejection
Sunitinib (Sutent)	2006	c-Kit, PDGFR, VEGFR	Renal cancer, imatinib-refractory gastrointestinal stromal tumors
Temsirolimus (Torisel)	2007	mTOR	Renal cell carcinoma

Abbreviations: Bcr-Abl, break point cluster-Abelson tyrosine kinase; EGFR, epidermal growth factor receptor; PDGFR, platelet-derived growth factor receptor; DDR1, discoidin domain receptor tyrosine kinase 1; VEGFR, vascular endothelial growth factor receptor; mTOR, mammalian target of rapamycin.

Ultimately, each experimental approach seeks to describe the same biological phenomena, post-translational modification of proteins through phosphorylation. While the terms kinome and phosphoproteome analysis are sometimes used interchangeably we have proposed a stricter nomenclature whereby phosphoproteome analysis refers exclusively to investigations of phosphorylated proteins and kinome analysis is limited to descriptions of kinase activities. We believe that it is critical to specify which approach has been employed as the experimental techniques, the biological philosophy, and perhaps most importantly, the nature of emerging data, are quite distinct.

Phosphoproteome analysis seeks to define the subpopulation of the proteome that undergoes modification through phosphorylation. This faces many of the same challenges associated with conventional proteomic analysis, including the sensitivity and specificity required for detection and quantification of low-abundance proteins. Phosphoproteome analysis, however, is further complicated by: the relative scarcity of phosphoproteome members, the dynamic nature of protein phosphorylation and the tendency of the proteome to suppress detection of the phosphoproteome.[16–18] While a considerable fraction of the proteome (up to one third) undergoes phosphorylation, many of these proteins, and in particular those involved in signal transduction, are present at exceedingly low levels. Thus, many proteins that are of the greatest interest to signal transduction investigations are among the most difficult to characterize. These difficulties are further exacerbated by the fact that many of these low-abundance proteins are phosphorylated at substoichiometric levels (1–2%).[17,18] Thus, in addition to the theoretical limitations associated with phosphoproteome analysis, there are also a number of technical obstacles. Furthermore, phosphoproteome studies are often dependent upon mass spectroscopy, which can impose a technological barrier to this approach.

In light of these challenges, a promising alternative strategy is to characterize the kinome rather than the phosphoproteome. In particular, as investigations of enzymatic activity, as opposed to quantifications of particular biomolecules or metabolites, offer greater opportunity to amplify weak biological signals. Kinases are highly appropriate for this type of enzyme-based format as the well-defined and highly conserved chemistry of enzymatic phosphorylation permits rapid characterization of kinase activity, provided an appropriate substrate is available.

6.4   Peptide Arrays for Kinome Analysis

The nature of the substrate to be employed is a critical variable when developing kinase assays. Proteins represent the natural substrates for kinases but can be problematic as experimental substrates due to difficulties associated with their mass production/purification and instability on arrays. Fortunately, the substrate specificity of many kinases is dictated by the residues adjacent $(+/-4)$ from the phosphoacceptor site.[19] As such, it is possible to use short peptides (typically 15 residues in length) rather than full-length proteins, as surrogate kinase substrates. The appropriateness of synthetic peptides as kinase substrates is supported by the reports of kinases that recognize and phosphorylate peptide targets with comparable catalytic efficiency values as the natively structured protein.[20] Relative to full-length proteins peptides are easily synthesized, relatively inexpensive, highly stable and amenable to array technology.[21]

6.4.1 Generation and Application of Peptide Arrays

A considerable advantage of utilizing peptides as surrogate substrates for kinase assays is the ease with which peptides can be presented in an array format enables investigation of a large number of kinases, or phosphorylation events, within a single experiment. Currently, array-printing technology allows for construction of arrays of thousands to tens of thousands of peptides. Detailed information of the commonly employed methodologies for peptide synthesis and array spotting are available elsewhere.[22,23]

The number of unique peptides required on an array depends largely on whether the arrays are being applied as a form of kinome or phosphoproteome analysis. For example, working from the assumption that each peptide represents a substrate to investigate a specific kinase translates to the need for an array of just over five hundred peptides, which would enable characterization the entire human kinome, which is composed of just over 500 kinase members. In contrast, consideration of each peptide as a surrogate marker for a specific phosphorylation event translates into the requirement for tens of thousands of unique peptides for comprehensive description of human phosphorylation events.

Following array production, a biological sample, typically a cellular lysate, is applied to the slide. Within these samples active kinases phosphorylate their respective target peptides. The extent of phosphorylation of a particular peptide is used to quantify the activity of the corresponding kinase or is interpreted as a surrogate marker of the extent of phosphorylation of the corresponding protein within the cellular context. The data is generally considered in a relative fashion, *i.e.* levels of phosphorylation or kinase activity under one condition as compared to a control. As such, the objective is not to assign a specific level of activity to a particular kinase, or a degree of phosphorylation to a particular protein, but rather to assign a relative degree of activity or phosphorylation under different treatment conditions (disease *vs.* normal; treated *vs.* untreated; infected *vs.* uninfected, *etc.*) (Figure 6.1).

6.4.2 Peptide Arrays: Phosphoproteome or Kinome Analysis?

We certainly appreciate the challenge associated with labeling peptide arrays as tools of either kinome or phosphoproteome analysis.[15] This distinction is important, however, as it implies different approaches of data analysis and interpretation. For example, if increased phosphorylation of a peptide is interpreted as a marker of increased activity of a kinase then the investigation represents kinome analysis. In contrast, if the increased phosphorylation of the

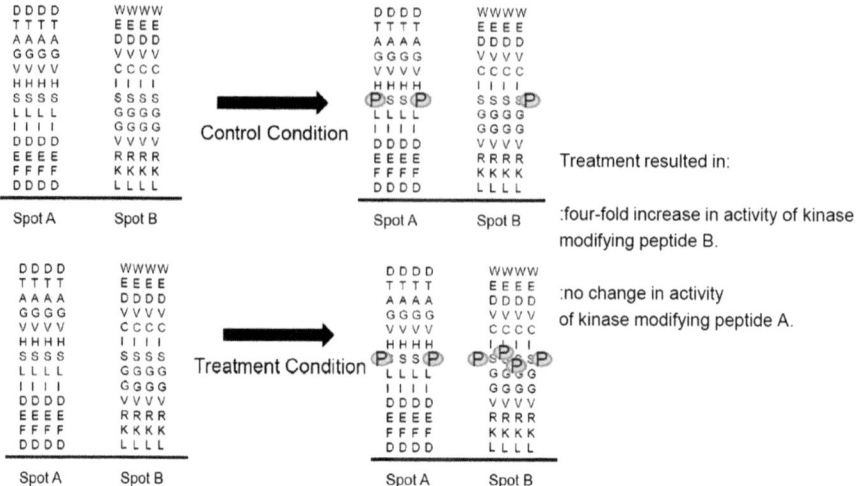

Figure 6.1   Evaluation of changes in kinase activity through peptide arrays. A hypothetical array of two peptides representing substrates for Kinase A and Kinase B. Under the control conditions there is greater baseline activity of Kinase A. However with the treatment the activity of Kinase B increases four-fold and Kinase A shows no further increases. This would be interpreted that the treatment condition results in a four-fold increase in activity of Kinase B but does not affect the activity of Kinase A.

peptide is used as a surrogate marker of increased phosphorylation of a protein then this would be consistent with a phosphoproteome investigation. This is an issue that remains to be fully addressed within the field.

Having presented these concerns we do believe that it is important not to exaggerate the significance of these differing nomenclatures. Within most of the labs employing peptide arrays, including our own, the philosophy behind application of the arrays is to identify potentially interesting, phosphorylation-associated biological events that are then validated through independent techniques. Thus far, when applied as a tool for either kinome or phospho-proteome analysis, the arrays have proven to be robust tools, which offers powerful insight into biology. While it is important to appreciate the issue, the debate of whether peptide arrays represent a form of phosphoproteome or kinome analysis is best left for another forum. In the interests of clarity and consistence we will refer to these peptide array-based investigations as representing kinome analysis but appreciate the complexities of this nomenclature.

6.5   Recent Advances

6.5.1   Customized Arrays

A large number of predesigned peptide arrays are available for purchase (Table 6.2). These arrays representing defined phosphorylation sites and/or kinase targets for dozens to thousands of peptides. These predesigned arrays

Table 6.2 Current kinome peptide array platforms.

Manufacturer	Peptide Sequences	Platform	Detection Method(s)
JPT	720	Spotted array	Radiolabeling; Phosphate-specific fluorescent stain
PepScan	1024	Spotted array	Radiolabeling; Phosphate-specific fluorescent stain
LC Sciences	1231	Microfluidics array	Phosphate-specific fluorescent stain
PamGene	144 Tyr kinase targets; 140 Ser/Thr kinase targets	Spotted array	Antibody-based

represent an effective way to perform kinomic investigations with minimal financial or training investment. An alternative approach is to create customized arrays based on peptides selected for their associated biological function. This approach allows for the development of tools that provide more specific and comprehensive description of priority phosphorylation events, pathways or processes. The cost of development of a customized array is not significantly greatly than for purchase of a pregenerated array, although customized arrays are typically ordered in batches (typically 100 arrays) to make the process of customization more cost effective.

Within our group, efforts to develop process-customized arrays begin with surveys of the phosphoproteome. There are extensive, publically available databases that contain extensive information on manually curated and literature-based serine, threonine and tyrosine phosphorylation sites. These include Phosphosite (www.phosphosite.org) and Phosphobase (phospho.elm.eu.org). A significant portion of the information available from these sites relates to phosphorylation events that have been characterized in humans or mice. From these databases information about the sequence contexts of phosphorylation sites are presented as peptides of 15 amino acids in length with the phosphoacceptor site centered in the peptide. This information can be rapidly translated into a customized array as peptides of this length and positioning of the phospho-acceptor site are perfectly complimentary with the requirements for the arrays. Utilized in this manner these databases essentially represent a shopping list of potential peptides which can be included on an array (Figure 6.2). These databases also include important information to guide peptide selection such as the identity of the corresponding kinase, associated biology and supporting publications.

6.5.1.1 Species-Specific Arrays

Peptide arrays were initially created based on phosphorylation events characterized for a particular species and then applied for investigations of that same species. For example, an array of peptides that represent sites of phosphorylation within the human phosphoproteome would be employed to

A

Overview

STAT5A transcription factor of the STAT family. Phosphorylated and activated by receptor-associated kinases downstream of many cytokines and growth-factor receptors. Activation of this protein in myeloma and lymphoma associated with a TEL/JAK2 fusion protein is essential for the tumorigenesis. Induces the expression of BCL2L1/BCL-X(L) in the mouse, suggesting an antiapoptotic function of this protein. Forms homo- or heterodimers that translocate into the nucleus where they regulate transcription. Two alternatively spliced isoforms have been described. Note: This description may include information from UniProtKB.

Protein type: Transcription factor; DNA binding protein

B

SS	MS		human
0	8	122	LRQNQVLyGQHPPIE
0	214	190	LKIELGByAIQLQRT
0	2	T92	IKLGByAIQLQRTyD
0	151	Y114	RCIBRILyNEQRLVR
1	1	S127	VREARMCSSPAGILV
1	1	S128	REARMCSSPAGILVD
0	2	171	KLQGTQByFIIQ;QE
0	2	S193	FAQLAQLsFQERLSR
0	1	T372	NFPQVRAtIIsBQQA
0	1	S375	QVRAtIIsBQQAESL
0	1	T647	LWNLKPttRDFSIR
0	1	T648	WNLKPPttRDFSIRS
0	4	1668	GDLsyLIyVFPDRPK
0	11	1682	KDEVFSKyTPVLAK
0	14	1683	DEVFSKyTPVLAKA
90	1877	1694	LAEAVDGyVFPQIRD
12	1	S726	TINDQAPsPAVCPQA
3	0	T757	GEFDLDETHDVARBV
0	1	S774	LLRAPMDsLDsRLsP
0	1	S777	RPMDsLDsRLsPPAG
7	6	S780	DsLDsRLsPPAGLFT

C

Site Information

DSLDSRLsPPAGLFT   **SwissProt Entrez-Gene**

Predicted information: Scansite

Orthologous residues: STAT5A (rat): **S779**, STAT5A (mouse): **S779**, STAT5A (sheep): **P780**, STAT5A (fruit fly): −, STAT5A iso2 (mouse): −

Blast this site against: NCBI SwissProt PDB

In vivo Characterization

Methods used to characterize site in vivo: mass spectrometry (2), mutation of modification site (1, 2, 3), phospho-antibody (1), phosphoamino acid analysis (2), western blotting (1)

Disease tissue studied: leukemia (1), acute lymphocytic leukemia (1), acute myelogenous leukemia (1), chronic myelogenous leukemia (1), lymphoma (1), Burkitt's lymphoma (1)

Relevant cell line – cell type – tissue: 293T (epithelial) (1), bone marrow (1), CCRF-CEM (T lymphocyte) (1), Daudi (B lymphocyte) (1), K562 (erythroid) (1), KU812 (myeloid) (1), lymphocyte-spleen [STAT5A (mouse), homozygous knockout) (2), MV4-11 (myeloid) (1), RAMOS (B lymphocyte) (1), Reh (B lymphocyte) (1), RL-7 (1), Y7 (2)

Controlled by

Putative upstream kinases: ERK1 (human) (3), ERK2 (human) (3)

Kinases, in vitro: ERK1 (human) (3), ERK2 (human) (3)

Treatments: IL-2 (2)

Downstream Regulation

Effects of modification on STAT5A: molecular association, regulation (3)

Inhibit interaction with: ERK1 (human) (3)

Disease / Diagnostics Relevance

Relevant diseases: acute lymphocytic leukemia (1), acute myelogenous leukemia (1), chronic myelogenous leukemia (1)

References

1. Friedbichler, K. et al. (2010) Stat5a serine 725 and 779 phosphorylation is a prerequisite for hematopoietic transformation. Blood **116**, 1548-58 20508164 **Curated Info**

2. Xue HH, et al. (2002) Serine phosphorylation of Stat5 proteins in lymphocytes stimulated with IL-2. Int Immunol **14**, 1263-71 12407017 **Curated Info**

3. Pircher TJ, Petersen H, Gustafsson H, Haldosén LA (1999) Extracellular signal-regulated kinase (ERK) interacts with signal transducer and activator of transcription (STAT) 5a. Mol Endocrinol **13**, 555-65 10194762 **Curated Info**

Figure 6.2 Publically accessible phosphorylation databases (phosphosite). (A) A brief summary of the physiological role of the protein. (B) A list of all the sites within the protein which have been suggested to undergo phosphorylation. This includes physiological function of the modification as well as the specific sequences surrounding the phosphacceptor site. (C) Each putative phosphorylation event can be examined in greater detail for kinases involved and supporting publications.

investigate human samples. The vast majority of information available within the phosphorylation databases relates to humans and mice. While investigations of humans and mice are of obvious priority for many labs, the inability to extend this technology beyond those species nevertheless represents a significant limitation. Kinases, and the phosphorylation events they catalyze, are well conserved across species. This conservation has significant value in: enabling more effective extrapolation of kinome research findings from model species to humans, testing of inhibitors of human kinases in animal models of disease and identification of biomarkers in animal models of disease that have applicability to human health. Also, given the current emphasis of the pharmaceutical industry on kinase inhibitors, there is a priority to conduct kinome investigations, including studies of disease mechanisms and their treatments, in animal models of human disease. The only, but often overlooked, caveat is the requirement for appropriate models of disease.

The opportunity for mouse kinomic investigations to advance human medicine is well recognized. Characterization of the mouse kinome has been heralded for its ability to "enhance the exploration of the roles of all kinases in mouse models of human diseases".[24] There are a number of factors that have contributed to the emergence of mice as a species-of-choice for research investigations includes their size, cost, availability of defined genetic strains and ease of experimental manipulation. While these are all valid considerations, these criteria do not consider whether mice serve as sufficiently accurate models of human disease. There have been a number of investigations demonstrating divergent biological mechanisms, responses and pathologies when comparing the responses of mice and humans to a spectrum of stimuli (pathogens, treatments, *etc.*).[25] This has prompted many groups to develop alternative, and likely more appropriate, animal models of human disease including large animal species such as pigs, cows, sheep and nonhuman primates.[25] The opportunity of these species to serve as more effective models of human disease, and to better facilitate translation of research, helps offset the additional costs and efforts associated with these species. A secondary consequence of the increased utilization of these nontraditional animal models is the concurrent push to develop research tools and reagents for these species.

The importance of these large-animal models, as well as the considerable economic value of many livestock species, prompted our group to expand the utility of this technology to species whose phosphoprotoeomes had yet to be characterized. As sites of phosphorylation, and subsequent biological consequences, are often conserved across species we hypothesized it would be possible to use bioinformatic approaches to predict the sequence contexts of phosphorylation events in proteins of other species based on genomic information. To test the feasibility of this approach we first investigated the extent of sequence conservation of phosphorylation sites between a species of interest, the cow (bovine), and humans. This comparison was done by using Blastp program to search nearly nine hundred peptides, representing human phosphorylation sites, against the NCBI-NR protein database to generate orthologous bovine peptides. From this we quantified the extent of sequence conservation surrounding sites of phosphorylation. Approximately half of the bovine sequences were perfect matches to their human counterparts.[26] The remaining peptides had limited sequence divergence, usually only 1 or 2 amino acids. An annotation comparison between the query and hit sequences was used to confirm that both referred back to the same protein identity.[26]

The results of this bovine-to-human comparison were highly encouraging as they supported the potential to use bioinformatic approaches to create species-specific arrays as well as highlighting the importance of customizing arrays to the species of interest. For example, the application of a human array to a bovine sample would generate data that would be problematic to interpret as nearly half of the peptides would be imperfect matches for the corresponding kinases. Interestingly, other labs have utilized peptides arrays based on sequences of human phosphoproteins to investigate species as evolutionarily distant as *Arabidopsis*.[27] Such efforts, in our opinion, speak more highly of the enthusiasm to obtain kinomic data for species of interest, as well as the absence of species-appropriate tools, than to the appropriateness of transcending species barriers with generic, noncustomized, peptide arrays.

The first generation of species-specific bovine array developed by our lab consisted of 300 unique peptides representing phosphorylation events involved in a spectrum of biological processes but with emphasis on proteins and pathways associated with innate immunity. Peptides were selected for the array based on the criteria of biological significance as well as the degree of sequence conservation surrounding the phosphoacceptor site. Notably, many proteins undergo phosphorylation at multiple sites to control discrete aspects of their function. Thus, for high-priority proteins, it is often of value to have multiple peptides that represent distinct phosphorylation events to provide more comprehensive insight into cellular responses.

In the earliest iterations of the arrays developed by our lab, each peptide was printed in triplicate to provide a measure of technical reproducibility. More recent versions involve printing each individual peptide in triplicate within a block, and then each block is subsequently printed in triplicate for a total of nine replicates of each peptide per array (Figure 6.3). As the technology advances it will become easier to make informed decisions about the appropriate number of technical replicates.

We have employed a similar bioinformatics approach to develop arrays for other species including, but not limited to, pigs, chickens, horses and honeybees. There is a similar degree of sequence conservation of phosphorylation sites between pigs and humans as observed for bovine. This likely reflects the comparable evolutionary distance. It appears that generation of arrays for more evolutionarily distant organisms may face more significant obstacles. In a preliminary investigation we calculated the degree of conservation of phosphorylation sites between a number of mammalian, plant and insect species. It is apparent from these efforts that conservation of phosphorylation sites reflects, at least in part, evolutionary distance. As such, development of arrays for evolutionarily distant organisms from the phosphodatabases of humans or mice will likely require significantly greater effort to

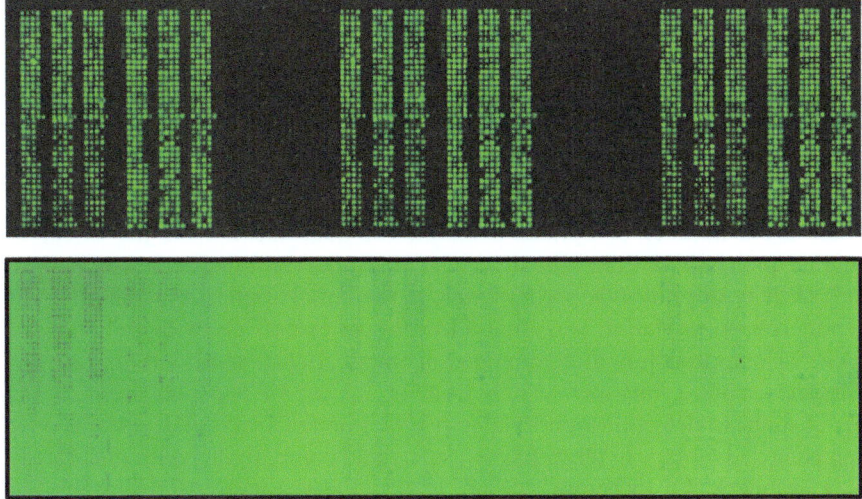

Figure 6.3 Development of customized arrays: Top image: Quality control slides to ensure proper spotting of all peptides. This particular array consists of 300 unique peptides, each printed in triplicate within each block and each block printed in triplicate for a total of 9 technical replicates per peptide per array. Bottom image: Data scan following incubation of the array with an appropriate cellular lysate. The relative intensity of each spot reveals the relative activity of the corresponding kinase.

take into account specific biology associated with these species as well as degrees of conservation of kinase with respect to their consensus sequences and phosphorylation targets. Databases are emerging describing the kinase complement for plant species, such as rice (phylomics.ucdavis.edu/kinase/). It may be possible to use this information as a more effective starting point to predict phosphorylation sites within other plants. Similarly well characterized insect species, such as Drosophila, may offer a more effective starting point for development of insect arrays. Given the economic importance of these organisms, in particular plants, and the potential for kinome analysis to offer functional insight into phenotypes and provide biomarkers for selective breeding, these efforts are certainly justified.

6.5.1.1.1 Interspecies Comparative Analysis. There is obvious and immediate value for developing research tools for species that are of economic importance or serve as models of human disease. Extending the range of species to which the peptide arrays can be applied may also offer longer-term opportunities by enabling comparative interspecies investigations. This type of information may be critical for the discovery of signaling-associated biomarkers as well as understanding the range of species to which such biomarkers can reliably be extrapolated. Understanding the conservation of signaling responses across species would also offer critical evidence for extrapolation of findings from animal models into clinical applications. Such activities benefit from knowledge regarding the degree of conservation, and specific points of divergence, of cell-signaling responses in model species.

6.5.2   Technical Advances

6.5.2.1   *Nonradioactive Stains*

There are a number of options available for quantification of the extent of peptide phosphorylation. These include phosphorylation-specific antibodies, radioactivity or phospho-specific stains. In the initial applications the extent of peptide phosphorylation was quantified almost exclusively with radiolabeled ATP. This approach is disadvantaged in terms of cost, sensitivity and safety. Published manuscripts increasingly describe the use of phosphospecific fluorescent stains to detect phosphorylated peptides on the arrays. These stains offer immediate advantage in terms of cost, safety and compatibility with microarray data scanners. Furthermore, in our hands, these nonradioactive stains, in particular Diamond ProQ phosphostain, offer greater sensitivity and reproducibility than radioactive protocols.[28]

6.5.2.2   *Bead-Based Solution-Phase Heterogeneous Kinase Assays*

In the earliest iterations of the arrays peptides were presented as immobilized targets affixed to a solid array surface. An interesting alternative that has recently emerged involves the covalent linkage of the peptides to beads through acrylamide linkers.[29] In this approach a panel of beads, each of which is decorated with multiple copies of a unique peptide, is incubated with a cellular lysate. The extent of phosphorylation of peptides of a particular bead is quantified with phosphorylation-specific, fluorescently labeled antibodies. Taken a step further, Sylvester and Kron[29] used Luminex beads and the Luminex flow system to measure two-channel fluorescence (fluorescent of the bead and the phospho-specific antibody) for identification of the bead (hence the identity of the peptide) as well as the extent of phosphorylation of the associated peptide. An immediate advantage of this approach, and its adaptation to 96-well plates, is the ability to screen a variety of conditions in parallel. It has also been suggested that presentation of the peptides on beads offers greater accessibility for recognition and modification by the corresponding kinase.[30] A marked disadvantage of this approach, however, is the requirement for phosphorylation-specific antibodies to quantify the extent of phosphorylation of each peptide.

6.6   Understanding Biology

With any emerging technology there are important questions that need to be addressed regarding the power and utility of the methodology. A number of examples are selected from the recent literature to illustrate the utility of the arrays for investigation of a spectrum of diseases. From these examples there is ample evidence of the value of peptide arrays in understanding biology as well as providing biomarkers and therapeutic targets.

6.6.1 Inflammation

Chronic inflammation is associated with a number of disorders including rheumatoid arthritis, inflammatory bowel disease, psoriasis, asthma and selected cancers.[31] It is generally accepted that a more comprehensive understanding of the molecular mechanisms of these conditions would assist in the development of effective, targeted treatments. The appreciation that dysregulation of the inflammatory cascade represents a crucial precipitating event of chronic inflammation is an important starting point to understanding these diseases.

Cytokines interleukin (IL)-17 and IL-32 serve critical roles in the pathology of immune-mediated chronic inflammatory diseases and autoimmune disorders. It has been hypothesized that IL-17 and IL-32 initiate cellular responses that are differentially dependent on the tumor necrosis factor (TNF) signaling pathway. This hypothesis, if correct, has immediate consequences for the use of TNF-based therapeutics for these disorders. Utilizing peptide-arrays Neeloffer *et al.*[32] demonstrated that a peptide representing TNF-receptor-1 (TNF-R1) underwent increased phosphorylation following treatment with IL-32 but not IL-17.[32] This was validated with a monoclonal antibody that specifically blocks TNF-R1-mediated responses; this antibody suppressed IL-32-induced, but not IL-17-induced responses. As such, peptide arrays were able to validate the hypothesis of differential TNF-pathway dependence of IL-32 and IL-17 (32).

Within the same study, peptide arrays suggested p300 (transcriptional coactivator) and death-associated protein kinase-1 (DAPK-1) as phosphorylation targets for both IL-32 and IL-17 induced signaling. Increased phosphorylation of p300 and DAPK-1 upon cytokine stimulation was confirmed with immunoblotting. Consistent with these results, targeted knock down of p300 and DAPK-1 suppressed the responses induced by both IL-32 and IL-17. These findings suggest that p300 and DAPK-1 represent shared nodes where the inflammatory networks of IL-32 and IL-17 overlap, such that p300 and DAPK-1 impact both TNF-dependent and -independent processes. This suggests that p300 and DAPK-1 may represent viable therapeutic targets for chronic inflammatory diseases. The authors praised the peptide arrays for their ability to "greatly accelerated discovery in contrast to using other techniques to investigate hypothesized targets one at a time".[32]

6.6.2 Infectious Diseases

The interaction between a pathogen and its target cell are often complex and multifaceted, which presents a considerable challenge for any experimental approach. The efforts required to understand these interactions are justified, in particular for pathogens that establish chronic infections as understanding the mechanisms used by a pathogen to subvert the host response may provides critical targets for therapeutic intervention. In particular, many pathogens, both viral and bacterial, have been shown to target host phosphorylation-mediated signal-transduction pathways including the production eukaryotic like kinase effector molecules, which are translocated into the host cell for direct subversion of host processes.[33–36] These eukaryotic-like kinases have been identified as highly attractive therapeutic targets.[37] Therefore, understanding host responses at the level of the kinome is highly appropriate.

6.6.2.1   *Monkeypox Virus*

Variola (VARV), the causative agent of smallpox, has been eradicated, however, incidences of *Monkeypox* virus (MPXV) infection, another member of the Poxviridae family, are increasing.[38] Due to the potential for the accidental, or intentional, release of this zoonotic virus to nonendemic regions, and the increasingly susceptible population resulting from the cessation of vaccination, MPXV is considered a Category C Priority Pathogen by the National Institute of Allergy and Infectious Diseases (NIAID).[39] Pathogens of this category are considered the third highest priority by NIAID, in conjunction with the Department of Homeland Security and the Centers for Disease Control and include emerging pathogens that could be manipulated for bioterrorism purposes. MPXV is comprised of two clades: Congo Basin MPXV, with an associated case fatality rate of 10%, and Western African MPXV, which is associated with less severe infection and minimal lethality.[40] Although there have been numerous animal model investigations into the disease pathologies of the two MPXV clades but no molecular investigations into this virulence disparity.

Kindrachuk *et al.*[41] employed peptide arrays to investigate the hypothesis that Congo Basin and West African MPXV differentially modulate host cell responses. Using this approach they demonstrated that compared to West African MPXV infection, Congo Basin MPXV downregulated a number of cellular responses including growth factor- and apoptosis-related signaling. The differential modulation of apoptosis was confirmed through fluorescence-activated cell sorting (FACS) as West African MPXV infection resulted in a significant increase in apoptosis in human monocytes as compared to Congo Basin MPXV. Interestingly, West African and Congo Basin MPXV share 99.4% amino acid sequence homology[42] with most of the differences being localized to proteins associated with modification of host responses. One of these proteins, BR-203, has been implicated in the subversion of apoptosis.[43] The BR-203 homolog of MPXV is severely truncated in West African MPXV but full length in Congo Basin MPXV.[42]

This investigation also identified a subset of kinases that were differentially phosphorylated by the two MPXV clades, including Akt and p53. These distinct patterns appear to be of functional significance as chemical inhibition of Akt phosphorylation resulted in a >250-fold decrease in Congo Basin

MPXV virus titers, while West African MPXV virus yield was not affected by these treatments.[41] Interestingly, chemical inhibition of cellular translocation of AKT with BML-257 resulted in similar reductions in viral titers for both MPXV clades suggesting that the modulation of AKT-mediated cell signaling by Congo Basin MPXV is related to the phosphorylation state of AKT rather than its cellular location. Overall, in this investigation the peptide arrays were able to shed light on the cellular mechanisms underlying the distinct pathologies of two closely related viruses as well as provide therapeutic targets that were functionally verified.

6.6.2.2 Pichinde Virus

Bowick *et al.*[44] compared host-cell signaling responses to virulent and avirulent *Arenaviridae* family members. Due to the experimental restrictions of maximum containment research, the authors employed a temporal systems kinomics investigation of *Pichinde* virus infection in guinea pigs that results in similar disease pathology as *Lassa* virus infection in humans but is not constrained by either maximum containment or biothreat classification.[45,46]

Guinea pigs were infected with either an attenuated *Pichinde* virus variant (P2) or a lethal disease variant (P18), and peritoneal macrophages were isolated 1 or 6 days postinfection. Kinome analysis demonstrated that infection with the attenuated *Pichinde* variant, resulted in a temporal increase in canonical signaling pathways (day 1 as compared to day 6). These included multiple growth factor signaling pathways [epidermal growth factor (EGF), insulin-like growth factor-1 (IGF-1), platelet derived growth factor (PDGF)], phosphoinositide 3 kinase (PI3K)/AKT signaling and extracellular signal-regulated kinase (ERK)/mitogen-activated protein kinase (MAPK) signaling. In contrast, P18 infection resulted in only modest perturbation of a subset of these canonical signaling pathways (EGF signaling; nuclear factor kappa beta (NF-κB) signaling; p38 MAPK signaling).

These results demonstrate that P18 infection results in a more limited activation of host responses than the attenuated P2 variant. These results are in agreement with previous *arenavirus* investigations in which pathogenic *Lassa* virus failed to activate human dendritic cells and monocytes, whereas the attenuated *Mopieia* virus resulted in activation of both cell types.[47,48] Interestingly, though *Arenaviridae* and *Poxviridae* are distinct viral families with no overlap there appears to be conservation in the mechanisms of disease pathogenesis during infection. Indeed, infection with pathogenic viral species from both families results in dampened host responses as compared to their attenuated variants and it will be prudent to further examine this phenomenon across a broader range of viral pathogens.

6.6.2.3 Streptococcus Pneumonia

The increasing medical and economic burden associated with multidrug-resistant bacterial strains, and the lack in development and approval of novel antibacterials, has prompted exploration of bacterial disease pathogenesis for identification of novel therapeutic targets.

Hoogendijk et al.[49] recently employed systems kinomics with peptide arrays to examine the host response to *Streptococcus pneumonia*, the causative pathogen of community-acquired pneumonia (CAP). The authors demonstrated that *Streptococcus pneumoniae* infection of mouse lung tissue activates a Th1 response with downregulation of B cell receptor, nuclear factor of activated T-cells, and v-abl Abelson murine leukemia viral oncogene homolog 1 phosphorylation. Chemotoxic stress also appeared to be upregulated during infection with increased phosphorylation of both DNA-dependent protein kinase and Ataxia telangiectasia mutated 6 h postinfection. In contrast, insulin receptor (INSR) and Wnt signaling also appeared to be downregulated during infection. Further, cell-cycle activity appeared to be downregulated during infection as numerous kinases involved in cell-cycle responses had reduced phosphorylation states. Although previous investigations of host-signaling responses to bacterial insult have primarily focused on TLR signaling cascades, Hoogendijk and colleagues have demonstrated that alternative pathways also play central roles in bacterial disease pathogenesis.

6.6.3   Stress

While there is considerable anecdotal evidence linking stress to increased risk of infection there is little information of the specific mechanisms through which stress may compromise immune function. In a cleverly designed experiment van Westerloo et al.[50] examined the consequences of acute stress (a first time bungee jump) on innate immune function. Innate immune function was evaluated by the capacity of leukocytes to release cytokines and mediate phagocytosis. The potential role of catecholamines, which are classically associated with stress responses, in suppression of immune function was considered with the inclusion of a subject group pretreated with the beta-receptor antagonist propanolol. Consistent with the hypothesis that acute stress compromises immune function, the jump resulted in an immune suppressive phenotype as determined by reduced release of the proinflammatory mediators in whole blood stimulated with lipopolysaccharide as well as reduced leukocyte phagocytic ability.

In contrast to the stated expectations, pretreatment of the subjects with propranolol did not influence the impact of the stressor on cytokine release nor phagocytic responses.[50] This challenged the researchers to seek alternate mechanisms of the stress-mediated immunosuppressive phenotype. Peptide arrays provided the opportunity to obtain an unbiased perspective on signaling events associated with the stress response. The pre- and postjump kinome profiles of the peripheral blood leukocytes indicated a strong noncanonical glucocorticoid receptor signaling response as well as downregulation of Lck/Fyn. This suggests that acute stress leads to a catecholamine-independent suppression of immune function and implicates noncanonical glucocorticoid receptor signal transduction as the link between acute stress and impaired innate immunity. This provides a powerful example of the ability to the arrays to facilitate novel discovery through a global, unbiased perspective of cell signaling.

That this model of acute stress was performed with human subjects lends considerable value to the importance of these findings. Furthermore, the physiological consequences of stress are not limited to increased vulnerability to infection but also associated with development of autoimmune disorders such as allergy[51] and inflammatory bowel disease.[52] The kinomic analysis of this study helps to define the mechanisms of regulation and interaction between stress and immune function. This provides useful biomarkers of stress-associated responses as well as greater understanding of the interplay between stress and immune suppression that may lead to the identification of potential therapeutic targets.

6.6.4 Impact of Glucocorticoids on Insulin Signaling

Glucocorticoids (GCs) are employed as therapies for anti-inflammatory and immunosuppressive disorders. These treatments are sometimes associated with side effects that include GC-induced insulin resistance.[53] Characterization of the molecular mechanism through which GC therapy induces insulin resistance may provide an opportunity to improve on existing therapeutic regimes.

Lowenberg *et al.*[54] utilized peptide arrays to investigate signaling responses of adipocytes and CD4 + T cells following treatment with dexamethasone (DEX). Decreased phosphorylation of peptides corresponding to regulatory events of INSR as well as 5′ adenosine monophosphate-activated protein kinase (AMPK), 70-kDa ribosomal protein S6 kinase (p70S6k), glycogen synthase kinase 3 (GSK-3) and Fyn kinase were observed in response to short-term DEX treatment. These results, which were confirmed with phosphorylation-specific antibodies and inhibitors, support a nongenomic GR-dependent mechanism of DEX-induced inhibition of insulin signaling activity. Similar effects on signaling were observed in two distinct, physiologically relevant, cell lines implying a generalized mechanism through which DEX impacts insulin sensitivity.[54]

In addition to proposing a cellular mechanism of GC-induced insulin resistance this study also highlights the tools and opportunity to address these challenges. Specifically, the array results of this study emphasize the importance of developing GC analogs that retain immunosuppressive and anti-inflammatory activities without having an accompanying effect on INSR tyrosine kinase activity. The arrays could be a valuable tool through which prospective analogs could be screened for these characteristics in a high-throughput fashion. The characterization of such drug-response biomarkers could represent a critical advancement toward the development of a new class of safer GC therapies.

6.6.5 Lupus

The immunologic defects and pathogenesis of lupus are thought to reflect altered signaling patterns in B lymphocytes. For example, deficiency of a number of proteins that regulate signaling, such as Lyn and CD22, can lead to production of anti-DNA auto-antibodies and onset of a lupus-like condition.[55–57] To provide a more comprehensive description of signaling events associated with lupus Taher *et al.*[58] utilized peptides arrays to investigate signaling responses in B lymphocytes from systemic lupus erythematosus (SLE) patients.

From these investigations it was observed that the activity of kinases involved in survival and differentiation, such as PI3K, as well as kinases involved in regulation of the cytoskeleton and migration, such as Rac and Rho, were markedly increased in SLE patients relative to healthy controls. In contrast, ataxia telangiectasia and Rad3-related protein (ATR), which regulates the cell cycle, had significantly reduced activity. This study demonstrated that in addition to signaling abnormalities already associated with SLE, including defects in B-cell receptor signaling, further signaling differences occur in the B lymphocytes from lupus patients. From these observations the authors proposed a new model of lupus-associated signaling highlighting the key pathways that appear to be altered in B cells of lupus patients. Specifically, the activities of phosphatidylinositol 3-kinase, which regulates survival and differentiation, and Rac and Rho, which regulate cytoskeleton rearrangement and migration, were upregulated.[58] Unexpected findings such as this are common outcomes in global investigations but, in our opinion, in particular of kinome investigations. Such findings present the opportunity, and challenge, to think outside the accepted paradigms of various models and mechanisms of particular disease states.

6.6.6 Angiotensin II-dependent Hypertensive Renal Damage

Hypertension can predispose for chronic kidney disease and potentially end-stage renal disease. Although current treatment regimens, including angiotensin-converting enzyme (ACE) inhibitors, are of value in the treatment and prevention renal disease progression, the prevalence of end-stage renal disease is nevertheless increasing.[59] As such, it is important to identify novel targets for intervention in hypertensive renal damage. In an attempt to identify such targets, de Borst *et al.*[60] utilized peptide arrays to profile the kinomes of renal lysates from a mouse model of angiotensin II-mediated renal fibrosis. Within this investigation the impact of treatment with angiotensin-converting enzyme inhibitor (ACEi) was also considered.

A central conclusion of this work was that p38 MAPK is involved in angiotensin II-dependent hypertensive renal damage, which is consistent with previous reports of increased p38 phosphorylation in experimental[61] and human renal disease.[62] Furthermore, inhibition of p38 is renoprotective in a number of animal models of this disease.[63] The arrays also indicated PDGFRβ activation that demonstrated increased activity in the mouse model of disease that was reversed by ACEi. Consistent with these array results, Imatinib treatment reduces renal perivascular fibrosis and microvascular hypertrophy in this mouse model of disease.[64] While Imatinib reduces tyrosine kinase activity of BCR/Abl it also inhibits other tyrosine kinases, including PDGFRβ and c-kit.[65]

Since pharmacological inhibition of either p38 or PDGFR is renoprotective, the authors concluded that the array identified kinases relevant to renal injury. The authors praise the arrays as "a powerful tool for identification of relevant kinase pathways *in vivo* and may lead to novel strategies for therapy. The most relevant property of this kinase array is its hypothesis-free character. This allows selection of a limited number of kinases with markedly altered activities under particular experimental conditions, which can then be studied in more detail. Ultimately, analysis of the renal kinome may provide novel targets for intervention to combat chronic renal disease." It is noteworthy that the renal samples of this study contain a variety of cell types and that the arrays proved sufficiently robust to identify relevant changes in kinase activity in these heterogeneous samples.

6.6.7 Cancer

The study of kinases, and the history of development and application of tools for kinome investigations, is closely intermingled with cancer research.[66] One of the exciting applications of peptide arrays to cancer research is the opportunity to survey cellular kinase activities in a global, high-throughput fashion. This offers tremendous potential for elucidating the cellular mechanisms of different cancers as well as for drug discovery where the ability to rapidly determine the specificity and potency of potential therapeutic agents is of tremendous value. A number of examples where peptide arrays have been applied to understand biology within different cancers are presented.

6.6.7.1 Hematopoietic Malignancies

Hematopoietic malignancies often treated with allogeneic bone marrow transplantation. In these treatments the graft-versus-leukemia response mediates elimination of malignant cells. At a cellular level, cytotoxic T lymphocytes and NK cells, which are responsible for the immune-elimination of malignant cells, express Fas ligand and depend on induction of Fas receptor-mediated apoptosis for this protective function. Graft-versus-leukemia reactions are often successful for the removal of cancer cells in myeloid malignancies but the efficiency of this treatment is low in T cell leukemias, perhaps as a result of the ability of malignant T cells to escape apoptosis.[67]

A study of Maddigan *et al.*[68] demonstrates that Eph family receptor EphB3 is consistently expressed by malignant T lymphocytes, often in combination with EphB6. Stimulation of these receptors with their respective ligands, ephrin-B1 and ephrin-B2, suppresses Fas-induced apoptosis in these cells. This effect is associated with Akt activation and subsequent suppression of the Fas receptor-initiated caspase proteolytic cascade. The hypothesis that Akt was critical for this prosurvival response was confirmed with kinase inhibitors. Inhibition of Akt, but not of other molecules central to T-cell biology, blocked the antiapoptotic effect. The peptide array analysis utilized in this investigation brings forth a new role for EphB receptors in the protection of malignant T cells from Fas-induced apoptosis in an Akt-dependent manner. These findings indicate that EphB receptors support the immunoevasivenes of T-cell malignancies and may represent targets for therapies administered to enhance immune-elimination of cancerous T cells.[66]

6.6.7.2   Barrett's Esophagus

Barrett's esophagus is a precursor metaplastic lesion of the mucosa that predisposes for esophageal adenocarcinoma.[69] van Baal et al.[70] utilizing peptide arrays to investigate cellular kinase activity in biopsies of Barrett's esophagus relative to the profiles of neighboring cardia and squamous epithelia.

This investigation revealed a number of kinotypic profiles that were associated with the diseased tissue. First, relative to healthy squamous and gastric cardia epithclia, MAPK and EGFR activity were significantly influenced in Barrett's esophagus; downregulation of activity of the MAPK signaling was the major reported event. The array data also suggested differences in glyolytic activity in Barrett's esophagus. These array results were validated with enzymatic assays for pyruvate kinase, which plays an important role in the final irreversible step in the glycolysis. Assays of patient biopsies verified that pyruvate kinase activity were indeed significantly upregulated in Barrett's esophagus compared with normal squamous and gastric cardia epithelium.[70]

This outcome is of particular interest, as often investigations of signal transduction activity are often associated with cellular responses such as growth and differentiation. The ability of the arrays to accurately predict and describe changes in metabolic activity emphasizes their utility in describing a wider spectrum of biological responses. Overall, this study offers critical insight into Barrett's esophagus pathogenesis which may assist in the treatment and prevention of esophageal adenocarcinoma. Another important aspect of this study is the use of full biopsies to conduct kinome analysis. These samples represent a heterogeneous cell population containing epithelial cells, stromal tissue and inflammatory cells. This offers a significant advantage over techniques that depend on homogenous populations of cells that fail to capture on the biological complexity associated with many, if not most, biological samples. The ability of the arrays to decipher signaling kinotypes in a complex population of cells indicates the robust nature of the technology.

6.6.7.3   Colon Cancer

The activities of receptor tyrosine kinases (RTK), as well as expression of cyclooxygenase-2 (COX-2), serve as important biomarkers for predicting tumor growth as well as overall prognosis in colorectal cancer (CRC). In CRC COX-2 also functions as a critical therapeutic target, although the mechanism of this anticarcinogenic action, in particular the relationship with RTK signaling, has yet to be elucidated.[71]

Tuynman *et al.*[72] applied peptide arrays to provide an unbiased assessment of the signaling consequences of a COX-2 inhibitor (celecoxib) in CRC cells. Within this investigation there was a general consensus that celecoxib impairs downstream signaling in the Gab/PI3K pathway by targeting c-Met auto-phosphorylation. Specifically, the arrays revealed that celecoxib treatment impeded phosphorylation of c-Met and IGFR substrates. The arrays also revealed an increase in GSK3β activity that was accompanied by increased phosphorylation of β-catenin. This result was supported by reduction of β-catenin-T-cell factor-dependent transcription following celecoxib treatment as well as chemical inhibition of c-Met phosphorylation. Collectively a major effect of celecoxib treatment of CRC appears to be the corepression of c-Met–related and β-catenin–related oncogenic signal transduction. This provides a rationale to use c-Met inhibitors and celecoxib analogous to target c-Met and Wnt signaling for treatment of CRC.[72]

Importantly, the mechanism of drug action identified by the arrays was not anticipated for celecoxib. Once suggested, however, there was considerable evidence within the literature to support this mechanism of action. One of the remarkable opportunities stemming from the peptide arrays investigations is the potential to represent a spectrum of biological pathways and processes to facilitate novel discovery. Thus, while the arrays are valuable tools for hypothesis validation, their true value and power may be in hypothesis generation.

6.6.7.4 Pediatric Brain Tumors

The growth and progression of pediatric brain tumors likely reflects signaling events regulating proliferation. Such events may be of significance in understanding mechanisms of disease progression as well as providing therapeutic targets.

With a stated objective of "defining new targets for treatment of human malignancies, without *a priori* knowledge on aberrant cell signaling activity" Sikkema *et al.*[73] utilized flow-through peptide microarrays to probe signaling events in pediatric brain tumors. The resulting kinome profiles demonstrated unique signaling patterns associated with EGFR, c-Met, and VEGFR. In addition, peptides representing phosphorylation targets of the Src family kinases showed increased levels of phosphorylation; 10 of 20 potential Src substrates present on the array showed significantly higher level of phosphorylation in pediatric brain tumors relative to healthy tissue. Increased Src kinase activity was confirmed with Phos-Tag SDS-PAGE as well as inhibitors; Src family kinase inhibitors PP1 and dasatinib induced tumor cell death in nine pediatric brain tumor cell lines but not in control cell lines.[73]

Collectively, this investigation demonstrates application of peptide array technology to generate kinomic profiles for pediatric brain tumors. This information was of considerable value for drug discovery as results suggested by the arrays were validated as potential targets for therapeutic intervention with pediatric brain tumors. The authors praise the arrays with the following statement: "In the era of a rapidly increasing number of small-molecule inhibitors, this tyrosine kinase activity profiling method will be of tremendous value by enabling us to rapidly screen for potential druggable targets in a broad range of malignancies".

6.6.7.5 *Validating* in vitro *Cancer Models*

Considerable effort and expense is expended to screen kinase inhibitors libraries for activities that are of potential therapeutic value in the treatment of cancer. While these cell-culture-based models provide an opportunity for high-throughput screening, it is critical that these inhibitors are evaluated in an appropriate biological context as the value of the results ultimately depends on the relevance of the biological system. Specifically, it is critical to evaluate whether cell-culture-based models accurately represent the biology of target cancer cells in the host.

Our group was involved in an investigation in which peptide arrays were used to monitor changes in signaling associated with transfer of transformed B cells from the host of a Bovine Leukemia Virus (BLV) ovine model system to tissue culture.[74] Our results revealed significant changes in kinome profiles when cancer cells were removed from host and passaged in tissue culture. These alterations related to pathways that define transformation and could likely influence conclusions if one was using such a host-derived cell line for therapeutic screens. This suggests, perhaps obviously, that the environment, or biological context, of the cells has a profound effect on cell signaling events. This finding could be of considerable significance as cell lines derived from various disease states are often employed as a convenient method to evaluate cellular mechanisms disease and screen potential therapeutics. The changes in signaling activity that occur as a result of the *in vivo* to *in vitro* transition may have significant impact on the translation of knowledge from such models to clinical applications.[74]

6.7   Validation of Results

Our labs, and most others, maintain a cautiously optimistic approach in the application of peptide arrays and interpretation of their data. The philosophy of our lab is that the most appropriate application of the peptide arrays is to suggest signal responses, and associated phenotypes, which are then validated through independent experimental approaches such as phosphorylation-specific antibodies and functional assays. It is important to not overinterpret array data. Sikkema *et al.*[73] perhaps stated this best: "generating hypotheses is the limit of what is possible at the current developmental stage of peptide microarray technology". The peptide arrays, in their current state, simply point us in directions of further study using other, more established, methodologies.

Every experimental approach has the potential to introduce biases or artifacts that cloud the biological significance of the data. Peptide arrays for kinome analysis are not unique in this regard. We believe the phosphorylation of peptides by kinases that are not physiologically relevant represents the greatest potential source of biological error that could emerge from a peptide array experiment. This could happen in a number of ways. First, there is still uncertainty as to how efficiently or specifically individual kinases recognize and phosphorylate a spotted peptide target sequence. Classic biochemical approaches have demonstrated that some kinases recognize linear substrates with similar kinetics as their protein substrates it must be appreciated that the specificity and efficiency of each individual kinase or each target sequence is unknown. This can greatly affect the data interpretation if one considers the magnitude of phosphorylation on the array as biologically significant. Analyzing the data such that degrees of phosphorylation are considered in a relative, rather than absolute, fashion may help to account for some of this background but this will nevertheless complicate extraction of biological information.

Secondly, the loss of cellular structure and organization within cellular lysates may place kinases in contact with substrates that they normally would not encounter due to compartmentalization or the absence of coincidental production of the substrate and kinase. For example, β-adrenergic receptor kinase translocates from the cytosol to the cell membrane to phosphorylate and desensitize G-protein-coupled receptors, Similarly, the subcellular localization of Erk defines its phosphorylation state and activation state.[75,76] Multipurpose kinases that regulate distinct and functionally unrelated processes within the same cell further underscore this complexity. For example, PKCδ regulates depolarization events, apoptosis and cell growth based on subcellular localization.[77,78] Depending upon the specific needs of a cell these distinct patterns of translocation and associated responses can be achieved simultaneously. Compartmentalization of regulatory small molecules, such as cAMP and Ca^{2+}, also contributes to the regulation of kinase activity.[79] Within a cellular lysate these finer points of control are compromised. This again highlights the importance of the overall philosophy is that the purpose and value of the arrays is to suggest biological responses that are validated through independent approaches. The loss of cellular organization likely represents a greater challenge for kinome rather than phosphoproteome investigations as kinome analysis depends on measurements of enzymatic activity that are taken postlysis, whereas phosphoproteome analysis characterizes events that occur prior to cell lysis.

6.8   Remaining Challenges

6.8.1   Data Statistics and Mining

Kinome analysis, like any global omics approach, faces the challenge of how to effectively extract meaningful biological information from the data.

Oligionucleotide arrays, which enjoy considerable seniority over peptide arrays, have had considerable opportunity to evolve and mature. As such, the tools available for management of gene expression data are well established. Given the apparently similar nature of the data which emerges from these approaches it was natural that early peptide array researchers borrowed heavily from the mathematical manipulations and statistical tools established for nucleotide arrays. While these initial efforts provided a reasonable foundation for analysis of kinome data it has become increasingly clear the simple exposition of the DNA array analytical methodologies has minimized the quality and breadth of data obtained from these investigations.

There are a number of reasons why the approaches for manipulation of gene-expression data may not be ideally suited of kinomic data. First, there is a considerable difference in the number spots presented on a typical peptide array as compared to current oligonucleotide or cDNA arrays. With current technologies the number of spots on a nucleic acid oligionucleotide array exceeds those that can be printed for a peptide array by approximately an order of magnitude. This difference means that each peptide spot, relative to each oligionucleotide spot, carries greater value such that there is greater potential for loss of valuable information with the omission or editing of data points from peptide array datasets. It is important that the statistical approaches used to process the data appreciate this distinction. Secondly, there are fundamental differences in how each approach generates signal. For nucleotide arrays, signal is achieved through hybridization of labeled cDNA to the array spots. Kinome arrays, in contrast, generate signals through the transfer of phosphoryl groups such that the scale of response ranges from no phosphorylation to many multiples of the control. Thus, whereas traditional microarray approaches have relied on standard fold-change cut-offs for establishing data significance the significance threshold for fold-changes associated with kinome arrays has yet to be established. Also, as the variance in signal reflects the dynamic range of phosphorylation it is more difficult to make statistical assumptions about the data. Data normalization is required before statistical tests can be conducted; however, normalization procedures are based primarily on nucleotide microarray data and may not be the most appropriate for kinome data.

6.9   Conclusions

Given the emerging importance of cellular kinases as causes, indicators and therapeutic targets of disease it seems certain that there will be ongoing effort to develop tools and strategies that enable characterization of the kinome in a high throughput, cost-effective fashion. In our opinion, peptide arrays represent the most practical and robust approach to date for achieving these goals. This technology continues to rapidly evolve with frequent advancements in the utility and scope to which the arrays can be applied. The examples presented in this chapter highlight the already existing potential of the technology to understand disease, discover biomarkers and evaluate therapeutics.

In the future, with the anticipated ongoing interest in kinase inhibitors, the arrays may be of tremendous value to ensure their appropriate application to enhance human health and minimize the risk of adverse reactions. A number challenges remain to be addressed for peptide-array kinome analysis. We believe that the development of customized software packages for the analysis and interpretation of kinome data will represent the next critical advance in the evolution of this technology. Software platforms that more effectively extract meaningful biological information from kinomic data, with fewer requirements for expertise in data manipulation, will greatly enhance the uptake of the technology into labs wishing to embark in signal-transduction investigations.

References

1. G. Manning, D. B. Whyte, R. Martinez, T. Hunter and S. Sudarsanam, The protein kinase compliment of the human genome, *Science*, 2002, **298**, 1912–1934.
2. M. E. Noble, J. A. Endicott and L. N. Johnson, Protein Kinase Inhibitors: Insights into Drug Design from Structure, *Science*, 2004, **303**, 1800–1805.
3. R. A. Arsenault, P. Griebel and S. Napper, Peptide Arrays for Kinome Analysis: New Opportunities and Remaining Challenges, *Proteomics*, 2011, **11**, 4595–4609.
4. G. Manning, S. Caenepeel, *Encyclopedia of Protein Kinases in Human Diseases, Catalog Reference Manual*, Cell Signaling Technology, Beverly, MA, 2005.
5. S. Knuutila, A. M. Bjorkqvist, K. Autio, M. Tarkkaqnen, *et al.*, DNA copy number amplifications in human neoplasms: review of comparative genomic hybridization studies, *Am. J. Pathol.*, 1998, **152**, 1107–1123.
6. P. Cohen, Protein kinases–the major drug targets of the twenty-first century? *Nature Rev. Drug Discov.*, 2002, **1**, 309–315.
7. A. L. Hopkins and C. R. Groom, The druggable genome, *Nature Rev. Drug Discov.*, 2002, **1**, 727–730.
8. M. Noble, J. Endicott and L. Johnson, Protein kinase inhibitors: insights into drug design from structure, *Science*, 2004, **303**, 1800–1805.
9. B. Druker, S. Tamura, E. Buchdunger, S. Ohno, *et al.*, Effects of a selective inhibitor of the Abl tyrosine kinase on the growth of Bcr-Abl positive cells, *Nature Med.*, 1996, **2**, 561–566.
10. B. Druker, M. Talpaz, D. Resta and B. Peng, *et al.*, Efficacy and safety of a specific inhibitor of the CR-ABL tyrosine kinase in chronic myeloid leukemia, *N. Engl. J. Med.*, 2001, **2**, 561–566.
11. R. P. Danis and M. J. Sheetz, Ruboxistaurin: PKC-beta inhibition for complications of diabetes, *Expert Opin. Pharmocther*, 2009, **17**, 2913–2925.
12. R. M. Eglen and T. Reisine, *The Current Status of Drug Discovery Against the Human Kinome. Assay Drug Dev. Technol.*, 2009, **7**, 22–43.

13. K. Yamashita, Y. Kotani, Y. Nakajima, M. Shimazawa, S. Yoshinmura, S. Nakashima, T. Iwama and H. Hara, Fasudil, a Rho kinase (ROCK) inhibitor, protects against ischemic neuronal damage in vitro and in vivo by acting directly on neurons, *Brain Res.*, 2007, **1154**, 215–224.

14. P. M. Reeves, B. Bommarius, S. Lebeis, S. McNutty, J. Christensen, A. Swimm, A. Chahraudi, R. Chavan, M. B. Feinberg, D. Veach, W. Bornmann, M. Sherman and D. Kalman, Diabling poxvirus pathogenesis by inhibition of Abl-family tyrosine kinases, *Nature Med.*, 2005, **11**, 731–739.

15. S. Jalal, J. Kindrachuk and S. Napper, Phosphoproteome and kinome analysis: unique perspectives on the same problem, *Curr. Anal. Chem.*, 2007, **3**, 1–15.

16. M. Mann, S. E. Ong, M. Gronborg and H. Stern, *et al.*, *Trends Biotechnol.*, 2002, **20**, 261.

17. D. E. Kalume, H. Molina, A. Pandey, Tackling the phosphoproteome: tools and strategies, *Curr. Opin. Chem. Biol.*, 2004, **7**, 64–69.

18. A. Gorg, W. Weiss and M. J. Dunn, Current two-dimensional electrophoresis technology for proteomics, *Proteomics*, 2004, **4**, 3665–3685.

19. A. Kreegipuu, N. Blom, S. Brunak and J. Jarv, Statistical analysis of protein kinase specificity determinants, *FEBS Lett.*, 1998, **430**, 45–50.

20. O. Zetterqvist, U. Ragnarsson, E. Humble, L. Berglund and L. Engstrom, The minimum substrate of cyclic AMP-stimulated protein kinase, as studied by synthetic peptides representing the phosphorylatable site of pyruvate kinase (type L) of rat liver, *Biochem. Biophys. Res. Commun.*, 1976, **70**, 669.

21. Z. Ouyang, Z. Takats, T. A. Blake, B. Gologan, *et al.*, Preparing protein microarrays by soft-landing of mass selected ions, *Science*, 2003, **301**, 1351–1354.

22. B. T. Houseman, J. H. Huh, S. J. Kron and M. Mrksich, Peptide chips for the quantitative evaluation of protein kinase activity, *Nature Biotech.*, 2002, **20**, 270–274.

23. S. Jala, J. Kindrachuk and S. Napper, Phosphoproteome and Kinome Analysis: Unique Perspectives on the Same Problem, *Curr. Anal. Chem.*, 2007, **3**, 1–15.

24. S. Caenepeel, G. Charydczak, S. Sudarsanam, T. Hunter and G. Manning, The mouse kinome: discovery and comparative genetics of all mouse protein kinases, *Proc. Natl. Acad. Sci. USA*, 2004, **32**, 11707–11712.

25. V. Gerdts, S. van den Hurk, P. Griebel and L. Babiuk, Use of animal models in the development of human vaccines, *Future Med.*, 2007, **2**, 667–675.

26. S. Jalal, R. Arsenault, A. Potter, L. Babiuk, P. Griebel and S. Napper, Genome to Kinome: Species-Specific Peptide Arrays for Kinome Analysis, *Science Signaling*, 2009, **54**, 1–11.

27. T. Ritsema, J. Joore, W. van Workum and C. J. M. Pieterse, Kinome profiling of Arabidopsis using arrays of kinase consensus substrates, *Plant Methods*, 2007, **3**, 3.
28. R. J. Arsenault, S. Jalal, L. A. Babiuk and A. Potter, *et al.*, Kinome analysis of toll-like receptor signaling in bovine monocytes, *J. Recept. Signal. Transduct. Res.*, 2009, **29**, 299–311.
29. J. E. Sylvester and S. J. Kron, A bead-based activity screen for small-molecule inhibitors of signal transduction in chronic myelogenous leukemia cells, *Mol. Cancer Ther.*, 2010, **9**, 1469–1481.
30. D. Wu, J. E. Sylvester, L. L. Parker, G. Zhou and S. J. Kron, Peptide reporters of kinase activity in whole cell lysates, *Biopolymers*, 2010, **94**, 475–486.
31. S. A. Doggrell, Inflammation, the key to much pathology, *Drug News Perspect.*, 2005, **18**, 531–539.
32. E. Turner-Brannen, K. G. Choi, R. Arsenault and H. El-Gagalwy, *et al.*, Inflammatory cytokines IL-32 and IL-17 have common signaling intermediates despite differential dependence on TNF-receptor 1, *J. Immunol.*, 2011, **186**, 7127–7135.
33. T. Alber, Signaling mechanisms of the Mycobacterium tuberculosis receptor Ser/Thr protein kinases, *Curr. Opin. Struct. Biol.*, 2009, **6**, 650–657.
34. A. M. Krachler, A. R. Woolery and K. Orth, Manipulation of kinase signaling by bacterial pathogens, *J. Cell. Biol.*, 2011, **195**, 1083–1092.
35. Y. Takeuchi, P. J. Birckbichler, M. K. Patterson, Jr., K. N. Lee, Measles virus V protein blocks interferon (IFN)-alpha/beta but not IFN-gamma signaling by inhibiting STAT1 and STAT2 phosphorylation, *FEBS Lett.*, 2003, **546**, 276–80.
36. M. W. Bahar, S. C. Graham, R. A. Chen, S. Cooray, G. L. Smith, D. I. Stuart and J. M. Grimes, How vaccinia virus has evolved to subvert the host immune response, *J. Struct. Biol.*, 2011, **175**, 127–34.
37. M. Schreiber, I. Res and A. Matter, Protein kinases as antibacterial targets, *Curr. Opin. Cell. Biol.*, 2009, **2**, 325–330.
38. A. W. Rimoin, P. M. Mulembakani, S. C. Johnston, J. O. Lloyd Smith, N. K. Kisalu, T. L. Kinkela, S. Blumberg, H. A. Thomassen, B. L. Pike, J. N. Fair, N. D. Wolfe, R. L. Shongo, B. S. Graham, P. Formenty, E. Okitolonda, L. E. Hensley, H. Meyer, L. L. Wright and J. J. Muyembe, Major increase in human monkeypox incidence 30 years after smallpox vaccination campaigns cease in the Democratic Republic of Congo, *Proc. Natl. Acad. Sci. USA*, 2010, **107**, 16262–16267.
39. P. B. Jahrling, E. A. Fritz and L. E. Hensley, Countermeasures to the bioterrorist threat of smallpox, *Curr. Mol. Med.*, 2005, **5**, 817–26.
40. S. Parker, A. Nuara, R. M. Buller and D. A. Schultz, Human monkeypox: an emerging zoonotic disease, *Future Microbiol.*, 2007, **73**, 17–34.
41. J. Kindrachuk, R. Arsenault, T. Kusalik, K. N. Kindrachuk, B. Trost, S. Napper, P. B. Jahrling and J. E. Blaney, Systems Kinomics Demonstrates Congo Basin Monkeypox Virus Infection Selectively Modulates

Host Cell Signaling Responses as Compared to West African Monkeypox Virus, *Mol. Cell. Prot. Published Dec*, 2011, **28**, 2011.

42. A. M. Likos, S. A. Sammons, V. A. Olson, A. M. Frace, Y. Li, M. Olsen-Rasmussen, W. Davidson, R. Galloway, M. L. Khristova, M. G. Reynolds, H. Zhao, D. S. Carroll, A. Curns, P. Formenty, J. J. Esposito, R. L. Regnery and I. K. Damon, A tale of two clades: monkeypox viruses, *J. Gen. Virol.*, 2005, **86**, 2661–72.

43. S. Hnatiuk, M. Barry, W. Zeng, L. Liu, A. Lucas, D. Percy and G. McFadden, Role of the C-terminal RDEL motif of the myxoma virus MT-4 protein in terms of apoptosis regulation and viral pathogenesis, *Virology*, 1999, **263**, 290–306.

44. G. C. Bowick, S. M. Fennewald, E. P. Scott, L. Zhang, B. L. Elsom, J. F. Aronson, H. M. Spratt, B. A. Luxon, D. G. Gorenstein and N. K. Herzog, Identification of differentially activated cell-signaling networks associated with pichinde virus pathogenesis by using systems kinomics, *J. Virol.*, 2007, **81**, 1923–33.

45. L. Borio, T. Inglesby, C. J. Peters, A. L. Schmaljohn, J. M. Hughes, P. B. Jahrling, T. Ksiazek, K. M. Johnson, A. Meyerhoff, T. O'Toole, M. S. Ascher, J. Bartlett, J. G. Breman, E. M. Eitzen Jr., M. Hamburg, J. Hauer, D. A. Henderson, R. T. Johnson, G. Kwik, M. Layton, S. Lillibridge, G. J. Nabel, M. T. Osterholm, T. M. Perl, P. Russell and K. Tonat, Working Group on Civilian Biodefense Hemorrhagic fever viruses as biological weapons: medical and public health management, *JAMA*, 2002, **287**, 2391–405.

46. P. B. Jahrling, R. A. Hesse, J. B. Rhoderick, M. A. Elwell and J. B. Moe, Pathogenesis of a Pichinde virus strain adapted to produce lethal infections in guinea pigs, *Infect. Immun.*, 1981, **32**, 872–80.

47. I. S. Lukashevich, R. Maryankova, A. S. Vladyko, N. Nashkevich, S. Koleda, M. Djavani, D. Horejsh, N. N. Voitenok and M. S. Salvato, Lassa and Mopeia virus replication in human monocytes/macrophages and in endothelial cells: different effects on IL-8 and TNF-alpha gene expression, *J. Med. Virol.*, 1999, **59**, 552–60.

48. D. Pannetier, C. Faure, M. C. Georges-Courbot, V. Deubel and S. Baize, Human macrophages, but not dendritic cells, are activated and produce alpha/beta interferons in response to Mopeia virus infection, *J. Virol.*, 2004, **78**, 10516–524.

49. A. J. Hoogendijk, S. H. Diks, T. van der Poll, M. P. Peppelenbosch and C.W. Wieland, Kinase activity profiling of pneumococcal pneumonia, *PLoS One*, 2011, **6**, e18519.

50. D. J. van Westerloo, G. Choi, E. C. Lowenberg, J. Truijen, A. F. de Vos, E. Endert, J. C. M. Meijers, L. Zhou, M. Pereira, Queiroz, S. H. Diks, M. Levi, M. P. Peppelenbosch and T. van der Poll, Acute Stress Elicited by Bungee Jumping Suppresses Human Innate Immunity, *Mol. Med.*, 2011, **17**, 180–183.

51. J. Montoro, *et al.*, Stress and allergy, *J. Investig. Allergol. Clin. Immunol.*, 2009, **19**(Suppl. 1), 40–47.

52. R. G. Maunder, Evidence that stress contributes to inflammatory bowel disease: evaluation, synthesis, and future directions, *Inflamm. Bowel Dis.*, 2005, **11**, 600–8.
53. G. Pagano, P. Cavallo-Perin, M. Cassader, A. Bruno, A. Ozzello, P. Masciola, A. M. Dall'omo and B. Imbimbo, An in vivo and in vitro study of the mechanism of prednisone-induced insulin resistance in healthy subjects, *J. Clin. Invest.*, 1983, **72**, 1814–1820.
54. M. Lowenberg, J. Tuynman, M. Scheffer, A. Verhaar, L. Vermeulen, S. van Deventer, D. Hommes and M. Peppelenbosch, Kinome Analysis Reveals Nongenomic Glucocorticoid Receptor-Dependent Inhibition of Insulin Signaling, *Endocrinology*, 2006, **147**, 3555–3562.
55. R. Cornall, J. Cyster, M. Hibbs, A. Dunn, K. Otipoby and A. Clark, *et al.*, Polygenic autoimmune traits: Lyn, CD22, and SHP-1 are limiting elements of a biochemical pathway regulating BCR signalling and selection, *Immunity*, 1998, **8**, 497–508.
56. T. L. O'Keefe, G. T. Williams, F. Batista and M. S. Neuberger, Deficiency in CD22, a B cell-specific inhibitory receptor, is sufficient to predispose to development of high affinity autoantibodies, *J Exp Med*, 1999, **189**, 1307–1313.
57. N. R. Pritchard, A. Cutler, S. Uribe, S. Chadban, B. Morley and K. G. Smith, Autoimmune-prone mice share a promoter haplotype associated with reduced expression and function of the Fc receptor Fc_RII, *Curr Biol.*, 2000, **10**, 227–230.
58. T. E. Taher, K. Parikh, F. Flores-Borja, S. Mletzko, D. A. Isenberg, M. P. Peppelenbosch and R. A. Mageed, Protein Phosphorylation and Kinome Profiling Reveal Altered Regulation of Multiple Signaling Pathways in B Lymphocytes from Patients with Systemic Lupus Erythematosus, *Arthritis and Rheumatism*, 2010, **62**, 2412–2423.
59. N. A. Meguid El and A. K. Bello, Chronic kidney disease: the global challenge. *Lancet*, 2005 365: 331–340.
60. de Borst, S. H. Diks, J. Bolbrinker, M. W. Schellings, M. B. van Dalen, M. P. Peppelenbosch, R. Kreutz, Y. M. Pinto, G. Navis and H. van Goor, Profiling of the renal kinome: a novel tool to identify protein kinases involved in aniotensin II-dependent hypertensive renal damage, *Am. J. Physiol. Renal Physiol.*, **293**, F428–F437.
61. C. Stambe, R. C. Atkins, P. A. Hill and D. J. Nikolic-Paterson, Activation and cellular localization of the p38 and JNK MAPK pathways in rat crescentic glomerulonephritis, *Kidney Int.*, 2003, **64**, 2121–2132.
62. C. Stambe, D. J. Nikolic-Paterson, P. A. Hill, J. Dowling and R. C. Atkins, p38 Mitogen-activated protein kinase activation and cell localization in human glomerulonephritis: correlation with renal injury, *J. Am. Soc. Nephrol.*, 2004, **15**, 326–336.
63. C. Stambe, R. C. Atkins, G. H. Tesch, T. Masaki, G. F. Schreiner and D. J. Nikolic- Paterson, The role of p38_ mitogen-activated protein kinase activation in renal fibrosis, *J. Am. Soc. Nephrol.*, 2004, **15**, 370–379.

64. M. W. Schellings, M. Baumann, R. E. van Leeuwen, R. F. Duisters, S. H. Janssen, B. Schroen, C. J. Peutz-Kootstra, S. Heymans and Y. M. Pinto, Imatinib attenuates end-organ damage in hypertensive homozygous TGR(mRen2)27 rats, *Hypertension*, 2006, **47**, 467–474.

65. D. G. Savage and K. H. Antman, Imatinib mesylate—a new oral targeted therapy, *N. Engl. J. Med.*, 2002, **346**, 683–693.

66. Z. A. Knight, H. Lin and K. M. Shokat, Targeting the cancer kinome through polypharmacology, *Nature Rev. Cancer*, 2010, **10**, 130–137.

67. J. A. Schetelig, M. Kiani, G. Schmitz, M. Ehninger and M. Bornhauser, T cell-mediated graft-versus-leukemia reactions after allogeneic stem cell transplantation, *Cance. Immunol. Immunother.*, 2005, **54**, 1043–1058.

68. A. Maddigan, L. Truitt, R. A. Arsenault, T. Freywald, O. Allonaby, J. Dean, A. Narendran, J. Xiang, A. Weng, S. Napper and A. Freywald, EphB receptors trigger Akt activation and suppress Fas receptor-induced apoptosis in malignant T lymphocytes, *J. Immunol.*, 2011, **187**, 5983–5994.

69. R. C. Haggitt, J. Tryzelaar, F. H. Ellis and H. Colcher, Adenocarcinoma complicating columnar epithelium-lined (Barrett's) esophagus, *Am. J. Clin. Pathol.*, 1978, **70**, 1–5.

70. J. W. van Baal, S. H. Diks and R. J. Wanders, *et al.*, Comparison of Linome Profiles of Barrett's Esophagus with Normal Squamus Esophagus and Normal Gastric Cardia, *Cancer Res.*, 2006, **66**, 11605–11612.

71. J. A. Meyerhardt and R. J. Mayer, Systemic therapy for colorectal cancer, *N. Engl. J. Med.*, 2005, **352**, 476–87.

72. J. B. Tuynman, L. Vermeulen, E. M. Boon and K. Kemper, *et al.*, Cyclooxygenase-2 inhibition inhibits c-Met kinase activity and Wnt activity in colon cancer, *Cancer Res.*, 2008, **68**(4), 1213–1220.

73. A. H. Sikkema, S. H. Diks, W. F. A. den Dunnen, A. ter Elst, *et al.*, Kinome profiling in pediatric brain tumors as a new approach for target discovery, *Cancer Res.*, 2009, **69**, 5987–5995.

74. A. Van den Broke, R. Arsenault, Y. Cleuter, C. Dehouck, *et al.*, Kinome profiling of bovine leukemia virus-indiced ovine leukemia: an approach for identifying altered signaling pathways and drugable targets in cancer, *JAIDS*, 2011, **56**, 98.

75. R. H. Strasser, J. L. Benovic, M. G. Caron and R. J. Lefkowitz, Beta-agonist and prostaglandin EI-induced translocation of the beta-adrenergic receptor kinase: evidence that the kinase may act on multiple adenylate cyclase-coupled receptors, *Proc. Natl. Acad. Sci. USA*, 1986, **83**, 6362–6366.

76. J. Pouyssegur and P. Lenormand, Fidelity and spatio-temporal control in MAP kinase (ERKs) signalling, *Eur. J. Biochem.*, 2003, **270**, 3291–3299.

77. P. R. Perillan, M. Chen, E. A. Potts and J. M. Simard, Transforming growth factor beta1 regulates Kir2.3 inward rectifier K1channels via

phospholipase C and protein kinase C gamma in reactive astrocytes from adult rat brain, *J. Biol. Chem.*, 2001, **227**, 1974–1980.

78. P. K. Majumder, N. C. Mishra, X. Sun, A. Bharti, *et al.*, Targeting of protein kinase C gamma to mitochondria in the oxidative stress response, *Cell Growth Differ.*, 2001, **12**, 465–470.

79. B. S. Skalhegg and K. Tasken, Specificity in the cAMP/PKA signalling pathway. Differential expression, regulation, and subcellular localization of subunits of PKA, *Front Biosci.*, 2000, **5**, d679–d693.

Bioinformatics and Statistics

CHAPTER 7

Bioinformatics and Statistics: LC-MS(/MS) Data Preprocessing for Biomarker Discovery

PÉTER HORVATOVICH,*[a,c,d] FRANK SUITS,[b]
BEREND HOEKMAN[a,c,d] AND RAINER BISCHOFF[a,c,d]

[a] Analytical Biochemistry, Department of Pharmacy, University of
Groningen, A. Deusinglaan 1, 9713 AV Groningen, The Netherlands; [b] IBM
T.J. Watson Research Centre, 1101 Kitchawan Road, Yorktown Heights,
10598, New York, USA; [c] Netherlands Bioinformatics Centre, Geert
Grooteplein 28, 6525 GA Nijmegen, the Netherlands; [d] Netherlands
Proteomics Centre, Padualaan 8, 3584 CH Utrecht, the Netherlands
*Email: p.l.horvatovich@rug.nl

7.1   Introduction

Computational methods form an important and integrated part of biomarker
research. Figure 7.1 presents the main components of a computational
workflow used for biomarker discovery. After the generation of raw data,
computational methods extract information for the identification and quantification of compounds and combine it into a form amenable to statistical
evaluation and validation. Statistical evaluation and validation seeks to identify
a list of compounds showing discrimination between groups of samples. This

RSC Drug Discovery Series No. 33
Comprehensive Biomarker Discovery and Validation for Clinical Application
Edited by Péter Horvatovich and Rainer Bischoff
© The Royal Society of Chemistry 2013
Published by the Royal Society of Chemistry, www.rsc.org

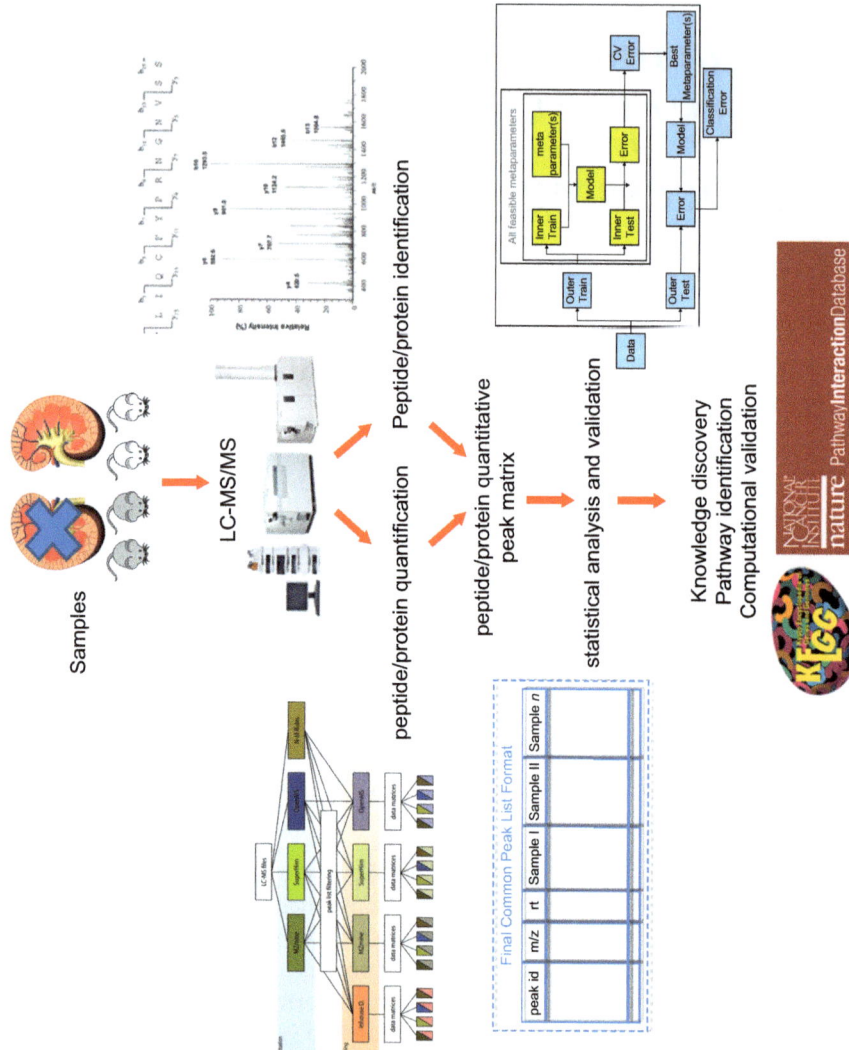

list, together with other elements of raw and partially preprocessed data, additionally serves to identify pathways involved in diseases and biological processes.[1,2] Advanced computational algorithms are used less intensively in this phase but play an important role in biomarker validation, especially when validation is performed using mass spectrometry with Multiple Reaction Monitoring (MRM) approach.[3–6]

The goal of data preprocessing is to extract quantitative and compound identity information from the raw, measured signal and convert this information into a form amenable to statistical analysis. Generally, this means providing a table containing quantitative information where rows (or columns) correspond to compounds (peptides or proteins in the case of proteomics analysis) and columns (or rows) correspond to measured samples, where compounds are annotated by their compound identities. Array-based profiling provides relatively simple data compared to mass-spectrometry-based proteomics, since array spots are already annotated with their compound identities and it is obvious from the raw data how the quantitative information maps to its compound identity. This is not the case for label-free LC-MS(/MS) data preprocessing of complex proteomics samples, where the link between quantitative information and compound identity must be extracted from raw data.[7] For that reason preprocessing of complex proteomics LC-MS(/MS) data is a challenging computational task that relies heavily on automatic algorithms due to the huge size and complexity of data. This chapter will present computational methods related to preprocessing and quality control of label-free LC-MS(/MS) data of complex proteomics samples.

The quantitative data produced by these experiments are generally related to compound concentration (proteins or peptides). However, protein or peptide concentrations do not always reflect biological activity or the specific role of a compound. In this case, analytical methods are needed to relate biological activity to the measured signal. An example of such a relation is the activity-based enrichment of metallo-proteases, which relies on an immobilized inhibitor to selectively extract only the active enzymes for subsequent analysis by LC-MS(/MS).[8,9] The biological samples contain a mixture of active metallo-proteases and nonactive ones, inhibited by natural inhibitors. The latter

---

Figure 7.1   Main parts of a generic bioinformatics workflow involved in biomarker discovery. Raw LC-MS(/MS) data are converted to HUPO/PSI standard format. The quantitative module extracts quantitative information using spectral counting from MS/MS data or using ion count intensity from MS1 data. The outcome of this identification is a table containing peptides or protein quantities, where rows (or columns) correspond to peptides/proteins and columns (or rows) correspond to samples. The identification module performs peptide/protein identification using MS/MS data. Then rows (or column) of the quantitative table are labeled with identification information. This table is then subjected to statistical analysis to select a set of discriminating features. The list of selected discriminatory features is then subjected to identify molecular pathways or protein interactions, which serves as a form of knowledge discovery.
Adapted from Christin, *et al.*[1]

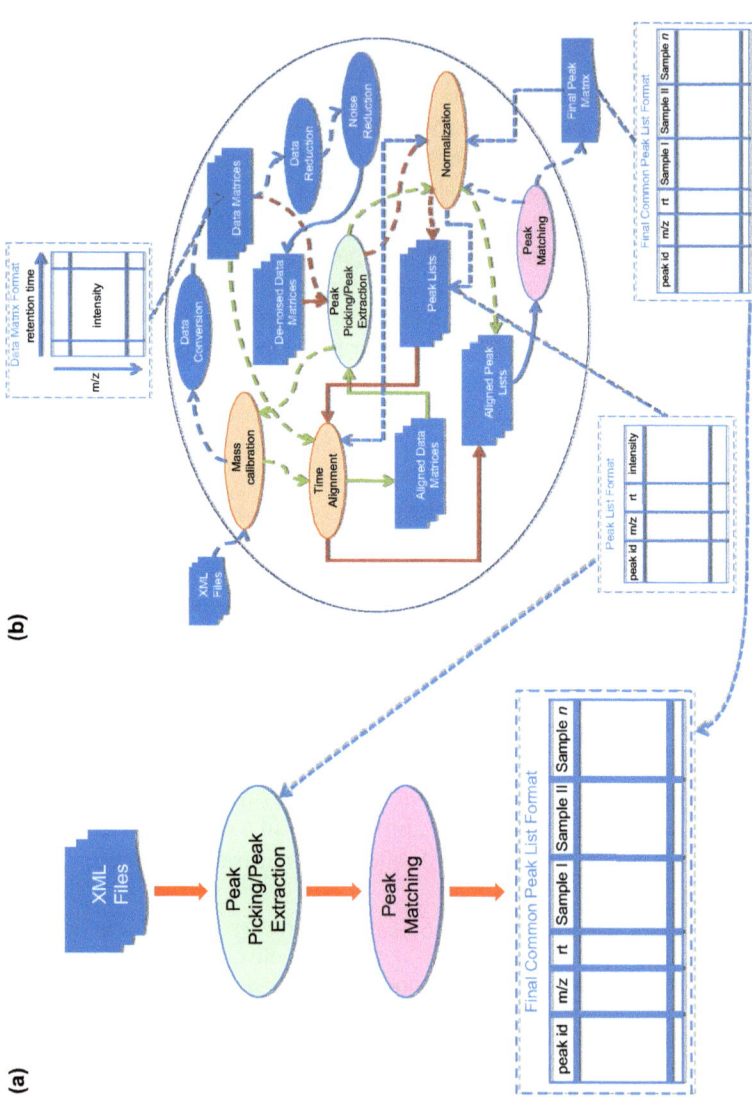

form is not retained on affinity columns and therefore the measured metallo-protease by LC-MS(/MS) reflects the biological activity of the enzyme, while the total concentration of metallo-protease would not be informative. In addition to extracting quantification and identification information, data preprocessing can also perform additional data transformations that help statistical analyzes such as logarithmic transformation, normalization, auto-scaling of quantitative data (*e.g.* mean subtraction and division by standard deviation of quantitative values for each compound) or methods dealing with missing values. Although we classify these operations as part of the prepro-cessing phase of analysis, other statistics books or manuscripts may view these steps as part of the statistical evaluation and validation phase.

7.2 Quantitative Preprocessing Workflow for Single-Stage LC-MS(/MS) Data

From raw label-free LC-MS(/MS) data it is possible to obtain quantitative information from single-stage mass spectra (MS1) using ion-count-based quantification, or from fragment mass spectra (MS/MS) using the frequency of MS/MS spectra recorded for a particular compound during Data-Dependent Acquisition. The latter quantification is called spectral counting. The dynamic quantification range of single-stage MS1-based quantification is generally higher by one or two orders of magnitude compared to spectral counting, which lends preference to perform quantification based on MS1 information. Figure 7.2a presents a simple LC-MS(/MS) quantitative preprocessing workflow using MS1 ion counts for quantification. This workflow has only one

---

Figure 7.2 Schematic workflow of quantitative LC-MS data-preprocessing workflows using MS1 information. (a) shows the minimal workflow containing one peak-picking module in green and one module in purple for peak-matching. (b) shows a comprehensive workflow, including multiple modules. Beside the main modules for peak picking in green and peak matching modules in purple it contains generally, data reduction and noise-reduction modules, modules for alignment in all 3 dimensions in orange such as m/z, ion intensity, and retention time (time alignment, normalization, mass (re)-calibration) and modules performing conversions between different file formats such as different peak-list formats. Workflows are reading raw data in HUPO/PSI XML format such as mzXML or mzML, generally remap the original raw ion count data to a grid where one dimension corresponds to retention time and the other to the m/z value. The peak-picking module provides the list of peaks present in one chromatogram represented by m/z, retention time and intensity value. There is no strict order of these modules, and multiple variations of module orders are possible, with few exceptions such as that peak-matching has to follow the peak-picking module, and that alignment in m/z and in retention-time dimension should precede the peak-matching module. The outcome of this workflow is a table containing the quanti-tative information on peptides or protein.
Adapted from Christin, *et al.*[1]

module for peak detection and quantification (peak picking) and one module for identifying the same compound in multiple samples (peak matching). However, this basic workflow can be completed by other algorithms as well (Figure 7.2b). For example, MS1 data are three-dimensional, with retention time, mass-to-charge ratio (m/z) and ion counts (cts) corresponding to each dimension. LC-MS(/MS) chromatograms of different samples may have a shift in each of these dimensions. The shifts in retention time are corrected with time-alignment algorithms and the shifts in m/z dimension are corrected with mass (re)calibration, while the adjustment of the ion counts is a form of normalization (rescaling) rather than shifting with a local offset. These corrections may be performed with respect to a null point. Mass calibration is always performed to reflect the exact mass and the calibration is performed using known compounds. Retention-time alignment in most cases is performed in a relative manner, by coaligning multiple LC-MS(/MS) chromatograms, since the time axis does not have a deterministic value as does m/z, but retention-time normalization methods to a null point do exist, such as those based on the accurate mass tag approach.[10,11] Normalization is also performed mainly in the context of relative quantification, but an absolute quantification method exists using the three most-abundant peptides of a protein.[12]

Of the three dimensions described above, the most challenging one to align is the elution time dimension.[13,14] Nonlinear retention time shifts are generally present even when particular attention is paid to maintain the same chromatographic conditions during the LC-MS(/MS) data acquisition. Acquisition temperature, changes in eluent composition or alterations of the chromatographic stationary phase cannot be or are not always strictly controlled, making it difficult to avoid retention-time shifts. Multiple retention-time alignment algorithms were developed during the last decade. These algorithms can be classified based on whether they use raw LC-MS data directly or a peak list obtained after the peak-picking procedure, the types of search algorithms they use, and how the local similarity between chromatograms is calculated.[1,2] Many algorithms based on raw LC-MS data and a two-dimensional signal such as total ion current (TIC) or base peak chromatogram (BPC) for alignment have been developed.[13,15,16] Recent time alignment algorithms, however, use all the three dimensions, including the m/z separation of compounds, and generally rely on noise filters to involve only the signal related to peaks of common compounds.[17,18] There are multiple approaches that use peak lists to perform time alignment, such as Warp2D[19] and others.[15,20,21] A plethora of search algorithms exist to find the optimal "warping" of the time dimension for retention-time alignment. The most popular are Correlation Optimized Warping,[16,18,19,22,23] Dynamic Time Warping,[17,23–25] Parametric Time Warping,[17,26] the Continuous Profile Model combined with a Hidden Markov Model[27] and nonlinear piece-wise polynomial-fit-based algorithms.[20,28] Multiple similarity measures exist that are used to govern the time-alignment procedure, such as the sum of overlapping peak volumes,[19] the sum of squares or correlation of full or part of data. Most of the time alignment approaches select one reference chromatogram and align pair-wise all the other chromatograms to this reference. If an alignment method is highly accurate and the chromatograms are highly similar, then the alignment accuracy should not depend on the chromatograms selected as the reference.[17,18] However, it is difficult to know *a priori* if these conditions are satisfied. Therefore, either the SIMA approach (simultaneous multiple alignment of LC/MS peak lists) can be used, which combines hierarchical pair-wise correspondence estimation with simultaneous alignment and global retention-time correction. Another solution is to perform all possible pair-wise alignments in a dataset and rearrange the similarity scores obtained after alignment using hierarchical clustering.[29] This method allows not only the selection of the best reference chromatogram showing the highest similarity to other chromatograms, but also it can discover, which groups of chromatograms can be aligned well to each other. Figure 7.3 shows an example dataset with 24 chromatograms acquired within an experimental design study aligned with this approach. In the experimental design the following factors were varied: LC-MS acquisition laboratory (2 levels), type of mouse (3 levels), disease/healthy state of the mouse (2 levels), type of mouse serum depletion (2 levels). This figure shows that all chromatograms within one laboratory can be aligned with sufficiently high similarity, however, there are no chromatographic pairs acquired between laboratories that can be aligned with sufficient quality. Misalignment in the m/z dimension generally indicates a need for mass recalibration, which is typically performed using known molecular mass of compounds either intentionally added or present as contaminant background ions, such as polysiloxanes, at the time of measurement.[1,2,30] Alignment in the ion-count (intensity) dimension is called normalization and is generally used to correct quantitative systematic nonlinearity or sensitivity variations at different ion-abundance levels.[31]

Other data preprocessing modules may improve the quality of LC-MS results as well. For example, noise reduction such as smoothing using a two-dimensional Gaussian kernel may improve the data performance of peak detection and quantification.[1] Different types of mass spectrometers provide different raw data characteristics, which then demand slightly different methods for preprocessing. For example, different mass analyzers such as time-of-flight, quadrupole, Fourier transform ion cyclotron resonance mass spectrometry or Orbitrap, provide mass spectra with a different dependency of resolution on the m/z value. For a quadrupole instrument the resolution is independent of m/z, but for time-of-flight instruments the resolution is linearly dependent on m/z. This means that the use of an equidistant sampling frequency in m/z would result in different peak picking quality at different m/z values. At low m/z the peaks may be undersampled, while at high m/z regions peaks would be oversampled. Figure 7.4 shows the effect of equidistant m/z sampling on isotopolog peaks of two double-charged peptides recorded at low and high m/z values using a QTOF instrument. Resolution at the low m/z value is high, therefore the equidistant space of 0.01 Th between sampling points in m/z provides a high-quality peak shape, which can be easily recognized by peak-picking algorithms. In contrast, double-charged isotopolog peaks at higher m/z values are much noisier and consist of more sampling points using the same grid

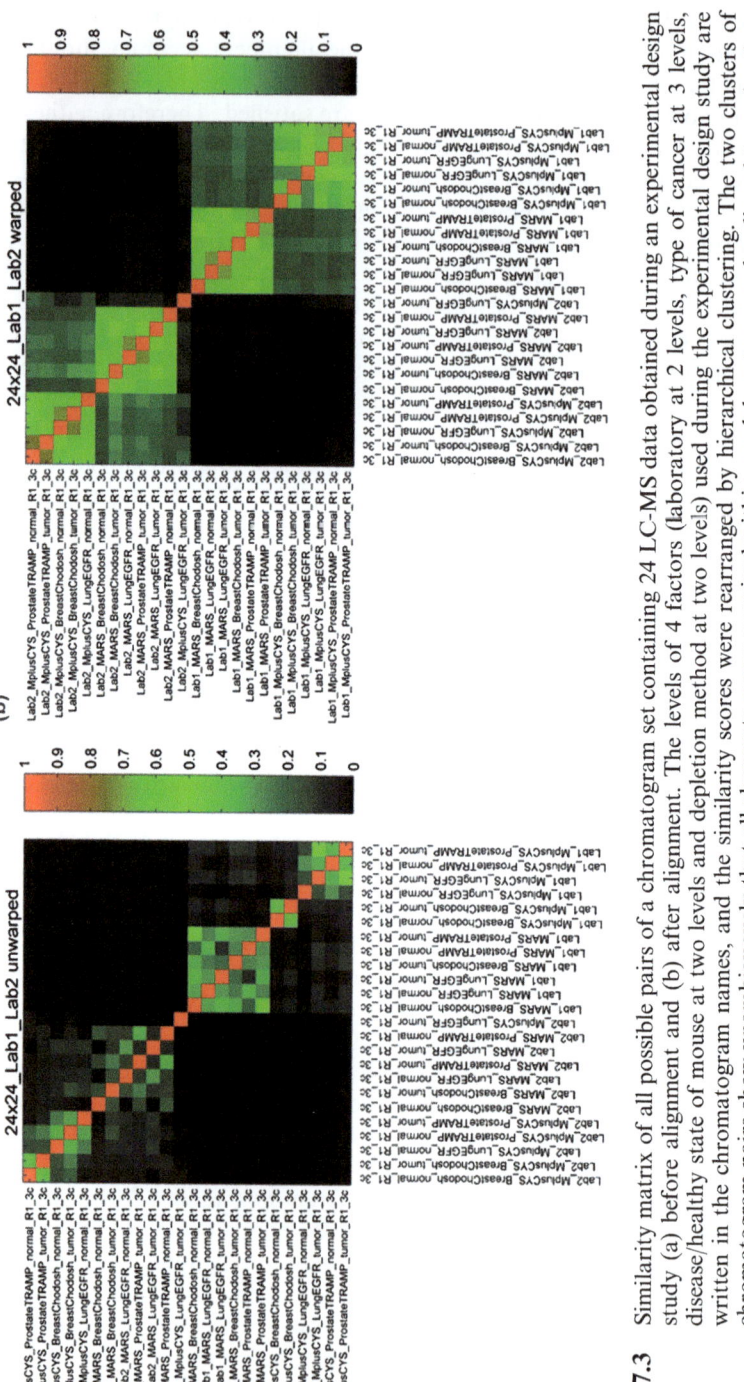

Figure 7.3   Similarity matrix of all possible pairs of a chromatogram set containing 24 LC-MS data obtained during an experimental design study (a) before alignment and (b) after alignment. The levels of 4 factors (laboratory at 2 levels, type of cancer at 3 levels, disease/healthy state of mouse at two levels and depletion method at two levels) used during the experimental design study are written in the chromatogram names, and the similarity scores were rearranged by hierarchical clustering. The two clusters of chromatogram pairs show unambiguously that all chromatograms acquired within one laboratory can be aligned to each other, while none of the chromatograms acquired in different laboratories are similar enough for an acceptable alignment. Adapted from Ahmad, *et al.*[29]

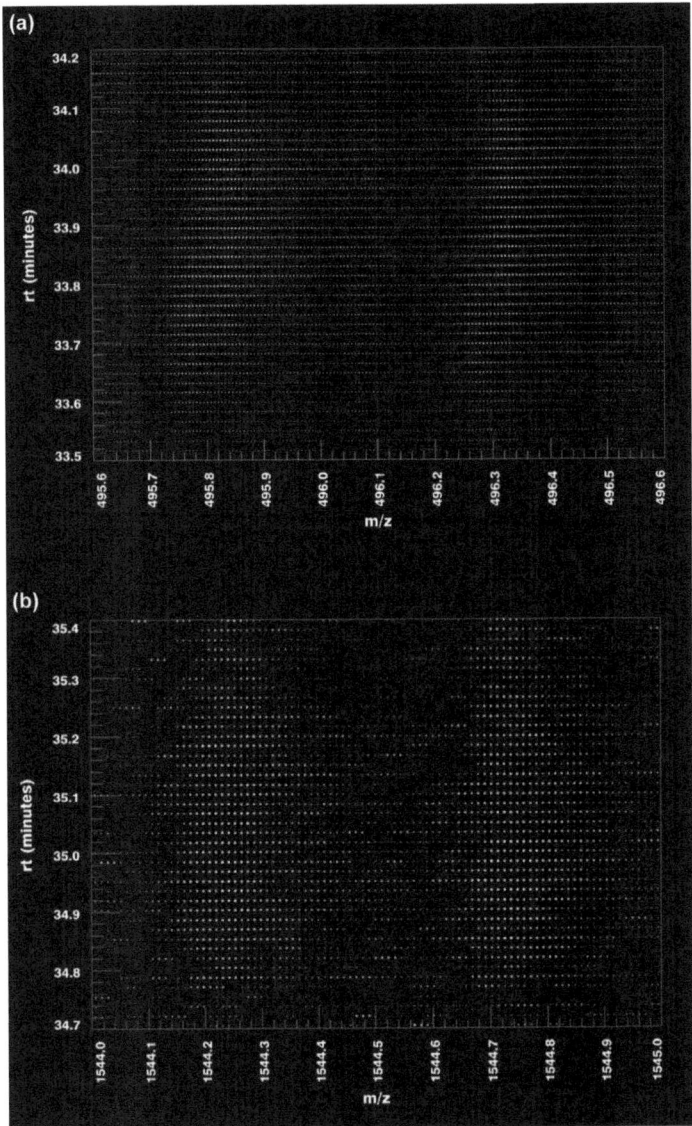

Figure 7.4 Effect of grid with equidistant *m/z* sampling on profile QTOF MS1 data. (a) double-charged isotopologs of a peptides at low *m/z* value are sampled with optimal number of points, while in (b) a doubly charged isotopolog peak of a different peptide at higher *m/z* value is more noisy due to the lower resolution in *m/z*. Lower-resolution results in more sampling points in *m/z*, therefore isotopolog peaks in (b) are much nosier and are more difficult to recognize and quantify by peak-picking algorithms than the isotopolog peaks in (a). Data was recorded in profile mode during label-free LC-MS analysis of a porcine cerebrospinal fluid sample, the sampling frequency in *m/z* direction is 0.01 Th. Adapted from.[32]

spacing in m/z due to lower resolution of the QTOF analyzer at higher m/z values. To provide the same peak-picking performance, the sampling frequency should be adopted to equal density resolution and not to equal density m/z values.[32] Another important aspect is the mass shift observed in the LC elution profile of a compound using a three-dimensional ion-trap mass analyzer. Compounds in the trap at high density may repel others with a similar m/z value, resulting in a slight change in their orbits, and therefore in their detected m/z values. This slight change in m/z value during an LC elution profile of a compound results in poor peak detection performance using algorithms that assume a normal Gaussian peak elution profile.[1] An additional factor affecting accuracy is the use of profile or centroid data for peak picking. Centroid mass spectra are obtained by applying a denoising algorithm on mass spectra by the acquisition software. This algorithm is limited by only considering information present in each spectrum and does not consider the whole LC elution profile of a compound, which may contain many consecutive spectra in time. This can lead to the rejection of small peaks and therefore to a reduced dynamic range of detected and extracted peaks.

The order of the preprocessing modules is not strictly fixed, however, there are a few rules that must be followed for certain tasks. For example, the peak-matching module must follow the peak-picking module, and alignment in m/z and retention-time dimensions should precede the peak-matching module, since generally peak matching is performed using the m/z and retention-time coordinates of peaks. Aside from these rules of precedence, there is allowed variability in the order of modules that results in multiple possible workflow designs. Regarding the used data format, the workflows generally read raw LC-MS(/MS) data in one of the HUPO-PSI standard formats such as mzXML, mzData.xml or the more recent mzML. Another important format allows the storage of peak lists produced by the peak-picking procedure. The final quantitative preprocessed data is stored in a table as mention above. This table or the peak lists can be submitted to further transformation steps, such as identifying the isotopic peak cluster of a compound in a given charge state (isotope deconvolution), combining peaks of multiple charge states (charge-state deconvolution) and, in conjunction with the identification results, combining peptide quantities to determine protein quantities. The latter transformation is, however, not unambiguous due to the well-known protein inference problem of shotgun proteomics.[33–35] Protein inference is the lack of an unambiguous combination of tryptic peptides to match the protein forms due to shared peptides and the fact that not all peptides are detected during a shotgun LC-MS/MS proteomics experiment. For this reason the application of much older methods such as the two-dimensional gel electrophoresis approach[36,37] and top-down protein characterization[36,37] are still popular since they can separately detect the different protein isoforms.

In this chapter only the quantitative aspects of label-free LC-MS data preprocessing are discussed, and statistical analysis with feature selection is the topic of Chapter 8, while Chapter 9 discusses applications of computational tools in verification and validation of functional biomarkers. The reader is invited to obtain a comprehensive overview on state-of-the-art approaches in topic-specific reviews on current shotgun MS/MS based peptides identification technology,[38–42] spectral counting quantification,[43] network analysis and knowledge discovery tools,[44–46] and MRM bioinformatics.[4,6]

7.3 Example Workflow: The Threshold Avoiding Proteomics Pipeline

In this section the main characteristics of the Threshold Avoiding Proteomics Pipeline (TAPP) as an example workflow for processing label-free LC-MS(/MS) data are presented.[32] The main steps of TAPP are presented in Figure 7.5. The TAPP pipeline has several innovations that distinguish it from conventional workflows. The first innovative element is that after reading the raw LC-MS data in HUPO-PSI format such as mzXML, TAPP resamples the

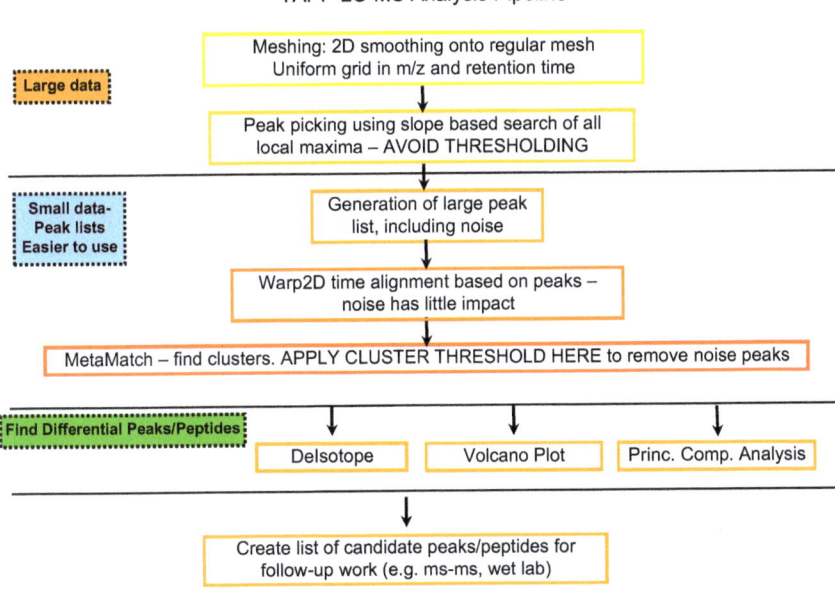

Figure 7.5 Main steps of TAPP pipeline.[32] TAPP reads a set of raw LC-MS data in HUPO-PSI format such as mzXML, and first performs signal smoothing with a three-dimensional Gaussian kernel, based on a resampling of the m/z and rt space onto a uniform grid. A slope-based geometric peak-picking algorithm detects and quantifies peaks. The generated peak list includes noise and is subjected to the Warp2D[19] time-alignment algorithm. Finally, the common peaks are identified in the set of peak lists using the MetaMatch tool, by identifying peaks in different LC-MS(/MS) chromatograms that cluster together tightly in retention time and m/z. The obtained quantitative peak table can be subjected to deisotoping or other statistical analyses such as construction of volcano plots, principal component analysis or selection of set of discriminatory features. Adapted from.[32]

ion abundance of the raw data into an uniform grid with m/z and retention time dimensions. Uniformity in the retention time dimension is represented by equidistant grid points in time, but in the m/z dimension grid points are equidistant with respect to mass resolution. This resampled grid provides the same sampling efficiency at all m/z values and therefore provides the same quality of data in the full m/z range for peak detection (Figure 7.3). The resampled data in a regular grid is subjected to a three-dimensional Gaussian filter, which preserves peak shape, and provides high-quality data for peak detection and quantification. This smoothing filter is particularly useful for processing centroid data obtained with a three-dimensional ion trap, which may contain shifts of peak maxima in the m/z direction. Using a three-dimensional Gaussian filter larger than the mass shift results in a more accurate peak centroid – both in its coordinates in (m/z, retention time) space, and in the quantification of the peak ion current.

The next step is peak detection and quantification. TAPP uses a slope-based geometric peak-picking algorithm, which first finds all local maxima of ion abundance in retention time and m/z space. Then, it identifies all sampling points that belong to this maximum with a descending slope. This approach is fast and does not require any assumed peak shape. In addition, this algorithm is able to resolve overlapping peak clusters as long as the peaks are separated by distinct local minima. Figure 7.6a shows the wide view of sampling points in an LC-MS map of a depleted human serum acquired using a three-dimensional ion-trap mass spectrometer. Figure 7.6b presents the close view of one part of the full LC-MS map showing the search path of sampling points from the local maxima of a peak until positions, where the descending ion intensities reach a minimum threshold on slope. The resolving power of two overlapping peaks is presented in Figure 7.6c. Sampling points for the left larger peak are represented by white spots, while central black spots correspond to the smaller peaks with the left maximum. The peripheral black spots represent the boundary of the larger left peak. The white spots are then used to determine the peak characteristics, such as position in m/z and retention time, peak height, peak volume, full width at half-maximum in retention time and m/z dimensions. It should be noted that the edge of the peaks (peripheral black dots) are used to calculate a three-dimensional background of the peak and provide local signal-to-noise estimate. These characteristics for all peaks are then summarized in a peak list without setting a threshold based on peak height or signal-to-noise ratio.

The generated peak lists are then subjected to time alignment using the Warp2D approach.[19] Warp2D is a pair-wise alignment algorithm based on correlation optimized warping, and uses the sum of peak overlap as the benefit function driving the time-alignment process. In Warp2D, peaks are modeled by three-dimensional Gaussians using peaks with peak location in m/z and retention-time space, peak height and full width at half-maximum (FWHM) of a peak in m/z and retention-time dimensions functioning as the $FWHM = \sigma \cdot \sqrt{2 \cdot \ln(2)} \approx 1.17741 \cdot \sigma$ in both dimensions (σ being the standard deviation of the Gaussian function matching each peak). The integration of two

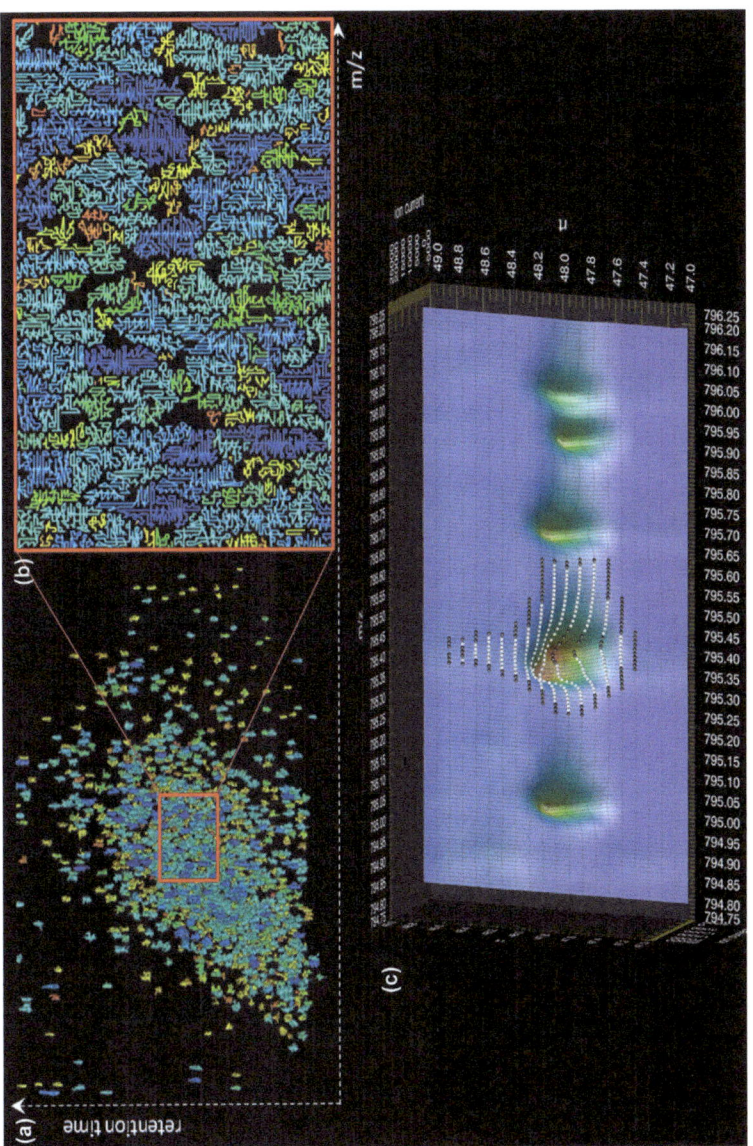

Figure 7.6 Slope-based geometric peak picking implemented in TAPP.[32] (a) shows the wide view LC-MS data, where sampling points corresponding to one peak in retention and m/z space are colored differently. (b) shows the close view of this map, which include the search path from local maximum until descending points reach a minimal slope. LC-MS data was acquired with a three-dimensional ion trap by analyzing a depleted human serum sample. (c) shows a three-dimensional plot of two overlapping peaks. Sampling points in m/z and retention-time space are represented by spots. White spots belong to the larger peak with right maximum, while central black spots belong to the smaller peak with left maximum. Black spots at the edge represent the boundary of the larger peak with left maximum. Plot (c) is adapted from.[32]

overlapping three-dimensional Gaussian peaks has an analytical solution, and therefore can be quickly calculated. The overlap of the most abundant peaks within reasonable distance is calculated in segments using the traditional dynamic programming approach of correlation optimized warping (COW) and drives the optimal segment stretching or shrinking to find an overall optimal time alignment. In the final result the alignment algorithm changes the retention time of peaks in one of the peak lists, referred to as a "sample", the peaks of which are aligned to the peaks of the "reference" chromatogram, which do not change at all during the alignment procedure. The algorithm also outputs a text file containing the retention-time transformation in the form of the segment end positions of correlation optimized warping of the sample chromatograms before and after alignment, which amounts to a piece-wise linear transformation of the sample's retention time coordinate to that of the reference. This transformation can be used to adjust the retention time of any location in the sample chromatograms corresponding to the reference chromatogram. This functionality can be used to change the retention time of mass spectra in the raw data followed by the construction of overlaid extracted ion chromatograms of a peak present in all chromatograms to assess the local quality of the time alignment (Figure 7.7a). For the alignment, one reference is selected and all the other chromatograms are aligned to this reference. As mentioned earlier, performing all possible pair-wise alignments allows the identification of groups of chromatograms that can be aligned with reasonable similarity, which can also help to select the best chromatogram to use as the reference.[29] Warp2D performs the alignment of LC-MS chromatograms automatically but the software requires some user parameters, such as the number of sampling points in the retention-time dimension, segment length, slack parameter, and the number of the most-abundant peaks per segment to use when calculating the sum of overlapping peak volume. Warp2D provides a similarity score before and after alignment based on the arithmetic mean of the sum of overlapping and sum of individual peak volume in the sample and reference chromatograms. This similarity score provides a global alignment quality and does not allow the evaluation of local misalignments. For that reason a visualization showing the overlaid extracted ion chromatograms of peaks present in all samples (*e.g.* spiked-in peptides or peptides originating from housekeeping proteins) should be performed before and after alignment as presented in Figure 7.7a. The visualization should include at least 2–3 peaks for each segment.

The final step of TAPP is to identify the same peaks in multiple chromatograms by finding peaks that cluster closely together in retention time and m/z. As mentioned before, peak list are made without the use of a threshold of peak height, or low signal-to-noise ratio. In fact, the goal of TAPP is to combine all available information within sample groups to discover differential behavior, and for that reason noise/signal discrimination is performed at the final step where peaks from multiple chromatograms are matched based on clustering rather than intensity. This matching is governed first by a user-defined maximum cluster size in retention time and m/z dimensions, and secondly by giving a threshold on minimal sample class occupancy of peaks. The

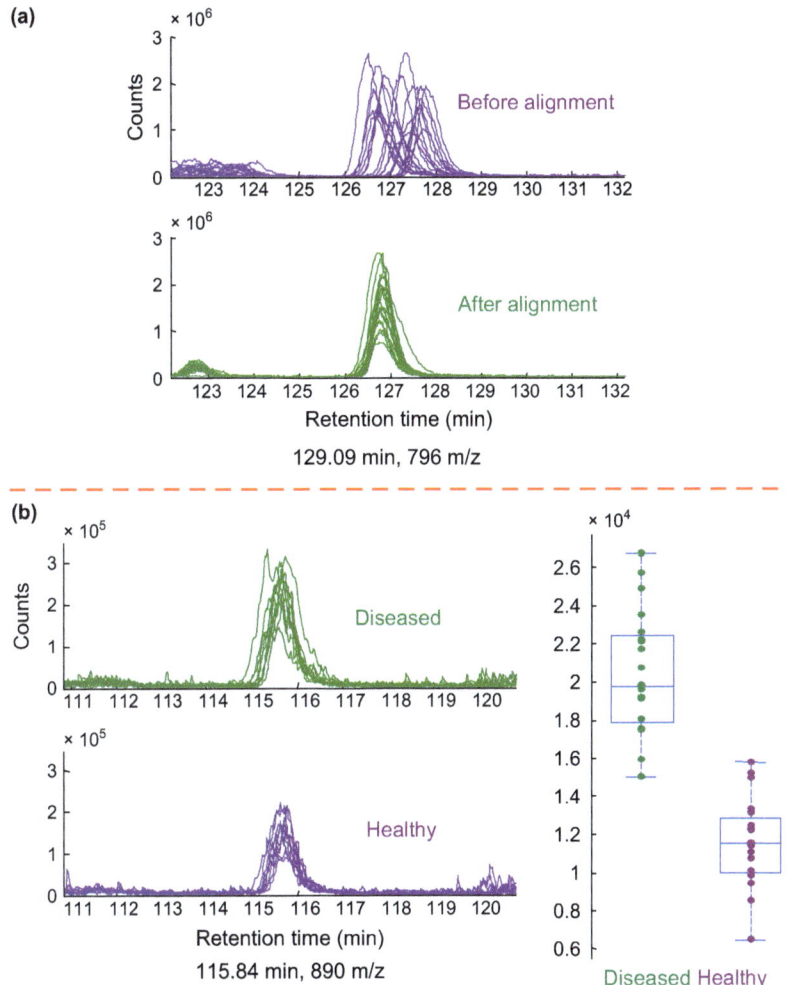

Figure 7.7 Quality assessment of label-free LC-MS data preprocessing pipeline parts using visualization. (a) assessment of local time alignment quality using overlaid extracted ion chromatograms of peak present in all chromatograms before (upper purple plots) and after application of a time alignment algorithm (lower green plots). (b) left plot shows overlaid extracted ion chromatograms of discriminatory isotopolog peaks of the same compound selected by statistical analysis in two sample groups of healthy and disease. Right box-plots showing measurement points by means of dots corresponding to isotopolog peaks of the same compounds in each chromatogram obtained with automatic LC-MS data preprocessing pipeline, which should be compared visually with the peaks heights in extracted ion chromatograms of the two sample groups.

noise/signal discriminatory power of the later parameter is based on the assumption that noise will appear in retention time and *m/z* space randomly, while true signal from multiple chromatograms of one sample group will consistently cluster closely around one position. This approach may capture very faint but true signal peaks that would otherwise be removed using conventional thresholds such as minimum allowed signal-to-noise ratio. The name of Threshold Avoiding Proteomics Pipeline originates from the application of a threshold for signal/noise discrimination in the final step, based on clustering rather than peak quality, making this decision based on the context of all chromatograms in a sample group. In contrast, the conventional application of a threshold at the peak list level uses only the information from a single chromatogram, and therefore extracts peaks with lower dynamic concentration range compared to TAPP.

The output of TAPP is a table containing quantitative information either based on peak height or peak volume, and the rows (or columns) correspond to peaks represented by retention time and m/z value, while columns (rows) correspond to LC-MS chromatograms. These data should be further subjected to isotope and charge-state deconvolution or other standard transformations on peak lists such as clustering peaks of the same compound with and without adducts. The peaks of interest should be annotated with peak identification results and peptide quantities can be further combined to estimate protein quantities. The annotated peptide or protein quantitative table is subjected to statistical analysis to select sets of discriminatory peptides or proteins. TAPP does not contain a normalization module or modules for mass recalibration, but this processing operation is considered independent and can be performed using algorithms based on either the raw or the processed data.

To verify the accuracy of the pipeline it helps to visualize some of the isotopologs of discriminatory peptides selected by the statistical analysis by means of overlaid extracted ion chromatograms showing separately chromatograms of different sample groups such as healthy and disease as shown in Figure 7.7b. The comparison of peak heights in these overlaid extracted ion chromatograms with box plots showing the quantification value obtained with TAPP helps verify the approximate accuracy of the automated quantitative LC-MS data preprocessing pipeline. These plots allow a visual confirmation that the peaks present in the overlaid extracted ion chromatograms reflect the quantification results found by the data preprocessing workflow. TAPP has been successfully used in several studies for processing LC-MS data for biomarker discovery such as to identify protein marker in rat cerebrospinal fluid modeling the development of experimental autoimmune encephalomyelitis[47] and to identify stability markers in human[48] and porcine[49] cerebrospinal fluids.

7.4 Performance Assessment of Quantitative LC-MS Data Preprocessing Workflows

During the last decade a large effort was made by the bioinformatics community to develop multiple software platforms for quantitative preprocessing of label-free LC-MS data. An incomplete enumeration of such open-source popular platforms include OpenMS,[50,51] mzMine,[52] SuperHirn,[52] Specarray,[53] and together they serve as a rich bioinformatics resource, but also make a difficult choice for the end user regarding which tools to use. A plethora of quality-control methods exist, some of which were presented earlier in this chapter such as use of overlay extracted ion chromatograms of common peaks to visualize local time alignment quality, or global quality metrics based on the sum of overlapping peak area. These quality metrics are perfectly adequate to assess part of the full data-processing pipelines, but they are not adequate to assess the overall quality of the complete, end-to-end pipeline performance. In this chapter a generic quality-assessment method is presented that enables the relative comparison of quantitative LC-MS data preprocessing pipelines based on a set of differential spiked complex LC-MS chromatograms. The various quantitative LC-MS data preprocessing pipelines are provided by the msCompare[54] tool. msCompare is a framework that integrates peak-picking and peak-matching modules from in house-developed, and from OpenMS,[50,51] mzMine,[52] SuperHirn[52] popular open-source label-free LC-MS data preprocessing pipelines (Figure 7.8a). In Figure 7.2a these two modules were shown as a minimal requirement in LC-MS data preprocessing workflows. The minimal workflow reads raw LC-MS data in HUPO-PSI format, outputs a series of peak lists after the peak picking modules, and provides the table containing quantitative information of compounds in different samples as the final outcome after the last peak-matching module. The other modules shown in the previous session can be easily combined with these two modules. For example, any alignment modules can be integrated as part of both modules, while data-quality-enhancing modules such as smoothing or noise filtering may be integrated as part of the peak-picking modules. After modularizing all integrated pipelines to peak-picking and peak-matching modules, an intelligent format conversion method enables the connection of all integrated peak-picking modules with all peak-matching modules. Therefore, besides assessing the quantification performance of the original pipelines msCompare makes it possible to assess pipelines with heterogeneous composition of modules from multiple pipelines as shown in Figure 7.8b.

The performance assessment of different data processing pipelines is performed in msCompare based on the rank of the features (peaks) of spiked-peptides amongst the most discriminatory features calculated between two spiking levels. A *t*-value calculated according to a two independent samples *t*-test is used to rank features according to their discriminatory power based on the following equation defining the quality score:

$$\text{Quality score}(p, x) = \sum_{\text{for all datasets}} \sum_{i=1}^{n} \frac{p}{p + \left(\sum_{j=1}^{i} NSF_j\right)^x} \tag{7.1}$$

where n is the number of all features in a dataset, NSF (Number of Nonspike-Related Features) is the number of features occurring between rank 1 and index i in the matched feature matrix that are not related to the spiked standard compounds, and p and x are constants influencing the degree of score

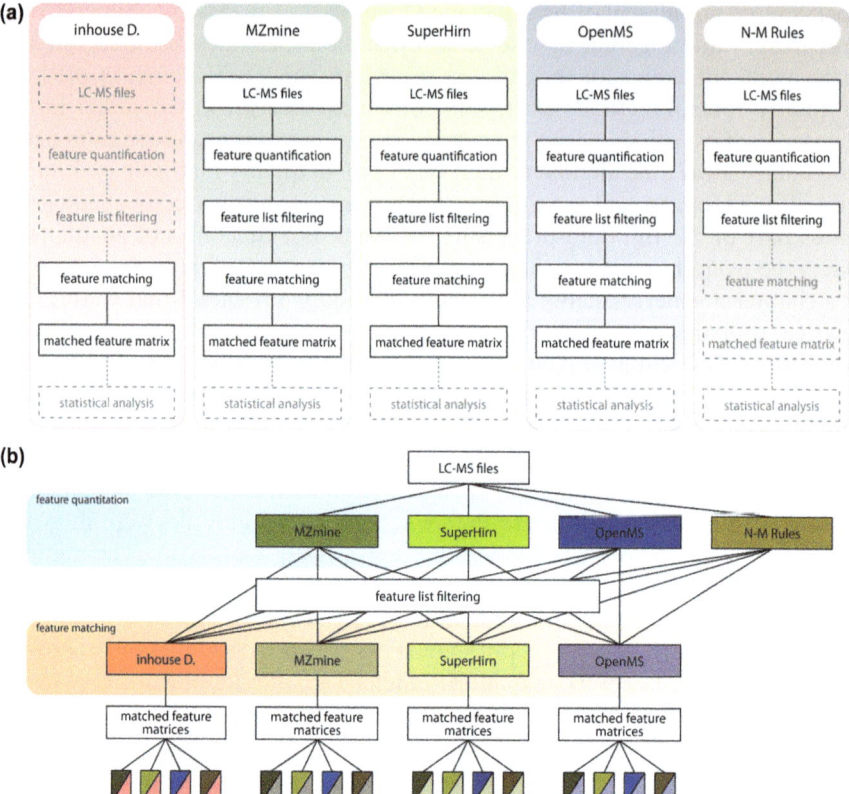

Figure 7.8 msCompare computational framework combining modules from different open-source data preprocessing workflows. (a) open-source data preprocessing workflows modularized to form peak-picking and peak-matching modules and integrated in msCompare. (b) computational framework, which allows execution of any combination of peak-picking and peak-matching modules of the original pipelines.
Adapted from.[54]

attenuation for nonspike-related features amongst the most discriminating features. A decoy approach is applied to calculate a quality score corresponding to a random distribution of the detected spiked-peptide related features in the complete feature list, and the decoy score is subtracted from the score obtained for a particular pipeline. The random distribution of features related to spiked peptides in the feature lists is obtained by randomly reshuffling the order of all features in the quantitative matched feature matrix. Example scenarios of quality scores and their dependence on values of p and x parameters, and the number and distribution of spiked-peptide-related features amongst the 8 most discriminating peaks, is presented in Figure 7.9a. Figure 7.9b shows the Venn diagram of the number of spiked-peptides related features amongst the 100 most discriminating features identified by three different homogenous pipelines of mzMine, SuperHirn and OpenMS. The

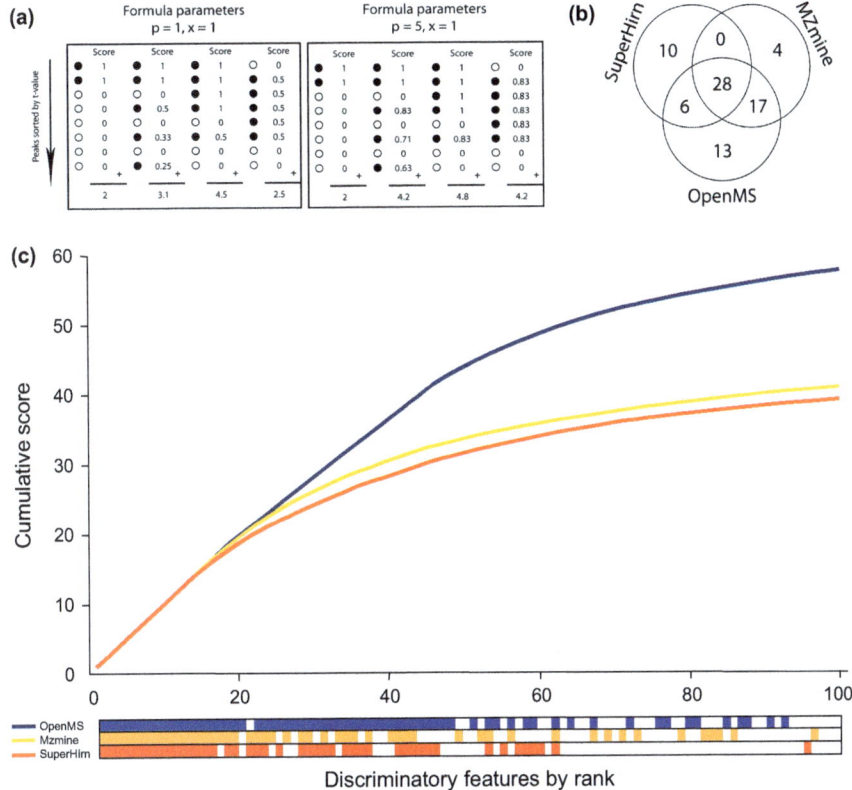

Figure 7.9 Assessment of different homogenous data preprocessing pipelines using a quality score (eqn (7.1)). (a) effect of p and x parameters on the quality score of individual feature and the sum of all features, for a few scenarios regarding the distribution of spiked-peptide-related features (●) amongst the most-discriminative ones. ○ shows features not related to spiked-in peptides. Features are sorted according to their t-value from top to the bottom. (b) shows the Venn diagram of the number of spiked-peptide-related features found amongst the 100 most discriminatory features. The upper plot in (c) shows the evaluation of the quality scores according to the discriminatory rank from the 1 to the top 100 most discriminating features. The upper part of (c) showing colored stripes indicates the presence (colored squares) or absence (empty white squares) of spiked-peptide-related features at given rank from 1 until the top 100 most discriminating features. (b) and (c) show clearly that OpenMS identified the most of the spiked-peptide-related features amongst the 100 most-discriminating features with lowest number of nonspiked-peptide related features. The analysis was performed using a set of label-free LC-MS data obtained from differentially spiked human urine samples with 175 spiked-peptide-related features. Adapted from.[54]

upper plot of Figure 7.9c shows the evaluation of the quality score in function of the discriminatory rank of features from rank 1 to 100, while the lower plot shows the presence (colored squares) and absence (empty white squares) of spiked-peptide-related features at a given rank in the three studied quantitative LC-MS data preprocessing pipelines. Figures 7.9b and c show clearly that there are large differences in the quantitative data preprocessing performances of various homologous pipelines, and that the OpenMS pipeline outperforms the two others and finds almost all (82%) of the spiked-peptide-related features identified by all the three workflows. In addition, the OpenMS pipeline makes fewer data processing errors compared to the other two pipelines. The bottom part of Figure 7.9c shows that the first nonspiked-peptides-related feature is identified at later rank compared to the other pipelines resulting in a higher-quality peak list. The close investigation of the most discriminating nonspiked-peptides-related features reveal the type of data preprocessing errors that may occur most frequently. The most often occurring data preprocessing errors providing highly discriminating features with low ranks are due to detection of one peak as two. Figure 7.10 shows this situation where the split of a peak results in two detected features, and one feature is matched to the features of the corresponding peaks in the same sample group (*e.g.* low spiking level), while the other feature is matched to the feature of the corresponding peaks in samples of the other sample group (*e.g.* high spiking level). This results in two features, with one being highly discriminatory between the two sample groups. Other errors such as the misalignment of one peak, or an error in peak quantification, result in specific features having lower discriminatory power since it affects only one or a few features, while the others remain accurate.

The application of the quality score to all possible combinations of peak-picking and peak-matching modules of popular and inhouse-developed LC-MS data preprocessing pipelines allows an assessment of the performance differences of these pipelines. Figure 7.11 shows the quality scores for 16 and 12 LC-MS data preprocessing pipelines used to preprocess a set of differentially spiked human urine (Figure 7.11a) and porcine cerebrospinal fluid (Figure 7.11b) LC-MS chromatograms with high, medium and low differences between the levels of spiked peptides. The red bars in the plots indicate the best-performing data-processing pipelines. For the human urine datasets with medium and high spiking level differences the best-performing pipelines were OpenMS to OpenMS, and OpenMS to inhouse developed with respect to peak-picking and peak-matching modules, while for low spiking level difference N-M rules to inhouse developed provided the best performance. The-best performing pipelines for the porcine CSF was OpenMS to Superhirn for high and OpenMS to OpenMS for low spiking level differences. In multiple cases, workflows with a heterogeneous combination of modules outperformed the original workflow. For example, note the better performance of the mzMine to SuperHirn combination compared to the SuperHirn to SuperHirn and mzMine to Mzmine combinations in the human urine dataset with high spiking level difference (upper bar plot in Figure 7.11a). It should be noted, however, that the performance of the different workflows also depends on the applied data preprocessing parameters and the type of LC-MS data.

Large differences exist between the best-performing and worst-performing pipelines independent of the spiking level difference during the processing of the

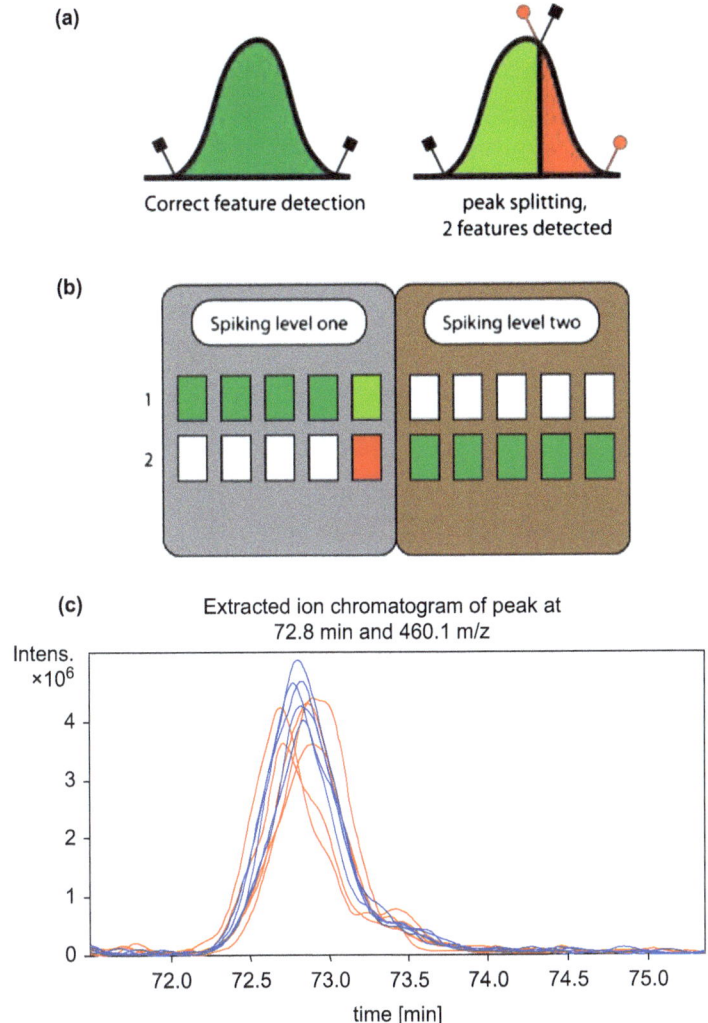

(a)

Correct feature detection

peak splitting,
2 features detected

(b)

Spiking level one

Spiking level two

1

2

(c) Extracted ion chromatogram of peak at
72.8 min and 460.1 m/z

Intens.
×10^6

4

3

2

1

0

72.0 72.5 73.0 73.5 74.0 74.5 75.0

time [min]

Figure 7.10 Most relevant quantitative LC-MS data preprocessing error providing highly discriminating features not related to spiked peptides. (a) one isotopolog peak is detected as two features in one chromatogram creating a peak-correspondence problem; (b) one part of the isotopolog peak is matched with the corresponding isotopolog peaks of the same compound in the same sample group, while the other part of this isotopolog peak is matched with the corresponding isotopolog peaks of the same compound in samples of the other sample group. This situation creates two sets of features, from which the upper one is highly discriminating between the two spiking levels. However extracted ion chromatograms in (c) of the isotopolog peaks colored according to low and high spiking levels shows no difference in peak intensities in the two sample groups.
Adapted from.[54]

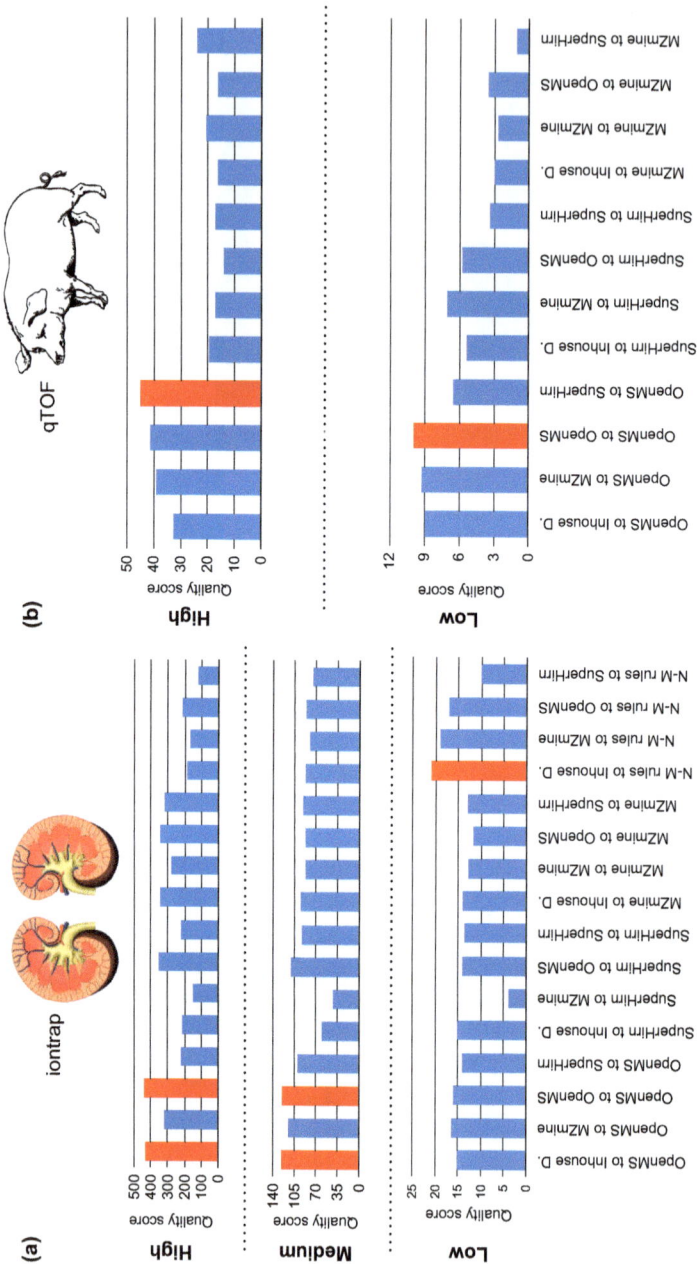

Figure 7.11 Comparison of the performance of peak-picking and peak-matching modules at high, medium and low concentration differences of spiked peptides using a human urine dataset (a) and the porcine CSF dataset (b), respectively. Labels of the hybrid workflows (x-axis) start with the name of the peak-picking module followed by the name of the peak-matching module. The best-performing workflows at each concentration level difference are highlighted in red. The homogeneous OpenMS workflow and combinations of the OpenMS peak-picking module with the inhouse-developed peak-matching module result in the highest scores when concentration differences are large or medium for the spiked human urine dataset (a), while the respective combination of the OpenMS-SuperHirn heterogeneous workflow provides the best performance for the porcine CSF dataset spiked with large concentration differences (b). Adapted from [54].

same LC-MS datasets. These results indicate that using the worst pipeline would provide multiple highly discriminatory features leading to, for example, the selection of many false-positive biomarker candidates. Therefore, when performing a large biomarker discovery study we advise the use of differentially spiked quality control (QC) samples, so that this QC dataset can empirically determine the best workflow. In addition, our scoring algorithm can be used to optimize parameters, but this is not a straightforward task due to the large number of parameters typically available in most of the data-processing workflows. The scoring method can also be used by a bioinformatician to evaluate novel developments as part of an existing or completely new data-processing workflow. The acquisition of differentially spiked QC samples allows not only the quantitative evaluation of the LC-MS data-preprocessing performance, but also an assessment of the performance of any subsequent statistical analysis to identify sets of discriminatory features.[55]

In a final note, msCompare and the quality score help determine the best workflow and provide a measure of relative global performance. However, the quality score does not replace the quality assessment of the preprocessed LC-MS datasets used for clinical biomarker discovery, for example to assess local and global time alignment quality and quantification accuracy as shown in Figures 7.3 and 7.7.

7.5 Summary and Future Trends

Accurate preprocessing and quality assessment of preprocessed LC-MS data are no simple tasks, and generally require the involvement of bioinformatics experts. However, proteomics biomarker discovery is still in a growing phase and currently there is an exploding amount of data generated by application-oriented clinical laboratories. It is therefore important to develop easy-to-use platforms with data-management and -processing capabilities that can comprehensively store raw and processed data with concomitant metadata, and that perform all bioinformatics processing and evaluation tasks starting from preprocessing and peptide/protein identification, and ending with network and pathway analysis. These platforms should satisfy the following criteria: (1) management (storage, retrieval and overview) of large data volumes, (2) storage of metadata such as experimental design, relevant clinical data or parameters applied for data processing and statistical analysis, (3) integration of massive parallel computation resources such as cloud, grid and large computational clusters, (4) integration of tools, building workflows and executing them on massive, parallel computational resources, (5) allowance for flexible and easy adaptation of workflows to different experimental designs and data structures, (6) extensive visualization for quality control and results assessment, preferably allowing user interaction. Finally, computational workflows for mass-spectrometry-based proteomics require flexible and a dynamic software infrastructure to adapt workflows to the high diversity in data structure and experimental design (*e.g.* label-free, various stable isotope labeling approaches, multidimensional chromatography, *etc.*).

The genomics community is already using such platforms to store and process genomics data. For example, the Molgenis platform offers UML model-based database design and an easy-to-use web interface to interact with stored data.[56–58] Molgenis was recently extended with a cluster and grid processing back end, and includes several imputation and sequencing workflows to process next-generation sequencing data. Other platforms use the Galaxy environment[59,60] to integrate tools, create and share workflows and data. A similar endeavor has emerged in the proteomics community: To facilitate the use of msCompare, this framework was integrated in the Galaxy environment and is available as source code at https://trac.nbic.nl/mscompare/ or as an online service at http://galaxy.nbic.nl/galaxy. The warp2D time alignment service presented earlier in Figure 7.3 is accessible at http://www.nbpp.nl/warp2d.html and serves as an additional example of an easy-to-use high-throughput data processing service. Recently, the Taverna workflow management[61] system was used to integrate popular proteomics tools, develop workflows, and execute them in a cloud environment.[62,63] The openBIS platform[64] provides a general purpose data-management system that can store a large volume of heterogeneous experimental data such as mass-spectrometry-based proteomics data, next-generation sequencing data or high-throughput microscopy images.

To alleviate the data-processing bottleneck it is necessary to involve massively parallel computation resources such as cloud, grid and large computational clusters. A few of the above-mentioned solutions already include the access to such resources. Recently, a huge community effort was made to develop gUSE and WS-Pgrade platforms,[65,66] which are generic workflow management systems. These platforms provide rich access to various massively parallel processing back ends and allow the execution of workflows remotely or internally.[67]

The OpenMS platform[50,51,68] is underway toward the realization of a comprehensive data-management and -processing platform for mass-spectrometry-based proteomics. In its current state it includes quantitative LC-MS data preprocessing for both label-free and stable-isotope labeled experiments, with rich visualization tools, peptides/protein identification[69] and the TOPPAS[70] workflow management system. The integration of OpenMS with openBIS and WS-Pgrade is currently ongoing, and has the potential to serve as a platform with a high level of data management, rich tools and workflows for LC-MS(/MS) preprocessing and evaluation, and easy access to multiple high-performance distributed computational resources. This platform, along with the other ongoing developments that offer high-throughput and easy-to-use data management and processing services will help researchers face the challenge created by current data explosion.

References

1. C. Christin, R. Bischoff and P. Horvatovich, *Talanta*, 2011, **83**, 1209–1224.
2. P. L. Horvatovich and R. Bischoff, *Eur J Mass Spectrom (Chichester, Eng)*, 2010, **16**, 101–121.

3. P. Picotti and R. Aebersold, *Nature Methods*, 2012, **9**, 555–566.
4. J. Lemoine, T. Fortin, A. Salvador, A. Jaffuel, J. P. Charrier and G. Choquet-Kastylevsky, *Expert Review of Molecular Diagnostics*, 2012, **12**, 333–342.
5. E. S. Boja and H. Rodriguez, *Proteomics*, 2012, **12**, 1093–1110.
6. M. Y. Brusniak, C. S. Chu, U. Kusebauch, M. J. Sartain, J. D. Watts and R. L. Moritz, *Proteomics*, 2012, **12**, 1176–1184.
7. W. Urfer, M. Grzegorczyk and K. Jung, *Proteomics*, 2006, **6**(Suppl 2), 48–55.
8. J. R. Freije, T. Klein, J. A. Ooms, J. P. Franke and R. Bischoff, *J Proteome Res*, 2006, **5**, 1186–1194.
9. R. Freije, T. Klein, B. Ooms, H. F. Kauffman and R. Bischoff, *J Chromatogr A*, 2008, **1189**, 417–425.
10. D. May, M. Fitzgibbon, Y. Liu, T. Holzman, J. Eng, C. J. Kemp, J. Whiteaker, A. Paulovich and M. McIntosh, *J Proteome Res*, 2007, **6**, 2685–2694.
11. A. D. Norbeck, M. E. Monroe, J. N. Adkins, K. K. Anderson, D. S. Daly and R. D. Smith, *Journal of the American Society for Mass Spectrometry*, 2005, **16**, 1239–1249.
12. J. C. Silva, M. V. Gorenstein, G. Z. Li, J. P. Vissers and S. J. Geromanos, *Mol Cell Proteomics*, 2006, **5**, 144–156.
13. K. M. Aberg, E. Alm and R. J. Torgrip, *Anal Bioanal Chem*, 2009, **394**, 151–162.
14. M. Vandenbogaert, S. Li-Thiao-Te, H. M. Kaltenbach, R. Zhang, T. Aittokallio and B. Schwikowski, *Proteomics*, 2008, **8**, 650–672.
15. K. Marc, S. Benjamin, S. Hanno, A. J. S. Judith and A. H. Fred, *Journal of Statistical Software*, 2007, **18**, 12.
16. G. Tomasi, F. van den Berg and C. Andersson, *Journal of Chemometrics*, 2004, **18**, 231–241.
17. C. Christin, H. C. Hoefsloot, A. K. Smilde, F. Suits, R. Bischoff and P. L. Horvatovich, *J Proteome Res*, 2011, **9**, 1483–1495.
18. C. Christin, A. K. Smilde, H. C. Hoefsloot, F. Suits, R. Bischoff and P. L. Horvatovich, *Anal Chem*, 2008, **80**, 7012–7021.
19. F. Suits, J. Lepre, P. Du, R. Bischoff and P. Horvatovich, *Anal Chem*, 2008, **80**, 3095–3104.
20. E. Lange, C. Gropl, O. Schulz-Trieglaff, A. Leinenbach, C. Huber and K. Reinert, *Bioinformatics*, 2007, **23**, i273–281.
21. K. Podwojski, A. Fritsch, D. C. Chamrad, W. Paul, B. Sitek, K. Stuhler, P. Mutzel, C. Stephan, H. E. Meyer, W. Urfer, K. Ickstadt and J. Rahnenfuhrer, *Bioinformatics*, 2009, **25**, 758–764.
22. N.-P. V. Nielsen, J. M. Carstensen and J. Smedsgaard, *Journal of Chromatography A*, 1998, **805**, 17–35.
23. A. M. van Nederkassel, M. Daszykowski, P. H. Eilers and Y. V. Heyden, *J. Chromatogr A*, 2006, **1118**, 199–210.
24. A. Kassidas, J. F. MacGregor and P. A. Taylor, *AIChE Journal*, 1998, **44**, 864–875.

25. A. Prakash, P. Mallick, J. Whiteaker, H. Zhang, A. Paulovich, M. Flory, H. Lee, R. Aebersold and B. Schwikowski, *Mol Cell Proteomics*, 2006, **5**, 423–432.

26. P. H. Eilers, *Anal Chem*, 2004, **76**, 404–411.

27. J. Listgarten, R. M. Neal, S. T. Roweis, P. Wong and A. Emili, *Bioinformatics*, 2007, **23**, e198–204.

28. E. Lange, R. Tautenhahn, S. Neumann and C. Gropl, *BMC Bioinformatics*, 2008, **9**, 375.

29. I. Ahmad, F. Suits, B. Hoekman, M. A. Swertz, H. Byelas, M. Dijkstra, R. Hooft, D. Katsubo, B. van Breukelen, R. Bischoff and P. Horvatovich, *Bioinformatics*, 2011, **27**, 1176–1178.

30. R. A. Scheltema, A. Kamleh, D. Wildridge, C. Ebikeme, D. G. Watson, M. P. Barrett, R. C. Jansen and R. Breitling, *Proteomics*, 2008, **8**, 4647–4656.

31. A. H. America and J. H. Cordewener, *Proteomics*, 2008, **8**, 731–749.

32. F. Suits, B. Hoekman, T. Rosenling, R. Bischoff and P. Horvatovich, *Anal Chem*, 2011, **83**, 7786–7794.

33. T. Huang and Z. He, *Bioinformatics*, 2012, **28**, 2956–2962.

34. T. Huang, J. Wang, W. Yu and Z. He, *Brief Bioinform*, 2012, **13**, 586–614.

35. A. I. Nesvizhskii and R. Aebersold, *Mol Cell Proteomics*, 2005, **4**, 1419–1440.

36. S. O. Curreem, R. M. Watt, S. K. Lau and P. C. Woo, *Protein Cell*, 2012, **3**, 346–363.

37. A. Gorg, W. Weiss and M. J. Dunn, *Proteomics*, 2004, **4**, 3665–3685.

38. M. R. Hoopmann and R. L. Moritz, *Curr Opin Biotechnol*, 2013, **24**, 31–38.

39. H. Lam, *Mol Cell Proteomics*, 2011, **10**, R111 008565.

40. J. K. Eng, B. C. Searle, K. R. Clauser and D. L. Tabb, *Mol Cell Proteomics*, 2011, **10**, R111 009522.

41. H. Johnson and C. E. Eyers, *Methods Mol Biol*, 2010, **658**, 93–108.

42. C. Hughes, B. Ma and G. A. Lajoie, *Methods Mol Biol*, 2010, **604**, 105–121.

43. D. H. Lundgren, S. I. Hwang, L. Wu and D. K. Han, *Expert Rev Proteomics*, 2010, **7**, 39–53.

44. P. Khatri, M. Sirota and A. J. Butte, *PLoS Comput Biol*, 2012, **8**, e1002375.

45. W. W. Goh, Y. H. Lee, M. Chung and L. Wong, *Proteomics*, 2012, **12**, 550–563.

46. G. L. Zhang, D. S. DeLuca and V. Brusic, *Methods Mol Biol*, 2011, **723**, 349–364.

47. T. Rosenling, M. P. Stoop, A. Attali, H. van Aken, E. Suidgeest, C. Christin, C. Stingl, F. Suits, P. Horvatovich, R. Q. Hintzen, T. Tuinstra, R. Bischoff and T. M. Luider, *J Proteome Res*, 2012, **11**, 2048–2060.

48. T. Rosenling, M. P. Stoop, A. Smolinska, B. Muilwijk, L. Coulier, S. Shi, A. Dane, C. Christin, F. Suits, P. L. Horvatovich, S. S. Wijmenga, L. M. Buydens, R. Vreeken, T. Hankemeier, A. J. van Gool, T. M. Luider and R. Bischoff, *Clin Chem*, 2011, **57**, 1703–1711.

49. T. Rosenling, C. L. Slim, C. Christin, L. Coulier, S. Shi, M. P. Stoop, J. Bosman, F. Suits, P. L. Horvatovich, N. Stockhofe-Zurwieden, R. Vreeken, T. Hankemeier, A. J. van Gool, T. M. Luider and R. Bischoff, *J Proteome Res*, 2009, **8**, 5511–5522.

50. O. Kohlbacher, K. Reinert, C. Gropl, E. Lange, N. Pfeifer, O. Schulz-Trieglaff and M. Sturm, *Bioinformatics*, 2007, **23**, e191–197.
51. M. Sturm, A. Bertsch, C. Gropl, A. Hildebrandt, R. Hussong, E. Lange, N. Pfeifer, O. Schulz-Trieglaff, A. Zerck, K. Reinert and O. Kohlbacher, *BMC Bioinformatics*, 2008, **9**, 163.
52. M. Katajamaa, J. Miettinen and M. Oresic, *Bioinformatics*, 2006, **22**, 634–636.
53. X. J. Li, E. C. Yi, C. J. Kemp, H. Zhang and R. Aebersold, *Mol Cell Proteomics*, 2005, **4**, 1328–1340.
54. B. Hoekman, R. Breitling, F. Suits, R. Bischoff and P. Horvatovich, *Mol Cell Proteomics*, 2012, **11**, M111 015974.
55. C. Christin, H. C. Hoefsloot, A. K. Smilde, B. Hoekman, F. Suits, R. Bischoff and P. Horvatovich, *Mol Cell Proteomics*, 2013, **12**, 263–276.
56. T. Adamusiak, H. Parkinson, J. Muilu, E. Roos, K. J. van der Velde, G. A. Thorisson, M. Byrne, C. Pang, S. Gollapudi, V. Ferretti, H. Hillege, A. J. Brookes and M. A. Swertz, *Hum Mutat*, 2012, **33**, 867–873.
57. D. Arends, K. J. van der Velde, P. Prins, K. W. Broman, S. Moller, R. C. Jansen and M. A. Swertz, *Bioinformatics*, 2012, **28**, 1042–1044.
58. M. A. Swertz, M. Dijkstra, T. Adamusiak, J. K. van der Velde, A. Kanterakis, E. T. Roos, J. Lops, G. A. Thorisson, D. Arends, G. Byelas, J. Muilu, A. J. Brookes, E. O. de Brock, R. C. Jansen and H. Parkinson, *BMC Bioinformatics*, 2010, **11**(Suppl 12), S12.
59. J. Hillman-Jackson, D. Clements, D. Blankenberg, J. Taylor and A. Nekrutenko, *Curr Protoc Bioinformatics*, 2012 **Chapter 10**, Unit 10 15.
60. J. Goecks, A. Nekrutenko and J. Taylor, *Genome Biol*, 2010, **11**, R86.
61. T. Oinn, M. Addis, J. Ferris, D. Marvin, M. Senger, M. Greenwood, T. Carver, K. Glover, M. R. Pocock, A. Wipat and P. Li, *Bioinformatics*, 2004, **20**, 3045–3054.
62. Y. Mohammed, E. Mostovenko, A. A. Henneman, R. J. Marissen, A. M. Deelder and M. Palmblad, *J Proteome Res*, 2012, **11**, 5101–5108.
63. J. S. de Bruin, A. M. Deelder and M. Palmblad, *Mol Cell Proteomics*, 2012, **11**, M111 010595.
64. A. Bauch, I. Adamczyk, P. Buczek, F. J. Elmer, K. Enimanev, P. Glyzewski, M. Kohler, T. Pylak, A. Quandt, C. Ramakrishnan, C. Beisel, L. Malmstrom, R. Aebersold and B. Rinn, *BMC Bioinformatics*, 2011, **12**, 468.
65. P. Kacsuk, *Concurrency and Computation: Practice and Experience*, 2011, **23**, 235–245.
66. P. Kacsuk, Z. Farkas, M. Kozlovszky, G. Hermann, A. Balasko, K. Karoczkai and I. Marton, *J Grid Computing*, 2012, **10**, 601–630.
67. M. Kozlovszky, K. Karoczkai, I. Marton, A. Balasko, A. Marosi and P. Kacsuk, *Computer Science*, 2012, **13**, 3.
68. A. Bertsch, C. Gropl, K. Reinert and O. Kohlbacher, *Methods Mol Biol*, 2011, **696**, 353–367.
69. S. Nahnsen, A. Bertsch, J. Rahnenfuhrer, A. Nordheim and O. Kohlbacher, *J Proteome Res*, 2011, **10**, 3332–3343.
70. J. Junker, C. Bielow, A. Bertsch, M. Sturm, K. Reinert and O. Kohlbacher, *J Proteome Res*, 2012, **11**, 3914–3920.

CHAPTER 8

Bioinformatics and Statistics: Statistical Analysis and Validation

HUUB C. J. HOEFSLOOT

Swammerdam Institute for Life Science, University of Amsterdam
Email: h.c.j.hoefsloot@uva.nl

8.1 Introduction

The aim of this chapter is to introduce some statistical techniques and validation strategies to extract information on possible biomarkers from a given data set. This chapter deals with the situation where a few biomarkers are to be selected from a large number of measured compounds. Because these measurements can be quite elaborate the number of samples is usually low. In this chapter the term biomarker will be used in a loose way. The result from a statistical analysis cannot be a biomarker, it should be tested in follow-up experiments. Despite this, the result of the statistical analysis will be called a biomarker. For the sake of simplicity only case versus control studies are discussed. So, one group consist of controls and the other group have an abnormality, if human are concerned most commonly a disease. The data is assumed to come from experiments with an adequate measurement – and a study design.

If the measurement design is inadequate statistical methods are not able to repair this. A classic example in a case control study is that all the controls are

RSC Drug Discovery Series No. 33
Comprehensive Biomarker Discovery and Validation for Clinical Application
Edited by Péter Horvatovich and Rainer Bischoff
© The Royal Society of Chemistry 2013
Published by the Royal Society of Chemistry, www.rsc.org

measured on one day and the cases are measured on another day. The two effects, case/control and measurement day, are then confounded and if a difference between case and control is found it is impossible to tell whether this is caused by the disease or by the measurement occasion.

In the case of an improper study design the differences found between two sample groups cannot be attributed to the disease under consideration or to the flawed design. To give an example, if in a breast cancer study the average age of the controls is 22 and that of the cases is 55, the differences found between the groups can be either due to age or to the cancer. Statistics cannot discriminate between the two possibilities.

The study design is crucial for the interpretation of the putative biomarkers. The markers are indicative of the differences between the groups. So, if the controls are healthy and the cases have lung cancer the markers found indicate a difference between healthy people and people with lung cancer. Whether these differences are specific for lung cancer or could also be found in many other diseases, statistics cannot decide. If the research question involves a specific marker for lung cancer, a control group of healthy people might not be the best option.

Statistical analysis is only one step in the pipeline to discover biomarkers. The chance of success depend on all previous steps, like sample handling, measurement quality and data preprocessing. If only one of these steps is not performed carefully then the statistical method will work according to the principle of garbage in means garbage out.

There are many methods to extract biomarkers. Many areas of research are contribution to this, *e.g.* machine learning, chemometrics and statistics. It is therefore impossible to discuss all the available methods. For clarity, the methods are split into three classes: (i) looking for a specific pattern, (ii) univariate, looking at each m/z value one at a time (iii) multivariate methods, looking at all variable simultaneously. Clear guide lines are given as to what class of methods should be used. Within each class, in particular the multivariate class, there are still many methods.[1] What method should be chosen heavily depends on the data at hand and is still a matter of research. Some suggestions on how to choose an appropriate method are made in this chapter.

The statistical validation of a biomarker is one of the key issues in this chapter. The reason that validation is a difficult issue, are the characteristics of a proteomics data set. The number of m/z values measured is usually large, in the thousands. The number of samples, *e.g.* the patient plus the controls is in most cases much smaller. This situation is called the curse of dimensionality.[2] If data suffers from this curse it is possible to find discriminative patterns between two groups that are in fact identical in a statistical sense, which means that the two groups come from the same population. Strategies to overcome this involve techniques like crossvalidation, permutation testing and the use of random data. These three methods are about the issue of finding a biomarker by chance and methods to control the extent of these findings. Another important aspect of the data analysis is how likely it is to find a true biomarker. Obviously this depends on the difference in this marker between the two groups. For univariate methods the analysis of how well a method picks up a difference is called power analysis. For multivariate methods there are no established methods that can perform a power analysis. In this chapter some *ad hoc* procedures are recommended to gain an insight into the ability of the used method to detect true biomarkers.

This chapter is organized in the following way. After this introduction some terminology will be introduced. This is followed by a section on validation strategies. Then, a section on different types of biomarkers and biomarker panels is presented. The next section describes statistical methods including statistical validation to find biomarkers. The final part consists of conclusions and recommendations.

8.2 Terminology

The methods for biomarker discovery come from different fields, *e.g.* statistics, machine learning, bioinformatics and chemometrics. These fields can have different terminology.

The data to be analyzed is in the form of a matrix, see Figure 8.1. One row of this matrix contain the measurement of one person and is called a sample or object. In the case of animal experiments one row generally contains the measurements values of one animal.

The number of samples is n. A column of the data matrix contains the values of the same entity, *e.g.* a protein or a peptide concentration, that are measured for all samples. A row in the matrix is called a variable, the number of variables is p. In proteomics a variable stands for a protein or a peptide, the number of variables is usually much larger than the number of samples, which means that p is much larger than n. Sometimes, a variable is also called a feature. Statistical methods that find variables/features that have good discriminative power between cases and controls are called feature extraction or feature selection.

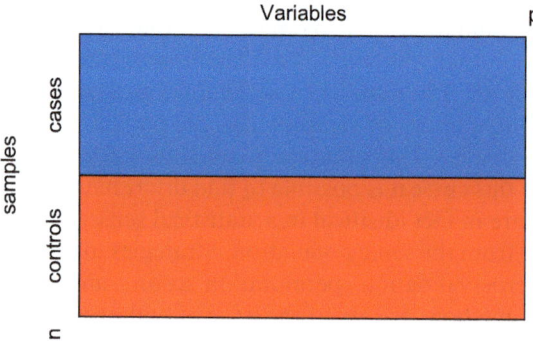

Figure 8.1 The data.

8.3 Validation Strategies

8.3.1 Crossvalidation

The statistical validation of a biomarker is one of the key issues in this chapter. Good markers should be useful in prediction. This leads to the idea that the found marker should be judged on how well it predicts new samples as being a member of the case or control sample group. To make this judgment new data is needed. Using the same data to find biomarkers and judge the predictive power leads to an overestimation of the predictive properties, or in statistical terms a bias. This has been shown many times in literature.[1,3] The new data preferably comes from a source other than the data used to find the biomarkers. If this is not possible, splitting the data into two data sets is an option.

One data set is used to find biomarkers and the second one is used to evaluate the performance. Taking this idea one step further leads to the concept of crossvalidation, see Figure 8.2. In crossvalidation a data set is split into a part used for model building and a part to evaluate prediction several times. This procedure is continued until every sample in the data has been in the test set ones. The number of samples used for training set and test set can be varied. As a rule of thumb the test data should be about 10% of the training data.[2] This leads to a so-called 10-fold crossvalidation. Often, the ratio of the class sizes is preserved in the training and test sets, making them accurate representations of the original data. This is called stratified crossvalidation.

There are many ways to split the data into different parts in k-fold crossvalidation. The performance may depend on the choice of split.[4] It is recommended to repeat the crossvalidation several times with different splits of the data. Kohavi and John[5] let the number of repeats depend on the standard deviation of the performance estimate. They repeated until the standard deviation becomes sufficiently small.

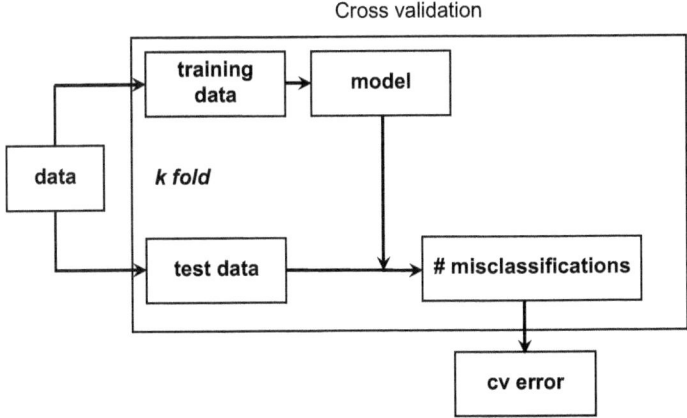

Figure 8.2 The crossvalidation scheme. In the loop a k-fold crossvalidation is used to determine the classification error rates.

8.3.1.1   *Permutation Testing*

One type of permutation test is repeatedly removing and randomly reassigning class labels to samples to create an uninformative data set of the same size as the data under study. An application of permutation tests is determining the relevance of a model. The model should perform much better on the original data than on the permuted data.

The rationale behind the use of the permutation test in this manner is that with uninformative data that is divided into two groups, a classifier on average assigns 50% to the wrong class. Thus, a model that returns an error rate that is on average deviating from the expected 50% error rate is biased. Permutation tests thus answer two questions: whether the information in the data is truly relevant and whether the model is unbiased.

From permutation tests a *p*-value for the relevance of the model can be calculated. This *p*-value is the ratio of the number of permutations that give an equal or better result than the original data and the total number of permutations. The number of permutations determines accuracy and the lower bound of the *p*-value. Using 100 permutations the lowest possible *p*-value is 0.01. Since the variance of the performance in permutations can be very large, a large number of permutations are needed to obtain a reliable result. In practice, this means that at least 10 000 permutations should be performed.

8.3.2   Model Assessment Using Random Data

A test for the performance of the method is to feed it with randomly generated data of the same size as the experimental data. The simplest option is to draw all numbers from a normal distribution but also drawing from other distributions, *e.g.* log-normal or uniform, is possible. A classification method should on average give 50% misclassification. If this is not the case, the implementation of the method is wrong. Because there are no biomarkers in random data all the findings of possible biomarkers are false-positives. This gives an indication of the amount of false-positives to expect when the real data is used.

8.3.3   Multiple Testing Corrections

Significance tests, such as the *t*-test, compare differences in means between the case and the control groups. The *t*-test assumes normality of the data. The Wilcoxon–Mann–Whitney test assesses differences between two groups without making this assumption.

These significance tests are designed to deal with univariate data, and a variable is considered to differ significantly when its test statistic is smaller than some value for α (generally, $\alpha = 0.05$ or $\alpha = 0.01$). Since proteomics data involves testing many individual variables corresponding to peptides or proteins simultaneously, applying the same value for α leads to many false-positives.[6] The Bonferroni correction sets an α-value for the entire set, so that the test statistic for each individual variable is compared to a value of α/(number of variables) and the false-positive rate or family wise error rate (FWER) is controlled.

A less-conservative correction for multiple testing is controlling the false discovery rate (FDR): the number of false-positives among all positives.[7,8] An observation that can be made is that multiple testing correction decreases the number of false-positives but increases the number of false-negatives. This means that a possible biomarker could be missed due to the correction.

8.3.4 Data Augmentation

The methods presented above are concerned with false-positives, which are variables that are marked as possible biomarkers but in fact they are not. It might also be useful to know how good is the method to find a real biomarker. From the measured data this is not possible because the markers are not known. But a new variable can be introduced, either of type I (see Figure 8.1) or type II, in the data as an extra column. Now the question can be asked does the method find this marker. A similar approach can be used to augment the data with a biomarker panel. In that case obviously more than one column has to be augmented to the data.

8.4 Types of Biomarkers and Biomarker Panels

In this chapter three different types of biomarkers will be distinguished, see Figure 8.3. The first one are the biomarkers that are present in one group and are not present in the other. With not present it is not meant that the concentration is below the detection limit but that it is really not present. If the biomarker is present in the case group and it is not present in the controls, it is easy to develop a clinical test; because the identified biomarker indicates that patient has the disease. This is a favorable situation as the prevalence of the disease is relatively low in the population compared to the healthy controls. In the data, this situation is characterized with zeros for the controls and positive values for the cases. Also, the opposite situation, zeros for the cases and values for the controls yields a biomarker of interest. We call this biomarker a type I marker.

The second type is where variables have different levels in the two groups. Thus, the controls have a higher value for the marker than the cases or *vice versa*. In the reminder of the document this will be called a type II marker. In the case that for one of the groups the biomarker is below the detection limits it might appear to be a type I biomarker, but if it can be detected by a more sensitive method it is actually a type II biomarker.

The third situation is where a group of markers is needed to discriminate between cases and controls. In that case the set of markers is called a biomarker panel.

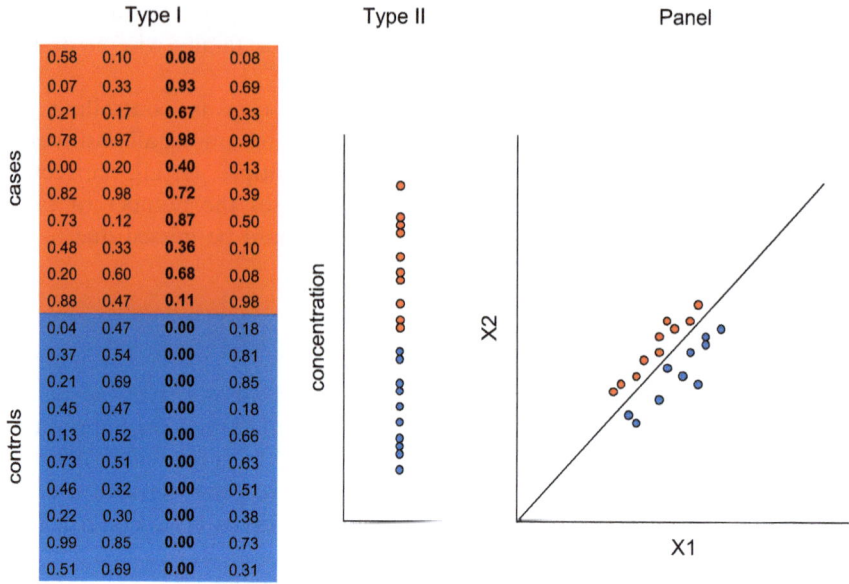

Figure 8.3   Different types of biomarkers. For the type I marker it holds that it is absent in one of the two groups. In the case of a type II marker the levels of the marker differ between groups. A biomarker panel is necessary if more than one variable is needed to reach adequate discrimination. In the case of a panel biomarker both variables X1 and X2 are needed for proper discrimination.

8.5   Discovery and Statistical Validation of The Biomarkers

8.5.1   How to Find a Type I Biomarker

This is a relative simple task. Look for every variable where one group has only zeros and the other group has only values not equal to zero. In any programming language this is only a few lines of code. To obtain confidence that the finding is really a type I marker the raw data should be inspected. This inspection should make sure that the variable has really zero values in one group. It is common in the preprocessing to set a threshold, most often considering the noise level. It should be made plausible that the zeros are due to the absence of the marker and not due to a threshold or any other analytical procedure. If, from visual inspection of the raw data, one is uncertain about the zeros, the biomarker probably is of type II, that means values below the set detection limit and values above this limit. To be more certain about a type I biomarker, a more sensitive analytical method should be used to confirm the absence of the biomarker in one of the two groups.

8.5.2   Validation for Type I Biomarker

If a type I biomarker is found the question of how valid it is or is it possible that it is just a chance result arises. The validation strategy for a type I biomarker is permutation tests. The two classes are randomly permuted so each sample has an equal chance of being in two classes. Then the search procedure for a type I biomarker is performed. Because the class assignment is random it is expected that there is on average no difference between the classes. So all biomarkers found are there just because of chance. If the permutation is performed many times a *p*-value for the number of type I biomarker in the original sample classes can be calculated. The *p*-value is just the number of times that a biomarker is found in the permuted data divided by the total number of permutations. To give an example, if the number of permutations is 100 000 and the number of times a type I biomarker is found is 23 then the *p*-value is 0.00023, which is a statistically significant result considering an usual threshold of 0.05. If in 100 000 permutations 20 000 times a biomarker is found the *p*-value is 0.2. This means that the marker that is found in the original data has a chance of 0.2 of being a false-positive.

8.5.3 How to Find a Type II Biomarker

The most used method to find differences between groups is the *t*-test. However, it should only be used if the measurement values are normally distributed. If this is not the case a Wilcoxon–Mann–Whitney test can be used. For both tests a multiple testing correction should be applied.

8.5.4 How to Find a Biomarker Panel

To find a biomarker panel usually classification methods are used. In the context of a two-class problem a classification method assigns an object in one of the two classes. These methods have to be trained on a set of data where the class is known. Then it should be able to predict the class for a sample where the class in unknown. In classification methods not all variables play the same role, some are important, while others are unimportant. Only keeping important variables can lead to a set of variables with which it is possible to obtain a good classification. This set of variables is the biomarker panel that we are looking for.

Multiple classification methods are available to determine the biomarker panel providing the best discrimination between class of samples. The classical approaches like linear discrimination analyses (LDA) and quadratic discriminant analysis (QDA) only work when the number of samples are exceeding the number of variables (compounds) such as $p < n$. So these methods cannot be used in proteomics data because this type of data has a much smaller number of samples than the number of variables. But there are many methods that can work with omics type of data where the number of variables exceeds by a large extent the number of samples. Some of these methods will be discussed below and references for these methods will also be given.

8.5.4.1   Embedded Methods

First, some methods are discussed that classify and do variable selection composing the biomarker panel simultaneously. These methods are called embedded methods.[3] Some of them are regression methods where the number of parameters is penalized. These methods are: Penalized Logistic Regression (PLR), LASSO, Elastic Nets.[9–11] The larger the penalty chosen in these methods the smaller the biomarker panel will be. Another embedded method is nearest shrunken centroids (NSC).[9,12,13,30] In the nearest centroid classification, a sample is assigned to the class with the nearest class mean. The class centroids are shrunken towards the overall centroid, thereby selecting variables. Embedded methods give directly a set of variables that is used in the classification. This set of variables can be used as a biomarker panel to classify new set of samples.

8.5.4.2   Wrapper Methods

In this section some methods are briefly mentioned that perform the classification on all available variables. This enumeration of methods is far from complete. A selection was made including methods that are actually uses in classification problems for proteomics data. Principal Component Discriminant Analysis is a method that combines Principal Component Analysis and Linear Discriminant Analysis (PCDA). It is described in detail in many papers and is used regularly in the classification of proteomics data.[14–18,30] Partial least squares discriminant analysis (PLSDA) is a method that is popular in the metabolomics field,[15,19] but can also be used for proteomics data.[30] PLSDA is a multivariate regression method in which classes are encoded numerically, e.g. all controls get a 1 and all cases a –1. Both methods, PCDA and PLSDA, come from the field of chemometrics. The next method stems from the area of machine learning. Support Vector Machines (SVM) tries to find a plane in high-dimensional space that separates the sample groups. In a linear SVM this is performed in the original variable space but in nonlinear variants variable transformations are performed. SVM are often used successfully in the proteomics fields, e.g.[9,10,17,18] In random forest classifiers many weak classifiers are combined to get a better performing classifier.[10] A k-nearest-neighbor classifier is a classification algorithm that is related to k-nearest-neighbor clustering. Roy et al.[20] provide a good example for applications of k-nearest-neighbor classifier in proteomics. Decision trees are tree-like structures that by rules at the junctions give an answer to a classification problem, and Wegdam et al.[18] give an example of the use of this algorithm for classification in proteomics.

All of the above methods only classify a new set of samples, but they do not include variable selection at all. Therefore, for biomarker discovery these methods have to be combined with variable selection approaches (see Figure 8.4.). The methods that have to be combined with a variable selection approaches are known as wrapper methods.

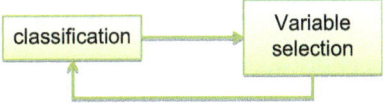

Figure 8.4   Structure of wrapper method.

How to select variables in the variable selection step of a wrapper? Wrapper classification method tests which variables are important in the classification. Variables that lead to good classification performance are selected, while variables that do not contribute to the classification are no longer considered. Forward selection starts with the variable that gives the best classification result. Given this first variable, another variable is added that gives the largest improvement of performance.[2] Variables are added until the performance no longer improves or a set criterion is met. Backward elimination works similarly, starting with all features and sequentially remove the less-discriminative features.[2] Genetic algorithms create many feature sets that are tested simultaneously, given a classification method. The best sets are recombined to make a new generation of improved feature sets. The algorithm is stopped when the performance does not improve over several generations or when a set performance measure is achieved.[21]

Recursive Feature Elimination (RFE) is used in combination with linear SVMs.[22,30] RFE starts with the set of all variables. The variables are ranked in decreasing order according to their weights in the dividing hyperplane obtained from a linear SVM. The lowest-ranked variables are eliminated. The process is continued until a predefined number of variables is reached.

Rank products[23] have been employed for selection of proteins using a PCDA classification rule.[14] Rank products were also used with a PLSDA classification in Van Velzen *et al.*[24] to obtain a set of relevant metabolites from a human nutritional intervention study that used NMR-based metabolomics to assess the metabolic impact of grape/wine extract consumption. So, the important metabolites that differed between the intake of a placebo and the extract were detected.

Bijlsma *et al.*[25] show the use of PLSDA to select potential metabolite biomarkers on the basis of regression coefficient. Another feature-extraction method for PLSDA is variable importance in the projection (VIP).[26] It has been used for selection of relevant metabolites in several studies: liver function in Hepatitis B,[27] intestinal fistulas[28] and metabolites in apple.[15] A good overview of variable selection methods is given in Hillario and Kalousis.[3]

8.5.4.3   Filter Methods

Filters form one other category of biomarker panel selection and are used together with classification algorithms.[3] In filter methods, variables are selected by a statistical test, *e.g.* a *t*-test or a Wilcoxon–Mann–Whitney test. Then, the classification method makes the rule of how to use the selected variables in the classification. So basically, the statistical test decides what variables are in the biomarker panel.

8.6 Selection of the Right Classification Method

A large number of papers discuss and compare classification algorithms on a single data set or on multiple data sets. From these papers no winner seems to emerge. As is stated by Hand[10] it is unsafe to make assertions about the superiority of any classification algorithm. This paper mentions in addition, that it is not only the algorithm that is compared but also how these algorithms are implemented by the user. If the user is not skillful with a method the performance of this method might be negatively influenced by the user. To avoid this issue it would be desirable to have publicly available software made by experts and publically available data to test algorithms. This situation has not been reached yet but progress towards this situation is being made.[30]

Therefore, the chosen method should be a method that the user or his/her collaborator should be skilful in and be an experienced user of the algorithm. It could be misleading to use a new method as a black box. Using algorithms that the user is not familiar with may provide a numerical outcome that is suboptimal without the user realizing the cause and the inexperienced user will certainly not be able to solve the problem and provide optimal results. For example, in almost all the classification methods that are discussed here, some parameters have to be set. This is by no means a simple task. In order to obtain the optimal parameters for your algorithm given your data is a problem in itself. Sometimes methods like crossvalidation are used to obtain the proper parameters for a method.[14,19,30] The situation where crossvalidation is used to obtain an estimate of the number of misclassifications and it is also used to estimate optimal parameters in the method is called double cross-validation.[1,9,14,16,18,19,30] Double crossvalidation is depicted in Figure 8.5.

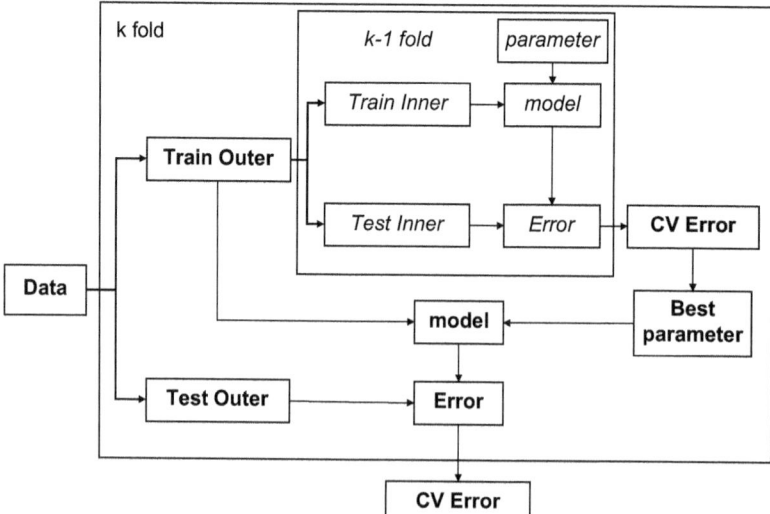

Figure 8.5   Double crossvalidation where crossvalidation is used to estimate the error and to obtain optimal parameters.

The default values that are proposed by the software usually do not give optimal results for the specific problem at hand. The best way, however, to exploit the full benefit of a new complex statistical method is to establish collaboration with the chemometrics and statistics laboratories who have deep understanding of the field. This usually creates a win-win situation. The experimentalist gets access to advanced algorithms and the data analyst gets access to new data.

8.7 How to Put the Pieces Together?

The statistical validation strategy that should always be applied is the use of a test set, independent from the training set, to see how well the combination of classifiers and biomarker panel work in reality. Using the same data to extract the biomarkers and to judge their quality gives overoptimistic results.

If the data set has a large sample size, which means that the number of samples is of the order of hundred it is always suggested to use an independent test set. As a rule of thumb about one third of the data should be used as a test set, while two thirds should be used to identify the biomarkers. After the biomarkers are found the test set can be used to judge the performance of the classifier using the identified biomarker panel.

For smaller data sets crossvalidation (cv) seems to be the method of choice. But in cv there is an extra complication. Because the data is different at every cv loop the results are also different for every loop of the cv, see Figure 8.6. This means that in a 10-fold cv 10 different models are generated. It also

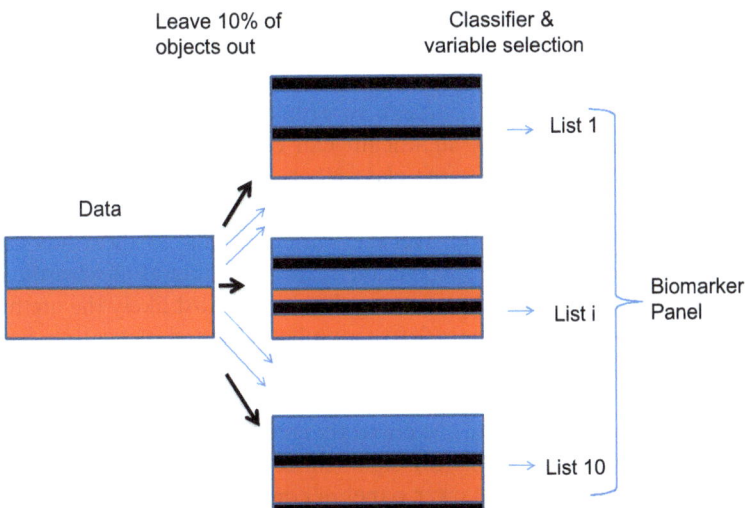

Figure 8.6 Combining lists of markers resulting from different crossvalidating loops.

means that 10 lists of potential biomarkers coming from the 10 different models are calculated. Because only one prediction and one biomarker panel is desired, the multiple models as well as the potential biomarker lists have to be combined.

For the classifier itself this problem can be dealt with the concept of aggregation as presented by Xu *et al.*[17] This paper describes a method where all the classifiers are used to build one overall classifier. The question remains how to come up with a single list representing a consensus selection of a biomarker panel? One possible solution could be to determine the consistency of the different biomarker lists. To give an example, in order to get into the biomarker panel a variable should be in at least a certain percentage of the lists as described by Wehrens *et al.*[15] Also, approaches like rank products or rank sums provide an encouraging solution to combine biomarker candidate lists.[30] Combining lists can be problematic because the existence of multiple sets of true biomarkers is possible.[29] If two variables are highly correlated one of them can be selected in the variable selector. Incorporating the correlated variable into the panel does not increase the classification rate and is therefore not selected in a forward selection algorithm. Because of the variability in the cv data sometimes one of the highly correlated variables will be selected while in the other instances of cv loops one other could be selected. This effect gets worse if the number of highly correlated markers increases. For processing proteomics data taking into account the biological function in combining biological data might be a good idea, because highly correlated variables are likely to have similar biological function.

8.8 Assessing the Quality of the Biomarker Panel

If a separate test set is present it can be used to assess the quality of the biomarker panel. The test samples can be classified with the classification method used in the training phase that has been used to identify the biomarker panel. The results should be compared with the true class of the samples. In this way the number of misclassification can be obtained. Whether this number of misclassifications is statistically significant can be assessed with permutation tests[14] (Figure 8.7). For this test the class labels of the samples are randomly permuted.

For such a permutation test the average number of misclassifications should be about half of the samples, so the average relative number of misclassification is 0.5. If in a permutation test this does not hold then most likely the method is not correctly implemented. The relative number of misclassification for the correctly labeled data should be better than for the permuted data. The statistical significance can be calculated as the number of times the permuted data gives equal or better results than the correctly labeled data divided by the total number of permutations. Statistical significance is not the same as clinical relevance. For example, a biomarker panel in conjunction with a classifier can give a performance of 85% correct, with a corresponding $p < 0.01$. But for a clinical application like screening in a large population the selectivity of the

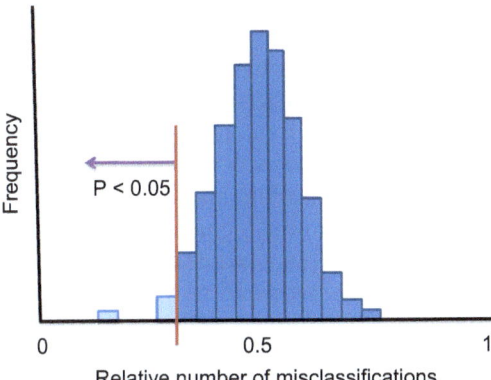

Relative number of misclassifications

Figure 8.7 Assessing the statistical significance to assign the right class labels for an independent test set by permutation testing. The blue bars give the frequency of the relative number of misclassification in a permutation test. The light blue bars are 5% of the total. Therefore, any result on the left of the red line has a significance of $p < 0.05$.

method could be very high in order to avoid large numbers of false-positives. To give an example of a disease like ovarian cancer with an incidence of about 40 per 100 000 per year, a sensitivity of 99.6% is necessary to fall within the acceptable limits for a screening.[31]

If the biomarker panel is obtained without a separate test set using a cross validation approach it is not possible to obtain results on how good it works on new samples. This fact shows the importance of having a separate test set. So, the number of samples should preferably be large enough to have a separate test set. But some assessments about the quality can be made even without disposal of an additional test set, because in every cv-loop an error rate of samples that were left out can be calculated. These error rates should be well above 50% otherwise the method is just as good as randomly classifying new set of samples. If the error rate is similar in all the cv loops and also the lists of selected biomarker from the cv loops are highly similar then it can be concluded that the performance of the derived biomarker panel is similar to the performance within the cv loops. Whether this within cv-loop performance is statistically significant can be assessed with permutation tests (see Figure 8.7). To assess if results could be obtained by chance, random data could be used. For example, a data matrix of the same size as the experimental data can be made by filling this matrix with random numbers from a given statistical distribution. In this simulated data set there is no difference between the sample groups, therefore no biomarkers signal is present. The total procedure to find a biomarker panel should be repeated many times. How often a biomarker panel does pop up when the total procedure is performed, is providing information on how likely it is that a biomarker panel is found by change alone.

8.9   What to Do if no Biomarker Panel is Found?

There are two reasons for not finding biomarkers in a data set. The first reason is that there is no marker present. The second reason is that the method used cannot find it. To see whether the second option is true the data could be augmented with a number of columns in which a biomarker (*e.g.* differential signal between sample groups in the data matrix) is present. If the method with augmented data found the introduced signal then returned, is most probably adequate to identify an unknown biomarker in the original data. If not, then the used statistical method does not perform well and another method should be tried.

Another option is to use data in which known biomarker signals are present[15,30] and assess how well these biomarkers are found by the used method.

Unfortunately, the first reason can also be true, there are no biomarkers in the data. Obviously statistics cannot change this situation and new experiments have to be performed in order to find biomarkers.

8.10   Conclusions and Recommendations

Discovery of biomarkers of type I is relatively straightforward. Using a permutation test the statistical significance of the finding can be calculated. A type II biomarker can be identified by using classical statistical methods, *e.g.* the *t*-test. However in this case, multiple testing correction is necessary in order to avoid large numbers of false-positives. Much work was performed on multiple testing in the field of transcriptomics. The generated knowledge in this field should be applied in proteomics as well.

Finding biomarker panels is a task for experienced users in the field of classification. This chapter provides a detailed summary of current state-of-the-art statistical methods used to identify biomarkers during biomarker discovery and concomitant statistical validation. The provided rich list of references may serve to acquire the knowledge and the experience required in the proper application of classification algorithms and the validation strategies such as crossvalidation and permutation testing as well the proper inter-pretation of the obtained data. Another option is the establishment of collaborations with statistics or chemometrics groups having experience in these fields. Some of the provided references provide links to freely accessible computer programs written for example in R[15] or MATLAB[24,30] to perform classification, biomarker selection and statistical validation. On the web there is also data available for algorithm testing such as: http://archive.ics.uci.edu/ml/ and in supplementary material.[15,30] Obviously this is not an exhaustive list of software and data. Development of statistical methods for biomarker selection and statistical validation strategies is currently a hot topic, and new algorithms and quality-control methods are entering the field. The list of available software and data is growing rapidly because in most newly published papers software and data is made publically available. This trend will help in the development of robust algorithms identifying true biomarkers in proteomics data.

Statistical analysis of a data set accompanied with proper validation tools can give a set of potential biomarkers. But it is never the final step. Findings should always be carefully evaluated using independently acquired data using the same analytical method used for biomarker discovery. The statistical power of selected biomarkers should be preferably validated using targeted quantification of selected biomarkers in large number of independent samples.

References

1. S. Smit, H. C. J. Hoefsloot and A. K. Smilde, Statistical data processing in clinical proteomics, *Journal of Chromatography B*, 2008, **866**(1–2), 77–88.
2. T. Hastie, J. Friedman, R. Tibshiranie, The Elements of Statistical Learning. *Data Mining, Inference and Prediction*, Springer, New York, 2001.
3. M. Hilario and A. Kalousis, Approaches to dimensionality reduction in proteomic biomarker studies, *Briefings in Bioinformatics*, **9**(2), 102–118.
4. L. F. A. Wessels, M. J. T. Reinders, A. A. M. Hart, C. J. Veenman, H. Dai, Y. D. He and L. J. van't Veer, *Bioinformatics*, 2005, **21**, 3755.
5. R. Kohavi and G. H. John, *Artif. Intell.*, 1997, **97**, 273.
6. D. I. Broadhurst and D. B. Kell, *Metabolomics*, 2006, **2**, 171.
7. Y. Benjamini and Y. Hochberg, *J. R. Statist. Soc. B*, 1995, **57**, 289.
8. J. D. Storey, *J. R. Statist. Soc. B*, 2002, **64**, 479.
9. M. M. W. B. Hendriks, S. Smit, W. L. M. W. Akkermans, T. H. Reijmers, P. H. C. Eilers, H. C. J. Hoefsloot, C. M. Rubingh, C. G. de Koster, J. M. Aerts and A. K. Smilde, How to distinguish healthy from diseased? Classification strategy for mass spectrometry-based clinical proteomics, *Proteomics*, 2007, **7**, 3672–3680.
10. D. J. Hand, Breast Cancer Diagnosis from Proteomic Mass Spectrometry Data: A Comparative Evaluation, *Statistical Applications in Genetics and Molecular Biology*, 2008, **7**(2), 15.
11. L. Waldron, M. Pintilie, M.-S. Tsao, A. Shepherd, C. Huttenhower and I. Jurisica, *Optimized Application of Penalized Regression Methods to Diverse Genomic Data*, 2011, **27**(24), 3399–3406.
12. R. Tibshirani, T. Hastie, B. Narasimhan and G. Chu, *Proc. Natl. Acad. Sci. U. S. A*, 2002, **99**, 6567.
13. U. T. Shankavaram, W. C. Reinhold, S. Nishizuka, S. Major, D. Morita, K. K. Chary, M. A. Reimers, U. Scherf, A. Kahn, D. Dolginow, J. Cossman, E. P. Kaldjian, D. A. Scudiero, E. Petricoin, L. Liotta, J. K. Lee and J. N. Weinstein, *Mol. Cancer Ther.*, 2007, **6**, 820.
14. S. Smit, M. J. Van Breemen, H. C. J. Hoefsloot, A. K. Smilde, J. M. F. G. Aerts and C. G. de Koster, Assessing the statistical validity of proteomics based biomarkers, *Analytica Chimica Acta*, 2007, **592**, 210–217.
15. R. Wehrens, P. Franceschi, U. Vrhovsek and F. Mattivi, Stability-based biomarker selection, *Analytica Chimica Acta*, 2011, **705**, 15–23.

16. H. C. J. Hoefsloot, S. Smit and A. K. Smilde, A classification model for the Leiden proteomics competition, *Statistical Applications in Genetics and Molecular Biology*, 2008, **7**(2), 8.
17. C.-J. Xu, H. C. J. Hoefsloot and A. K. Smilde, To aggregate or not to aggregate high dimensional Classifiers, *BMC Bioinformatics*, 2011, **12**, 153.
18. W. Wegdam, P. D. Moerland, M. R. Buist, E. V. L. van Themaat, B. Bleijlevens, H. C. J. Hoefsloot, C. G. de Koster and J. M. F. G. Aerts, Classification-based comparison of pre-processing methods for interpretation of mass spectrometry generated clinical datasets, *Proteome Science*, 2009, **7**, 19.
19. J. A. Westerhuis, H. C. J. Hoefsloot, S. Smit, D. J. Vis, A. K. Smilde, E. J. J. van Velzen, J. P. M. van Duijnhoven and F. A. van Dorsten, Validation of Plsda Models in Metabolomics, *Metabolomics*, 2008, **4**(1), 81–89.
20. P. Roy, C. Truntzer, D. Maucort-Boulch, T. Jouve and N. Molinari, Protein mass spectra data analysis for clinical biomarker discovery: a global review, *Briefings in Bioinformatics*, **12**(2), 176–186.
21. R. Wehrens and L. M. C. Buydens, TrAC, *Trends Anal. Chem.*, 1998, **17**, 193.
22. I. Guyon, J. Weston, S. Barnhill and V. Vapnik, *Machine Learning*, 2002, **46**, 389.
23. R. Breitling, P. Armengaud, A. Amtmann and P. Herzyk, *Febs Letters*, 2004, **573**.
24. E. J. J. van Velzen, J. A. Westerhuis, J. P. M. van Duynhoven, F. A. van Dorsten, H. C. J. Hoefsloot, D. M. Jacobs, S. Smit, R. Draijer, C. I. Kroner and A. K. Smilde, Multilevel Data Analysis of a Crossover Designed Human Nutritional Intervention Study, *Journal of Proteome Research*, 2008, **7**, 4483–4491.
25. S. Bijlsma, L. Bobeldijk, E. R. Verheij, R. Ramaker, S. Kochhar, I. A. Macdonald, B. van Ommen and A. K. Smilde, *Anal. Chem.*, 2006, **78**, 567.
26. I. G. Chong and C. H. Jun, Performance of some variable selection methods when multicollinearity is present, *Intell. Lab. Sys.*, 2005, **78**, 103–112.
27. J. Yang, X. J. Zhao, X. L. Liu, C. Wang, P. Gao, J. S. Wang, L. J. Li, J. R. Gu, S. L. Yang and G. W. Xu, *J. Proteome Res.*, 2006, **5**, 554.
28. P. Y. Yin, X. J. Zhao, Q. R. Li, J. S. Wang, J. S. Li and G. W. Xu, *J. Proteome Res.*, 2006, **5**, 2135.
29. Z. He and W. Yu, Stable feature selection for biomarker discovery, *Computational Biology and Chemistry*, 2010, **34**, 215–225.
30. C. Christin, H. C. J. Hoefsloot, A. K. Smilde, B. Hoekman, F. Suits, R. Bischoff and P. Horvatovich, A critical assessment of feature selection methods for biomarker discovery in clinical proteomics, *Mol. Cell Proteomic*, 2013, **12**(1), 263–76.
31. I. J. Jacobs and U. Menon, Progress and Challenges in Screening for Early Detection of Ovarian Cancer, *Molecular & Cellular Proteomics*, 2004, **3**, 355–366.

Bioinformatics and Statistics: Computational Discovery, Verification, and Validation of Functional Biomarkers

FAN ZHANG[a,b] AND RENEE DRABIER*[a]

[a] Department of Academic and Institutional Resources and Technology, University of North Texas Health Science Center, Fort Worth, USA;
[b] Department of Forensic and Investigative Genetics, University of North Texas Health Science Center, Fort Worth, USA
*Email: Renee.Drabier@unthsc.edu

9.1 What is a Biomarker?

A biomarker as defined by the National Cancer Institute is "a biological molecule found in blood, other body fluids, or tissues that is a sign of a normal or abnormal process, or of a condition or disease."[1] It is a characteristic that is objectively measured and evaluated as an indicator of normal biological processes, pathogenic processes, or pharmacologic responses to a therapeutic intervention.[2] The field of biomarkers has grown extensively over the past decade in many areas such as medicine, cell biology, genetics, geology, astrobiology, and ecotoxicology, *etc.* and biomarkers are currently being studied in many academic centers and in industry.

RSC Drug Discovery Series No. 33
Comprehensive Biomarker Discovery and Validation for Clinical Application
Edited by Péter Horvatovich and Rainer Bischoff
© The Royal Society of Chemistry 2013
Published by the Royal Society of Chemistry, www.rsc.org

9.2 What is Biomarker Research?

Biomarker research includes quantitative characterization of gene/protein mixtures in order to understand complex biological systems at the molecular level and determine relationships between the functions of genes/proteins across a large population of samples. Biomarker research improves the ability to predict, detect, and monitor diseases and promotes understanding of how individuals respond to drug intervention. It underlies personalized medicine: a tailored approach to patient treatment based on the molecular analysis of genes, proteins, and metabolites.

The goal of biomarker research is to discover and validate markers that can be used in clinical research applications such as patient stratification, diagnosis and therapeutic monitoring, or in pharmaceutical development to fully characterize the behavior of candidate drugs. The current challenge is not only to identify potential biomarkers, but also to validate them, ascertaining that they are, in fact, accurate and predictive for certain conditions. For example, in the past decade, more than 150 000 publications have documented thousands of newly discovered candidate biomarkers representing hundreds of millions of dollars in research investment, yet only about 100 biomarkers have demonstrated potential clinical utility. Additionally, although large-scale and high-throughput technologies such as proteomics and genomics have held special promise for the discovery of novel biomarkers that can be used clinically, these promises cannot be fulfilled without well-designed validation methods.

9.3 Biomarker Workflow Overview

In general, biomarker research process follows a continuum that begins with the study design and proceeds through validation to the eventual implementation of the biomarkers in a clinical assay development. We present a general biomarker workflow that comprises study design, discovery, verification, validation, and clinical application (Figure 9.1).

Study design includes determining biomarker discovery objectives, assessing the current clinical and molecular situation, establishing data management, exploration, and mining goals, and developing a project plan.

Biomarker discovery requires high-confidence identification of candidate biomarkers with simultaneous quantitative information to indicate which genes/proteins are changing to a statistically relevant degree in response to a given environmental change, drug treatment, or disease. One of the most important challenges in proteomics biomarker discovery is the difficulty of identifying medium- to low-abundance proteins in complex biological samples. A robust biomarker discovery workflow must be capable of uncovering a panel of biomarkers in samples such as human plasma, serum, erythrocytes and urine, with unambiguous identification and quantitative characterization.

Candidate biomarkers identified in the discovery stage need to be validated using larger sample sets covering a broad cross section of patients or populations because of normal clinical or biological variability. To avoid a potential

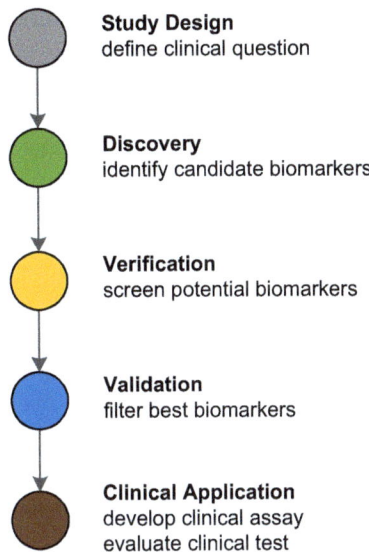

Figure 9.1   Biomarker development workflow overview.

bottleneck of taking a large number of candidates to validation, a biomarker candidate verification step can screen potential biomarkers to ensure that only the most promising putative biomarker candidates from the discovery phase are taken into the costly validation stage. The ability to narrow the number of candidates and take fewer, higher-quality leads into the validation stage will have significant cost- and time-saving benefits. A verification stage contains computational verification, analytical verification, and biological verification. An analytical verification is the process of establishing performance characteristics, such as limit of detection, sensitivity, specificity, and reproducibility, where a large number of candidates discovered using pooled set of samples are further filtered by quantification results of targeted MRM of these candidates on individual samples composing the discovery pooled set. A biological verification requires a high-throughput workflow that provides high specificity, sensitivity, requires minimal sample preparation, and can provide identification confirmation. A computational verification requires data modeling such as Neural Network, Support Vector Machine, Logic Regression and other methods; statistical analysis such as hypothesis testing, clustering analysis, variance analysis, survival analysis; and performance measuring such as Receiver Operating Characteristic (ROC), sensitivity, specificity and other characteristics. Additionally, the computational and analytical verification stage can result in measurement methods that can be used in validation.

Potential biomarkers identified in the verification stage need to be taken forward to the validation stage that can filter the best biomarkers for the clinical application. Validation of potential biomarkers requires targeted, multiplexed assays to screen and quantify proteins in patient plasma samples with high sensitivity, absolute specificity and sufficient throughput.

The validation stage also requires simultaneous measurement of large numbers of potential biomarkers in complex biological samples that precisely reflect the full variation of the targeted population. Multiple reaction monitoring (MRM) coupled with stable isotope dilution mass spectrometry (SID-MS) using a triple quadrupole mass spectrometer is a powerful method for quantitative measurement of target proteins. It provides superior sensitivity and selectivity for targeted compounds in complex samples. MRM also offers high precision and rapid MS cycle time, which makes it an ideal technology for validating biomarkers in a high-throughput fashion. It has been a principal tool for quantification of small molecules in clinical chemistry for a number of decades. MS-based quantitative assays have the necessary characteristics required for validation studies, namely: high specificity, sensitivity, large dynamic measurement range, multiplexing capability, and precision.

After a successful validation phase, the best biomarkers are selected for clinical assay development and clinical test evaluation in the clinical application stage.

9.4   Types of Biomarkers in Clinical Application

Biomarkers can be used clinically to screen for, diagnose or monitor the activity of diseases and to guide molecularly targeted therapy or assess therapeutic response. In the biopharmaceutical industry, biomarkers define molecular taxonomies of patients and diseases and serve as surrogate endpoints in early-phase drug trials. Biomarkers may have a variety of functions, which correspond to different stages (Table 9.1[3]) in disease development, such as in the progression in cancer or cardiovascular disease. Biomarkers can be used to detect and treat early-state (pre) cancers in the asymptomatic patients (screening biomarkers), to definitively establish the presence of cancer for those who are suspected to have the disease (diagnostic biomarkers), or to portend disease outcome at the time of diagnosis without reference to any specific therapy for those with overt disease (prognostic biomarkers) for whom therapy may or may not have been initiated. Biomarkers can also be used to predict the

Table 9.1   Clinical application of biomarkers: Rationale and objectives for use of biomarkers.

Type of biomarker	Objective for use
screening	to detect and treat early state (pre) cancers
diagnostic	to definitively establish the presence of cancer or other disease
prognostic	to portend disease outcome at time of diagnosis without reference to any specific therapy
predictive	to predict the outcome of a particular therapy
monitoring	to measure the response to treatment and early detect disease progression or relapse
risk profiling	to determine the risk profile
companion	to lead to companion diagnostic development
toxicity	to provide important, compound-specific information regarding toxic drug side effects

outcome of a particular therapy (predictive biomarkers) or to measure the response to treatment and early detect disease progression or relapse (monitoring biomarkers).[4] In addition, Biomarkers can be used to evaluate the disease risk profile[5] (risk profiling), to lead to companion diagnostic development (companion biomarker), and to provide important, compound- specific information regarding toxic drug side effects (toxicity biomarker).

9.5 Computational Methods for Biomarker Discovery, Verification, and Validation

9.5.1 Performance Measurements

The following measurements are generally involved in the biomarker discovery, verification, and validation. (1) Sensitivity (also called recall) is the proportion of actual positive pairs that are correctly identified; (2) Specificity measures the proportion of negative pairs that are correctly identified; (3) Precision is the probability of correct positive prediction; and (4) Accuracy is the proportion of correctly predicted pairs.

$$\text{Sensitivity} = \frac{TP}{TP + FN}$$

$$\text{Specificity} = \frac{TN}{TN + FP}$$

$$\text{Precision} = \frac{TP}{TP + FP}$$

$$\text{Accuracy} = \frac{TP + TN}{TP + TN + FP + FN},$$

where TP are true-positives, TN are false-negatives, TN are true-negatives, and FP are false-positives.

9.5.2 Receiver Operating Curve

A receiver operating characteristic (ROC), also receiver operating curve, is a graphical plot of the sensitivity (true-positive rate) versus 1-specificity (false-positive rate) for a binary classifier system as its discrimination threshold is varied. ROC curve displays the relationship between sensitivity (true-positive rate) and 1-specificity (false-positive rate) across all possible threshold values that define the positivity of a disease or condition. It shows the full picture of trade-off between the true-positive rate and the false-positive rate at different levels of positivity. A summary measure of ROC curve, such as the Area Under the Curve (AUC) is used to summarize the inherent capacity of a test or biomarker for discriminating a diseased from a nondiseased sample across all possible levels of positivity into a single statistic.

9.5.3 T-test

The *t*-test is a statistical hypothesis test in which the test statistic follows a Student's t distribution if the null hypothesis is supported, and assesses whether the means of two groups are statistically different from each other. Two-sample *t*-statistics and Welch's *t*-Test statistics are used to calculate the *p*-value of the null hypothesis that the means of two normally distributed populations are equal, for equal variance and unequal variance, respectively.

For equal variance of the two samples, two-sample *t*-test statistics is calculated as

$$t = \frac{\overline{X_1} - \overline{X_2}}{S_{X_1 X_2} \cdot \sqrt{\dfrac{1}{n_1} + \dfrac{1}{n_2}}} \tag{9.1}$$

where $S_{X_1 X_2} = \sqrt{\dfrac{(n_1 - 1)S_{X_1}^2 + (n_2 - 1)S_{X_2}^2}{n_1 + n_2 - 2}}$. $S_{X_1 X_2}$ is an estimator of the pooled standard deviation of the two samples. The degree of freedom for this test is $n_1 + n_2 - 2$.

For unequal variance, Welch's *t*-test statistics is calculated as

$$t = \frac{\overline{X_1} - \overline{X_2}}{s_{\overline{X_1} - \overline{X_2}}} \tag{9.2}$$

where

$$s_{\overline{X_1} - \overline{X_2}} = \sqrt{\frac{s_1^2}{n_1} + \frac{s_2^2}{n_2}}$$

For use in significance testing, the distribution of the test statistic is approximated as being a Student's *t* distribution with the degrees of freedom calculated using the following calculation:

$$df = \frac{\left(s_1^2/n_1 + s_2^2/n_2\right)^2}{\left(s_1^2/n_1\right)^2/(n_1 - 1) + \left(s_2^2/n_2\right)^2/(n_2 - 1)} \tag{9.3}$$

9.5.4 Fisher's Exact Test

Fisher's exact test is a statistical test used to determine if there are nonrandom associations between two categorical variables. The null hypothesis is that the relative proportions of one variable are independent of the second variable. Fisher's exact test is more accurate than the chi-squared test or goodness-of-fit test of independence when the expected numbers are small. The most common use of Fisher's exact test is for 2×2 tables. Given a 2×2 contingency table $\begin{bmatrix} a & b \\ c & d \end{bmatrix}$, the Fisher's exact test's p-value is the probability of obtaining any such set of values, which is given by the hypergeometric distribution:

$$p = \frac{\begin{pmatrix} a+b \\ a \end{pmatrix} \begin{pmatrix} c+d \\ c \end{pmatrix}}{\begin{pmatrix} a+b+c+d \\ a+c \end{pmatrix}} = \frac{(a+b)!\,(c+d)!\,(a+c)!\,(b+d)!}{a!\,b!\,c!\,d!\,(a+b+c+d)!}$$

where $\begin{pmatrix} n \\ k \end{pmatrix}$ is the binomial coefficient and the symbol ! indicates the factorial operator.

For example, in human genome background (43 895 genes total), 68 genes are involved in the p53 signaling pathway. A given gene list has found that 3 out of 300 belong to the p53 signalling pathway. Then we ask the question if 3/300 is a more than random chance comparing to the human background of 68/43 895. Here, the 2×2 contingency table is $\begin{bmatrix} 3 & 68 \\ 297 & 48\,827 \end{bmatrix}$ and the Fisher exact p-value $= 0.0126$, calculated by the above equation. Since the p-value > 0.01, we fail to reject the null hypothesis that this user gene list is not specifically associated in the p53 signaling pathway than random chance.

9.5.5 Neural Network

Neural Networks have several unique advantages and characteristics as research tools for cancer-prediction problems.[6–10] A very important feature of these networks is their adaptive nature, where "learning by examples" replaces conventional "programming by different cases" in solving problems.

A generalized "feedforward neural network" has three layers: input layer, hidden layer, and output layer and is trained using a backpropagation supervised training algorithm. The input is used as activation for the input layer and is propagated to the output layer. The received output is then compared to the desired output and an error value is calculated for each node in the output layer. The weights on edges going into the output layer are adjusted by a small amount relative to the error value. This error is propagated backwards through the network to correct edge weights at all levels. For example, Figure 9.2 described a feedforward neural network with an input layer of 5 nodes (corresponding to a five-marker panel), a hidden layer of 3 nodes, and an output layer of two-variable encoding scheme (healthy $= (0,1)$, cancer $= (1,0)$).

9.5.1.1 *Feedforward Neural Network Algorithm*

Phase 1: Propagation

Each propagation involves the following steps:

1. Forward propagation of a training pattern's input through the neural network in order to generate the propagation's output activations.
2. Backpropagation of the propagation's output activations through the neural network using the training pattern's classification target in order to generate the deltas of all output and hidden neurons.

Phase 2: Weight update

For each weight-synapse:

1. Multiply its output delta and input activation to get the gradient of the weight.
2. Bring the weight in the opposite direction of the gradient by subtracting a ratio of it from the weight.

This ratio influences the speed and quality of learning; it is called the *learning rate*. The sign of the gradient of a weight indicates where the error is increasing, this is why the weight must be updated in the opposite direction.

Repeat the phase 1 and 2 until the performance of the network is good enough, that is, a certain tolerance error value is reached.

Modes of Learning

There are basically two modes of learning to choose from, one is online learning and the other is batch learning. In online learning, each propagation is followed

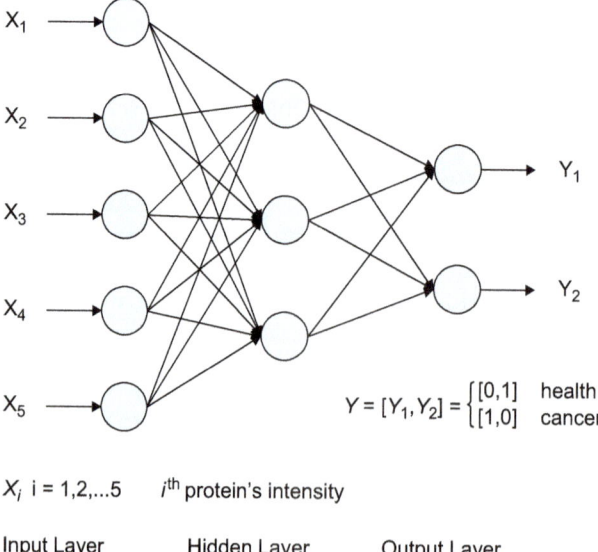

$$Y = [Y_1, Y_2] = \begin{cases} [0,1] & \text{health} \\ [1,0] & \text{cancer} \end{cases}$$

X_i i = 1,2,...5 i^{th} protein's intensity

Input Layer   Hidden Layer   Output Layer

Figure 9.2 Artificial Neural Network for five-biomarker panel.

immediately by a weight update. In batch learning, many propagations occur before weight updating occurs. Batch learning requires more memory capacity, but online learning requires more updates.

Algorithm
The following is an example of an actual algorithm for a 3-layer network (only one hidden layer):

      Initialize the weights in the network (often randomly)
      Do
                For each example e in the training set
                     O = neural-net-output(network, e) ; forward pass
                     T = teacher output for e
                     Calculate error (T − O) at the output units
                     Compute delta_wh for all weights from hidden layer to output layer ; backward pass
                     Compute delta_wi for all weights from input layer to hidden layer ; backward pass continued
                     Update the weights in the network
      Until all examples classified correctly or stopping criterion satisfied
      Return the network

9.5.5.2 Bayesian Classification

A simple Bayesian classifier is the naive Bayes classifier that is a probabilistic classifier based on applying Bayes' theorem with strong (naive) independence assumptions.

Assuming in general that Y is any discrete-valued variable, and the attributes $X_1,...,X_n$ are any discrete or real valued attributes, the Naive Bayes classification rule is

$$Y^* = \underset{y \in \{y_1,...,y_m\}}{\arg\max} \ P(Y|X_1,...,X_n). \tag{9.4}$$

Using Bayes' theorem and assuming the attributes $X_1,...,X_n$ are all conditionally independent of one another given Y, eqn (9.4) can be rewritten as

$$
\begin{aligned}
Y^* &= \underset{y \in \{y_1,...,y_m\}}{\arg\max} \ P(Y|X_1,...,X_n) \\
&= \underset{y \in \{y_1,...,y_m\}}{\arg\max} \ \frac{P(X_1,...,X_n|Y)P(Y)}{P(X_1,...,X_n)} \\
&\ \vdots \\
&= \underset{y \in \{y_1,...,y_m\}}{\arg\max} \ \frac{1}{P(X_1,...,X_n)} P(Y) \prod_{i=1}^{n} P(X_i|Y).
\end{aligned}
\tag{9.5}
$$

where $P(Y)$ is the prior probability, $\prod_{i=1}^{n} P(X_i|Y)$ is likelihood, and $P(X_1, ..., X_n)$ is evidence.

9.5.6 Bayesian Network

Bayesian networks are a probabilistic graphical model that represents a set of random variables and their conditional independencies *via* a Directed Acyclic Graph (DAG) whose nodes represent random variables (observable quantities, latent variables, unknown parameters or hypotheses) and edges represent conditional dependencies.

Let $G = (V, E)$ be a DAG, and let $X = \{X_1, X_2, ...X_n\}$ be a set of random variables. Suppose that each variable is conditionally independent of all its nondescendants in the graph given the value of all its parents. Then X is a Bayesian network with respect to G. Its joint probability density function (with respect to a product measure) can be written as a product of the individual density functions, conditional on their parent variables as follows:[11]

$$P(X_1, ..., X_n) = \prod_{i=1}^{n} P(X_i|\text{parents}(X_i)), \tag{9.6}$$

where $\text{parents}(X_i)$ is the set of parents of X_i.

For any set of random variables, the probability of any member of a joint distribution can be calculated from conditional probabilities using the chain rule as follows:[11]

$$P(X_1 = x_1, ..., X_n = x_n) = \prod_{i=1}^{n} P(X_i = x_i | X_{i+1} = x_{i+1}, ..., X_n = x_n) \tag{9.7}$$

9.5.7 Support Vector Machine

Support Vector Machine (SVM), a supervised learning method that analyzes data and recognize patterns for classification and regression analysis, performs classification by constructing an N-dimensional hyperplane that optimally separates the data into two categories. For example, the classification problem in biomarker can be restricted to consideration of the two-class problem without loss of generality (health and cancer).

Support Vector Machine (SVM) has several unique characteristics as a research tool for prediction in cancer classification applications. One unique characteristic as a specific type of learning algorithm is that it is characterized by the capacity control of the decision function, the use of the kernel functions, and the sparsity of the solution.[12] The second unique characteristic of SVM is that it is established on the unique theory of the structural risk minimization principle to estimate a function by minimizing an upper bound of the generalization error and therefore very resistant to the overfitting problem, eventually achieving a high generalization performance. The third unique characteristic of SVM is that training SVM is equivalent to solving a linearly constrained quadratic programming problem so that the solution of SVM is always unique and globally optimal, unlike neural networks training, which requires nonlinear optimization with the danger of getting stuck at local minima.

Consider the problem of separating the set of training patterns belonging to two separate classes (1, cancer; –1, health),

$$D = \{(x_1, y_1), \ldots, (x_1, y_1)\}, x \in \mathbb{R}^n, y \in \{-1, 1\} \tag{9.8}$$

with a hyperplane

$$< w, x > + b = 0 \tag{9.9}$$

The set of patterns is said to be optimally separated by the hyperplane if it is separated without error and the distance between the closest pattern to the hyperplane is maximal. Without loss of generality it is appropriate to consider a canonical hyperplane,[13] where the parameters w, b are constrained by,

$$\min_i | < w, x_i > + b | = 1 \tag{9.10}$$

That is, the norm of the weight vector should be equal to the inverse of the distance, of the nearest point in the data set to the hyperplane. A separating hyperplane in canonical form must satisfy the following constraints,

$$y_i[< w, x_i > + b] \geq 1 - e_i, i = 1, \ldots, l \tag{9.11}$$

Therefore, according to the structural risk minimization inductive principle, the training of a SVM is to minimize the guaranteed risk bound,

$$\min_{w,b,s} \phi(w, b, e) = \frac{1}{2} w^T w + \frac{1}{2} C \sum_{i=1}^{l} e_i^2 \tag{9.12}$$

subject to the constraints

$$y_i[< w, x_i > + b] \geq 1 - e_i, i = 1, \ldots, l \tag{9.13}$$

The above optimization problem can be used in a linear recognition problem, but in this case, the classification problem is nonlinear. To solve the nonlinear classification problem, we can map first the training data to another dot product space (called the feature space) F via a nonlinear map $\varphi : \mathbb{R}^n$ and then perform the above computations in F. For example, we can use Gaussian radius basis function (RBF) kernels function for SVM.

9.5.8 Crossvalidation

K-fold crossvalidation was used to increase the number of estimates and improve the accuracy of the prediction model by avoiding the overfitting. In k-fold crossvalidation, the original sample is randomly partitioned into k

subsamples. Of the k subsamples, a single subsample is retained as the validation data for testing the model, and the remaining $k–1$ subsamples are used as training data. The crossvalidation process is then repeated k times, with each of the k subsamples used exactly once as the validation data. The k results from the folds then can be averaged to produce a single estimation. The advantage of this method over repeated random subsampling is that all observations are used for both training and validation, and each observation is used for validation only once. If k equals the sample size, this is called leave-one-out crossvalidation.

9.6   System Biology Approaches for Biomarker Discovery, Verification, and Validation

System biology approaches for biomarker discovery, verification, and validation include: literature search, crossvalidation of multiple studies, pathway analysis and interassociation analysis. These system biology approaches can address the initial challenge of discovering, verifying, and validating biomarkers. At the end of the verification and validation, the knowledge base of biomarkers can be shared with the biological research community, along with peer-reviewed publications. This data sharing will enable the research community to perform subsequent experimental validations such as MRM or PCR with the aim of working towards clinical application.

9.6.1   Literature Search

A literature search is a systematic and explicit approach to the identification, retrieval, and bibliographic management of independent studies for the purpose of verifying and validating biomarkers. Literature search can evaluate if a result is consistent with previous reports simply by analyzing recall and precision. A more advanced method is to calculate the connectivity score for biomarkers and query components.

9.6.1.1   Recall and Precision

Recall and precision are more useful measures that have a fixed range and are easy to compare across queries and engines. Precision is a measure of the usefulness of a hitlist; recall is a measure of the completeness of the hitlist.

Recall is defined as:

recall = # relevant hits in hitlist / # relevant documents in the collection

Recall is a measure of how well the engine performs in finding relevant documents. Recall is 100% when every relevant document is retrieved. In theory, it is easy to achieve good recall: simply return every document in the collection for every query. Therefore, recall by itself is not a good measure of the quality of a search engine.

Precision is defined as:

precision = # relevant hits in hitlist / # hits in hitlist

Precision is a measure of how well the engine performs in avoiding returning nonrelevant documents. Precision is 100% when every document returned to the user is relevant to the query. There is no easy way to achieve 100% precision other than in the trivial case where no document is ever returned for any query!

For example, say that in order to evaluate the effectiveness of three search algorithms, we assemble a collection of 100 documents, 30 of which are considered relevant to our test query. Search algorithm #1 retrieves all 100 documents, so we can calculate Recall $= 30/30 = 100\%$ and Precision $= 30/100 = 30\%$.

9.6.1.2 Connectivity Score for Biomarkers and Query Components

A connectivity score Θ for each possible pair of biomarkers $\{p_1, p_2, \ldots, p_k\}$ and query components $\{d_1, d_2, \ldots, d_g\}$ can be assigned using a regularized log-odds function as following. The log-odds framework is able to qualify association strengths, in particular, facilitated the handling of words for which only sparse scientific literature existed.[14]

$$\Theta_{pd} = \ln(df_{pd} \times N + \lambda) - \ln(df_p \times df_d + \lambda)$$

where df_p and df_d are the total number of documents in which biomarker p and component d are mentioned, respectively, df_{pd} is the total number of documents in which biomarker p and component d are comentioned in the same document. N is the size of the entire PubMed abstract collection. λ is a small constant ($\lambda = 1$ here) introduced to avoid out-of-bound errors if any of df_p, df_d, or df_{pd} values are 0. The resulting Θ_{pd} is positive for when the biomarker–component pair is over-represented and negative when the biomarker–component pair is under-represented. The higher the Θ_{pd} is, the more significant the over-representation of connection between biomarker and query component becomes.

9.6.2 Crossvalidation of Multiple Studies

For novel biomarkers, crossvalidation of multiple studies can be used to test whether results are reproducible across multiple studies. The assumption is that the more reproducible across multiple studies, the more likely the newly characterized biomarker is to be real. Multiple hypothesis testing and multivariate analysis of variance are generally used for the crossvalidation of multiple studies. For biomarkers that cannot be verified in multiple studies, pathway analysis and interassociation analysis can be used for their verification.

9.6.3 Pathway Analysis

A biological system is very dynamic and complex. Systems biology results show that genes and proteins do not function in isolation;[15] instead, they work in interconnected pathways and molecular networks.[16] A mere study of individual molecules such as genes or proteins (which has been the traditional approach for many years) cannot help in decoding the mystery of an entire system.

With the advent of high-throughput technology over the last decade, vast amounts of data has been generated that has led to the development of systems-biology approaches that interrelate the elements of biological processes, such as mRNAs, micro-RNAs, and proteins, revealing higher-level pathways and networks of organization. Systems biology first describes the elements of the system, then the biological networks that interrelate these elements, and finally characterizes the flow of information that links these elements and their networks to an emergent biological process. A pathway network can be defined as a set of molecular interactions and in this set, a subset of genes that coordinate to achieve a specific task forms a pathway. Interactions among several genes and gene products lead to biological processes. Among these networks and pathways, some patterns recognized can be applied for biomarker discovery, verification and validation. In order to analyze these patterns, we need to understand their dynamic nature of how they evolve and change. It is also important to understand the transition from DNA sequence to disease symptoms and to their therapeutic targets.

A biological pathway seeks to understand a specific biological process. A pathway is a series of interactions leading to a cellular process/molecular function. Mathematically, pathways are described as graphs consisting of node and edges. A node represents a biological entity, which can be a gene, protein, or any compound, while edges reflect the relationship between two entities that it connects. The relationship can be activation, inhibition, chemical modification or undefined. Also, these edges can be directional or nondirectional. Pathways can be classified into several categories. Some of the important biological categories are: metabolic, regulatory, protein function and disease.

Biological pathway construction either follows a Data-Driven Objective (DDO) or Knowledge-Driven Objective (KDO). DDO is derived from experimental data such as genomic or proteomic data while KDO is constructed by considering a particular domain of interest such as disease, system, *etc.* The literature is the main source of information in KDO.

There are some pathway visualization tools such as CellDesigner, Cytoscape, and Ingenuity among others. There are various tools available to for manually curating the pathways including Pathway Editor, Knowledge Editor, and MapEditor. Several tools such as Pathway Studio, Pathway Finder, and Pubgene use Natural Language Processing (NLP) to identify associations from the literature that can be built into pathways. Some important pathway databases are described below.

KEGG (Kyoto Encyclopedia of Genes and Genomes): It is a free, online, open-source pathway database. Developed by Kanehisa laboratory, it is an integrated data resource consisting of 16 main databases that contain systems, genomic and chemical information. KEGG consists of pathways stored as pathway maps. The maps cover various domains such as metabolism (the most popular domain in KEGG), genetic information processing (transcription, translation, *etc.*), environmental information processing (cell growth, cell motility, *etc.*), cellular processes (immune system, nervous system *etc.*), and human diseases (neurodegenerative, circulatory, *etc.*). It contains 343 pathway maps, 114 human diseases, 9149 drugs, 5 135 391 genes in high-quality genomes and 16 055 metabolites and other small molecules.[17]

Reactome: It is also a free, online, curated, open-source pathway database. It crossreferences several databases such as UniProt, NCBI and others. It contains information about 23 species. For humans, it contains information on 3916 proteins, 2955 complexes, 3541 reactions and 1045 pathways. Reactome is an effort of collaboration between Cold Spring Harbor Laboratory, Ontario Institute for Cancer Research, European Bioinformatics Institute, Gene Ontology Consortium, and New York University. It also offers tools for pathway analysis.[18]

PathGuide: It is a meta database that contains information on various biological pathway resources. It contains information about 310 resources that include a listing of protein-protein interactions, metabolic pathway, signaling pathways, pathway diagrams, transcription factors/gene regulatory networks, protein compound interactions, genetic interactions, proteins sequences and some other resources. It contains this information for about 24 different species.[19]

BIOCYC: It is a collection of 505 pathway or genome databases and each database describes genome and metabolic pathways of a single organism. It contains many tools such as for a comparative analysis, visual analysis (editing of pathways, *etc.*), genome browser and display of individual metabolic maps. BIOCYC is organized into 3 tiers: the first tier contains an intensively curated database, tier 2 contains computationally derived databases subject to moderate curation and tier 3 contains computationally derived databases with no curation.[20]

NCI pathway interaction database: It is a very well-structured and curated collection of biomolecular interactions and key cellular processes pulled together into Signaling pathways. It is a collaborative effort between the National Cancer Institute (NCI) and the Nature Publishing Group (NPG). It contains 100 human pathways with 6298 interactions curated by NCl-nature, and 392 human pathways with 7418 interactions imported from Reactome/BioCarta.[21]

9.6.4 Interassociation Analysis

Systematic collection of pathway information not only in the form of pathway databases but also including interassociations between pathway, disease, drug, and organ specificity is crucial, because: 1) it provides a bridge between pathway, disease, drug and organ, and 2) this bridge can not only capture relevant biological pathways but also provide disease, drug target, and organ specificity information. For "interassociation", we refer to a biological connection between two or more biological components on the basis of inter-mediary genes. A component is a biomedical concept such as a pathway, disease, drug and organ. Some pilot studies about this kind of connections have been done in the past. For example, Li *et al.* investigated disease relationships based on their shared pathways.[22] First, they extracted disease associated genes by literature mining. Then, they connected diseases to biological pathways through overlapping genes. Lastly, they built a disease network by connecting diseases sharing common pathways. Smith *et al.*[23] combined pathway analysis and drug analysis to identify common biological pathways and drug targets across multiple respiratory viruses based on human host gene expression analysis. Their study suggested that multiple and diverse respiratory viruses invoked several common host response pathways. One study found that disease candidate genes were functionally related in the form of protein complexes or biological pathways and complex disease ensued from the malfunction of one or a few specific signaling pathways.[24] Another study aimed to explore complex relationships among diseases, drugs, genes, and target proteins altogether[25] and found that mapping the polypharmacology network onto the human disease-gene network revealed not only that drugs commonly acted on multiple targets but also that drug targets were often involved with multiple diseases. Berger and Iyengar[26] also discussed how analysis of biological networks had contributed to the genesis of systems pharmacology and how these studies had improved global understanding of drug targets. They described that an emerging area of pharmacology, systems pharmacology, which utilizes biological network analysis of drug action as one of its approaches, is becoming increasingly important in: 1) providing new approaches for drug discovery for complex diseases; 2) considering drug actions and side effects in the context of the regulatory networks within which the drug targets and disease gene products function; 3) understanding the relationships between drug action and disease susceptibility genes; and 4) increasing knowledge of the mechanisms underlying the multiple actions of drugs.[26]

The Integrated Pathway Analysis Database for Systematic Enrichment Analysis (IPAD) (http://bioinfo.hsc.unt.edu/ipad[27]) was developed for users to query information about genes, diseases, drugs, organ specificity, and signaling and metabolic pathways. The IPAD is a comprehensive database covering about 22 498 genes, 25 469 proteins, 1956 pathways, 6704 diseases, 5615 drugs, and 52 organs integrated from databases including the BioCarta,[5] KEGG,[28] NCI-Nature curated,[21] Reactome,[29] CTD,[30] PharmGKB,[31] DrugBank,[32] and Homer.[33] The database has a web-based user interface that allows users to perform enrichment analysis from genes/proteins/molecules and inter-association analysis from a pathway, disease, drug, and organ specificity.

The highly interassociated relationships between pathway, disease, drug and organ can be used to verify and validate the identified enriched pathway, disease, drug and organ candidates, which in turn verify and validate the biomarkers of the components. The more dense and complex the inter-association between the four components, the more reliable and robust are the identified biomarkers.

9.6.4.1 Similarity Measure for the Interassociation Analysis

The Jaccard Index measures similarity between pathways, diseases, drugs and organs, and is defined as the size of the intersection divided by the size of the union of the component sets. The component similarity measure can be defined as the extent of overlaps, *e.g.*, common number of genes/proteins, shared between two different components.[34]

The component–component similarity score $\text{JI}_{i,j}$ is defined as Jaccard Index,

$$\text{JI}_{i,j} = \frac{|p_i \cap p_j|}{|p_i \cup p_j|}; \; i = 1 \ldots N, \; j = 1 \ldots M$$

where, N, M denotes total number of components. P_i and P_j denote two different components, P_i and P_j can be the same or different type, while $|P_i|$ and $|P_j|$ are the numbers of molecules in these two components. Their intersection $P_i \cap P_j$ is the set of all molecules that appear in both P_i and P_j, while their union $P_i \cup P_j$ is the set of all molecules either appearing in the P_i or in the P_j. Duplicates are eliminated in the intersection set and union set.

Similarly, we define the left component–component similarity score $\text{LJI}_{i,j}$ as the Left Jaccard Index,

$$\text{LJI}_{i,j} = \frac{|p_i \cap p_j|}{min(|p_i|, |p_j|)} \; ; \; i = 1 \ldots N, \; j = 1 \ldots M$$

the right component–component similarity score $\text{RJI}_{i,j}$ as the Right Jaccard Index,

$$\text{RJI}_{i,j} = \frac{|p_i \cap p_j|}{max(|p_i|, |p_j|)} \; ; \; i = 1 \ldots N, \; j = 1 \ldots M$$

and the mean component–component similarity score $\text{MJI}_{i,j}$ as the Mean Jaccard Index,

$$\text{MJI}_{i,j} = \frac{LJI_{i,j} + RJI_{i,j}}{2} \; ; \; i = 1 \ldots N, \; j = 1 \ldots M$$

With the equations above, we can calculate similarity scores (Jaccard Index, Left Jaccard Index, Right Jaccard Index, and Mean Jaccard Index) for pathway–pathway, disease–disease, drug–drug, organ–organ, pathway–disease, pathway–drug, pathway–organ, disease–drug, disease–organ, and drug–organ associations.

9.6.4.2   Statistics for the Interassociation Analysis

In addition to similarity scores, we developed a statistic model based on Fisher Exact test[35,36] and number of genes involved in a component for systematic enrichment analysis. When members of two independent groups can fall into one of two mutually exclusive categories, the Fisher Exact test[35,36] is used to determine whether the proportions of those falling into each category differs by group. In interassociation analysis, the Fisher Exact test is adopted to measure the enrichment between components. Given p as the probability of success in a Bernoulli trial where one gene in component i falls in component j, the probability of x successes is

$$P(x) = C_L^x p^x (1-p)^{L-x}$$

where L is the total number of genes in component i, M is the total number of genes in component j, N is the total number of genes in the type of component, $p = M/N$, x is the number of genes corresponding to component i in component j, and C_L^x is the number of possible combinations of x genes from a set of L genes.

The p-value for component i in component j is the probability of obtaining a test statistic at least as extreme as the one observed, given that the null hypothesis that there is no enrichment between component i and component j is true, and calculated according to the following formula

$$P - value = \sum_x^M P(x)$$

To prevent multiple testing problems from happening, the p-values are then adjusted with the Benjamini and Hochberg method.[37]

The absolute enrichment value (AE) of component i in component j is defined as x, the number of genes corresponding to component i in component j. The expected enrichment value (EE) of component i in component j is defined as the expected number of genes of component i in component j under the null hypothesis that the component i and component j are independent of each other.

$$EE = L \cdot \frac{M}{N}$$

The relative enrichment value (RE) of component i in component j is defined as AE/EE.

9.7 Case Study: Breast Cancer Plasma Protein Biomarker Discovery and Verification by Coupling LC-MS/MS Proteomics and Systems Biology

Breast cancer is the second most common type of cancer worldwide after lung cancer. Excluding cancers of the skin, breast cancer is the most common cancer among women, accounting for nearly 1 in 3 cancers diagnosed in United States of America (USA) women. The American Cancer Society's most recent estimates for breast cancer in the USA for 2011 report: 230 480 new cases of invasive breast cancer, 57 650 additional cases of *in situ* breast cancer, and 39 520 deaths from breast cancer.

In the case study, we show how to apply a "systems-biology" approach to discovering, verifying and validating biomarker panels in breast-cancer proteomics. First, we used a *t*-statistics and permutation procedure to discover candidate protein biomarkers. Then an extensive literature-mining curation enabled us to determine the potential protein biomarkers. Lastly, focusing on these potential protein biomarkers, we used gene ontology analysis and Ingenuity pathway analysis to verify the list and unravel the intricate pathways, networks, and functional contexts in which genes or proteins function. Our results showed that the systems biology approach is essential to the understanding molecular mechanisms of panel protein biomarkers. In the future, MRM (Multiple Reaction Monitoring) is planned for validation of the ability of these potential biomarkers for early detection of breast cancer, and the best biomarkers will be selected for clinical assay development and clinical test evaluation in the clinical application.

9.7.1 Study Design

Plasma protein profiles were collected in two batches, referred to as Studies A and B. Studies A and B were processed in the same laboratory but at different times. Each sample was analyzed in a single batch by mass spectrometry. In each study, 80 plasma samples were collected (40 samples collected from women with breast cancer and 40 from healthy volunteer woman who served as controls). The demography and clinical distribution of breast-cancer stages/subtypes for studies A and B are comparable.

9.7.2 Biomarker's Statistical Discovery

4832 peptides in the Study A are mapped to 1422 proteins by searching against the International Protein Index (IPI) database. Using a *t*-Statistics and Permutation Process as described in the Method Section and setting a *p*-value cut-off (0.001) after initial ANOVA analysis of mass-spectra data, we identified 254 statistically significant differentially expressed proteins (per-family Type 1 error rate or PFER = 1.422, FDR = 0.0056), among which 208 are overexpressed and 46 are underexpressed in breast cancer plasma. Compared to the result of traditional statistical test (PFER = 2.5596, FDR = 0.01).

The test statistic is a mean of 40 values (protein intensities in health samples) minus the mean of another 40 values (protein intensities in cancer samples). A permutation procedure was used to determine the *p*-value for each protein, representing the chance of observing a test statistic at least as large as the value actually obtained. The 80 samples for each protein were permuted 100 000 times and the complete set of *t*-tests was performed for each permutation. The permutation *p*-value for a particular protein is the proportion of the permutations in which the permuted test statistic exceeds the observed test statistic in absolute values. We chose a significance level $\alpha = 0.001$ to select proteins where we estimated significant differences in the health and cancer sampled. The corresponding "per-family Type 1 error rate, PFER", that is, the expected number of false-positives for such a multiple test procedure is PFER = number of genes × 0.001. Alternatively, the nominal "false discovery rate, FDR", or expected proportion of false-positive among the genes declared differentially expressed, is FDR = PFER/number of genes declared differentially expressed.

The ANOVA analysis generalizes a linear mixed model that considers three types of effects when deriving protein intensities based on a weighted average of quantile-normalized peptide intensities: 1) the *group effect*, which refers to the fixed nonrandom effects caused by the experimental conditions or treatments that are being compared; 2) the *sample effect*, which refers to the random effects (including those arising from sample preparations) from individual biological samples within a group; 3) the *replicate effect*, which refers to the random effects from replicate injections from the same sample preparation.

9.7.3 Literature Search

We compared our results with 4 previously published proteomic studies of breast-cancer cell lines. Their methods and results presented in peer-reviewed journals[38-41] have established a higher reliability. 3085 protein biomarkers were identified from five breast cell lines, MCF-10A, BT474, MDA-MB-468, MD-MB-468, and T47D/MCF7 in their papers. A comparison of the set of 254 proteins with the published findings from proteomic analysis of human breast-cancer cell lines yielded 26 proteins with differentially expressed in human and cancer samples that were identified in breast-cancer cell lines.

9.7.4 Pathway Analysis and Gene Ontology Categorization of Significant Proteins

Ingenuity Pathway Analysis was used for building pathway and network. Top networks and canonical pathways were identified with Ingenuity Pathways Analysis (Table 9.2, Table 9.3 and Figure 9.3), based on computed score. The computed score (p-score $= -\log_{10}(p$-value)) was computed by IPA according to

Table 9.2 Top networks involved.

Primary Network Functions	Computed Score	Molecules in Network
Cancer, Cell-To-Cell Signaling and Interaction, Hepatic System Disease	**48**	**25**
Genetic Disorder, Hematological Disease, Ophthalmic Disease	43	24
Endocrine System Disorders, Skeletal and Muscular System Development and Function, Tissue Morphology	22	15
Cancer, Cell Cycle, Reproductive System Disease	22	14
Drug Metabolism, Small Molecule Biochemistry, Cancer	18	12

Table 9.3   Top pathways involved.

Pathway	−log(p-value)
Acute Phase Response Signaling	**1.20E+01**
Complement System	1.05E+01
Coagulation System	4.55E+00
PPAR Signaling	1.90E+00
Glutathione Metabolism	1.49E+00

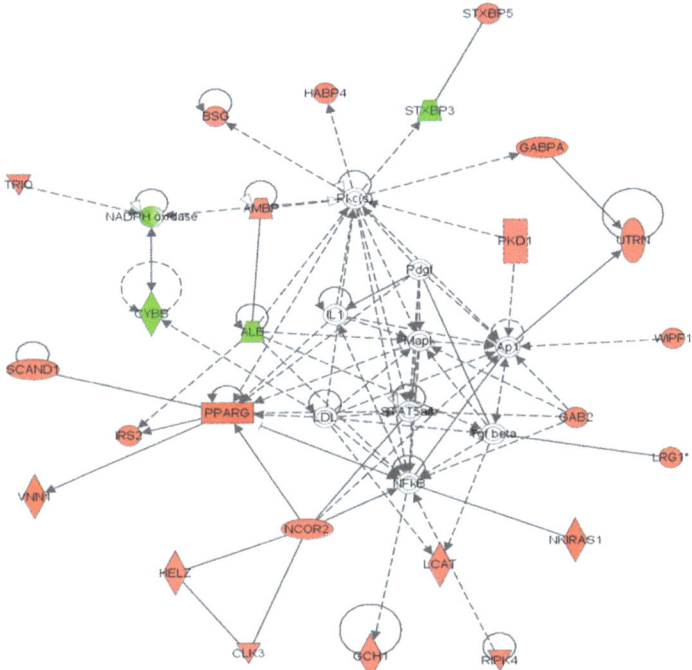

Figure 9.3   The 26 proteins are involved in a single cancer signaling network. (Red for overexpressed and green for underexpressed proteins).

the fit of the set of supplied genes and a list of biological functions stored in the IPKB. The score takes into account the number of genes in the network and the size of the network to approximate how relevant this network is to the original list of genes and allows the networks to be prioritized for further studies. The DAVID database was used to study level 2 and 5 of biological process in gene ontology. Table 9.4 shows the level 2 of biological process in gene ontology.

An interesting finding from Pathway Analysis is that those top networks and pathways shown in the Table 9.2 and Table 9.3, and Figure 9.3, especially the top 1 network (Cancer, Cell-To-Cell Signaling and Interaction, Hepatic System Disease) and top 1 pathway 9 Acute Phase Response Signaling), are validated by our study B dataset and the 26 candidate protein biomarkers, and are similar to previously reported works.[42–44] Another interesting finding from Gene

Table 9.4   Gene ontology biological processes enrichment analysis for 26
                   protein biomarkers.

GO Term	Percentage
response to external stimulus	50%
response to stress	46%
primary metabolic process	38%
defense response	38%
regulation of biological process	35%
establishment of localization	35%
regulation of biological quality	31%
anatomical structure development	27%
cellular metabolic process	27%
multicellular organismal development	27%
cell communication	27%
transport	27%
cellular component organization and biogenesis	23%
macromolecule metabolic process	23%
regulation of cellular process	19%

Ontology is the role of cellular metabolic process and response to external
stimulus (especially proteolysis and acute inflammatory response in level 5) in
Table 9.4 in breast cancer was also reported by other authors. For example,
cancer, like other diseases, is accompanied by strong metabolic disorders.[45] It
was also reported that stress and external stimulus such as microbial infections,
ultraviolet radiation, and chemical stress from heavy metals and pesticides
affect the progression of breast cancer.[46]

9.7.5   Crossvalidation of Multiple Studies

In order to validate the computational results, the same methods and
procedures as we used in Study A were applied to Study B. 48 candidate protein
biomarkers were identified, of which 13 were found in common with the 26
protein biomarkers we identified in study A. Fisher's Exact Test shows that our
methods are feasible and reliable (Fisher-s Exact Test, p-value $= 1.074 \times 10^{-09}$).
Using Ingenuity pathway analysis and DAVID GO analysis, we also found that
the 48 candidate protein biomarkers identified from the Study B have the
similar pathway, network and function as the 26 candidate protein biomarkers
identified from study A.

9.8   Conclusion

This chapter has discussed some of the technologies that are available for
computational methods for biomarker discovery, verification and validation,
has explored how these technologies can be applied to the discovery of different
types of biomarkers in genomics and proteomics, and finally, has described
system biology approaches for biomarker discovery, verification and validation

including literature search, crossvalidation of multiple studies, pathway analysis, and interassociation analysis. These approaches are illustrated through a case study of breast-cancer plasma protein biomarker discovery and verification. We hope that this will be useful in assisting the computational biomarker discovery efforts of current and future genomics and proteomics researchers.

Acknowledgements

We thank Brian Denton, Woody Hagar, Anthony Tissera, and Lynley Dungan for help with the IPAD's database design and web development. We also thank Dr. Peter Horvatovich for reviewing this chapter and giving invaluable comments.

References

1. National Cancer Institute Dictionary.
2. Biomarkers and surrogate endpoints: preferred definitions and conceptual framework, *Clinical pharmacology and therapeutics*, 2001, **69**(3), 89–95.
3. K. Soreide, Receiver-operating characteristic curve analysis in diagnostic, prognostic and predictive biomarker research, *J. Clin. Pathol.*, 2009, **62**(1), 1–5.
4. C. C. Jaffe, Pathology and imaging in biomarker development, *Arch. Pathol. Lab. Med.*, 2009, **133**(4), 547–549.
5. D. R. Rhodes, M. G. Sanda, A. P. Otte, A. M. Chinnaiyan and M. A. Rubin, Multiplex biomarker approach for determining risk of prostate-specific antigen-defined recurrence of prostate cancer, *J. Natl. Cancer. Inst.*, 2003, **95**(9), 661–668.
6. K. C. Lai, H. C. Chiang, W. C. Chen, F. J. Tsai and L. B. Jeng, Artificial neural network-based study can predict gastric cancer staging, *Hepatogastroenterology*, 2008, **55**(86-87), 1859–1863.
7. Z. Amiri, K. Mohammad, M. Mahmoudi, H. Zeraati and A. Fotouhi, Assessment of gastric cancer survival: using an artificial hierarchical neural network, *Pak. J. Biol. Sci.*, 2008, **11**(8), 1076–1084.
8. C. L. Chi, W. N. Street and W. H. Wolberg, Application of artificial neural network-based survival analysis on two breast cancer datasets, *AMIA Annu. Symp. Proc.*, 2007, 130–134.
9. I. Anagnostopoulos and I. Maglogiannis, Neural network-based diagnostic and prognostic estimations in breast cancer microscopic instances, *Med Biol Eng Comput.*, 2006, **44**(9), 773–784.
10. H. Q. Wang, H. S. Wong, H. Zhu and T. T. Yip, A neural network-based biomarker association information extraction approach for cancer classification, *J. Biomed. Inform.*, 2009.
11. C. J. Needham, J. R. Bradford, A. J. Bulpitt and D. R. Westhead, Inference in Bayesian networks, *Nature Biotechnol.*, 2006, **24**(1), 51–53.

12. V. N. Vapnik, An overview of statistical learning theory, *IEEE transactions on neural networks / a publication of the IEEE Neural Networks Council*, 1999, **10**(5), 988–999.
13. V. N. Vapnik, *Statistical Learning Theory*, Springer, NY 1998.
14. J. Li, X. Zhu and J. Y. Chen, Building disease-specific drug-protein connectivity maps from molecular interaction networks and PubMed abstracts, *PLoS Computational Biology*, 2009, **5**(7), e1000450.
15. P. Goymer, Cancer genetics: Networks uncover new cancer susceptibility suspect, *Nature Reviews Genetics*, 2007, **8**, 823–823.
16. A. Ergün, C. A. Lawrence, M. A. Kohanski, T. A. Brennan and J. J. Collins, A network biology approach to prostate cancer, *Molecular Systems Biology*, 2007, **3**, 82.
17. M. Kanehisa, S. Goto, S. Kawashima, Y. Okuno and M. Hattori, The KEGG resource for deciphering the genome, *Nucleic Acids Res.*, 2004, **32**(Database issue), D277–280.
18. L. Matthews, G. Gopinath, M. Gillespie, M. Caudy, D. Croft, B. de Bono, P. Garapati, J. Hemish, H. Hermjakob and B. Jassal, *et al.*, Reactome knowledgebase of human biological pathways and processes, *Nucleic Acids Res.*, 2009, (Database issue), D619–622.
19. G. D. Bader, M. P. Cary and C. Sander, Pathguide: a pathway resource list, *Nucleic Acids Res.*, 2006, **34**(Database issue), D504–506.
20. P. D. Karp, C. A. Ouzounis, C. Moore-Kochlacs, L. Goldovsky, P. Kaipa, D. Ahren, S. Tsoka, N. Darzentas, V. Kunin and N. Lopez-Bigas, Expansion of the BioCyc collection of pathway/genome databases to 160 genomes, *Nucleic Acids Res.*, 2005, **33**(19), 6083–6089.
21. C. F. Schaefer, K. Anthony, S. Krupa, J. Buchoff, M. Day, T. Hannay and K. H. Buetow, PID: the Pathway Interaction Database, *Nucleic Acids Res.*, 2009, **37**(Database issue), D674–679.
22. Y. Li and P. Agarwal, A pathway-based view of human diseases and disease relationships, *PloS one* 2009, **4**(2), e4346.
23. S. B. Smith, W. Dampier, A. Tozeren, J. R. Brown and M. Magid-Slav, Identification of common biological pathways and drug targets across multiple respiratory viruses based on human host gene expression analysis, *PloS one*, 2012, **7**(3), e33174.
24. M. Oti and H. G. Brunner, The modular nature of genetic diseases, *Clin. Genet.*, 2007, **71**(1), 1–11.
25. A. L. Hopkins, Network pharmacology, *Nature Biotechnol.*, 2007, **25**(10), 1110–1111.
26. S. I. Berger and R. Iyengar, Network analyses in systems pharmacology, *Bioinformatics*, 2009, **25**(19), 2466–2472.
27. F. Zhang and R. Drabier, IPAD: the Integrated Pathway Analysis Database for Systematic Enrichment Analysis, *BMC Bioinformatics*, 2012, **13**, 14.
28. K. G. Victor, J. M. Rady, J. V. Cross and D. J. Templeton, Proteomic Profile of Reversible Protein Oxidation Using PROP, Purification of Reversibly Oxidized Proteins, *PloS one*, 2012, **7**(2), e32527.

29. D. Croft, G. O'Kelly, G. Wu, R. Haw, M. Gillespie, L. Matthews, M. Caudy, P. Garapati, G. Gopinath and B. Jassal *et al.*, Reactome: a database of reactions, pathways and biological processes, *Nucleic Acids Res.*, 2011, **39**(Database issue), D691–697.

30. A. P. Davis, B. L. King, S. Mockus, C. G. Murphy, C. Saraceni-Richards, M. Rosenstein, T. Wiegers and C. J. Mattingly, The Comparative Toxicogenomics Database: update 2011, *Nucleic Acids Res.*, 2011, **39**(Database issue), D1067–1072.

31. E. M. McDonagh, M. Whirl-Carrillo, Y. Garten, R. B. Altman and T. E. Klein, From pharmacogenomic knowledge acquisition to clinical applications: the PharmGKB as a clinical pharmacogenomic biomarker resource, *Biomark Med.*, 2011, **5**(6), 795–806.

32. C. Knox, V. Law, T. Jewison, P. Liu, S. Ly, A. Frolkis, A. Pon, K. Banco, C. Mak and V. Neveu, *et al.*, DrugBank 3.0: a comprehensive resource for 'omics' research on drugs, *Nucleic Acids Res.*, 2011, **39**(Database issue), D1035–1041.

33. F. Zhang and J. Y. Chen, HOMER: a human organ-specific molecular electronic repository, *BMC Bioinformatics*, 2011, **12**(Suppl 10), S4.

34. X. Wu, S. R. Chowbina, P. M. Li, R. Pandey, H. N. Kasamsetty and J. Y. Chen, *Characterizing Mergeability of Human Molecular Pathways*, submitted.

35. R. A. FISHER and A. New, Test for 2×2 Tables, *Nature*, 1945, **156**, 388.

36. W. Huang da and B. T. Sherman, Lempicki RA: Bioinformatics enrichment tools: paths toward the comprehensive functional analysis of large gene lists, *Nucleic Acids Res.*, 2009, **37**(1), 1–13.

37. Y. Benjamini and Y. Hochberg, Controlling the False Discovery Rate: A Practical and Powerful Approach to Multiple Testing, *Journal of the Royal Statistical Society Series B (Methodological)*, 1995, **57**(1), 289–300.

38. P. J. Adam, R. Boyd, K. L. Tyson, G. C. Fletcher, A. Stamps, L. Hudson, H. R. Poyser, N. Redpath, M. Griffiths and G. Steers *et al.*, Comprehensive Proteomic Analysis of Breast Cancer Cell Membranes Reveals Unique Proteins with Potential Roles in Clinical Cancer, *J. Biol. Chem.*, 2003, **278**(8), 6482–6489.

39. V. Kulasingam and E. P. Diamandis, Proteomics Analysis of Conditioned Media from Three Breast Cancer Cell Lines: A Mine for Biomarkers and Therapeutic Targets, *Mol. Cell. Proteomics.*, 2007, **6**(11), 1997–2011.

40. F. Mbeunkui, B. J. Metge, L. A. Shevde and L. K. Pannell, Identification of Differentially Secreted Biomarkers Using LC-MS/MS in Isogenic Cell Lines Representing a Progression of Breast Cancer, *J. Proteome Res.*, 2007, **6**(8), 2993–3002.

41. R. Xiang, Y. Shi, D. A. Dillon, B. Negin, C. Horvath and J. A. Wilkins, 2D LC/MS Analysis of Membrane Proteins from Breast Cancer Cell Lines MCF7 and BT474, *J. Proteome Res.*, 2004, **3**(6), 1278–1283.

42. M. Berishaj, S. P. Gao, S. Ahmed, K. Leslie, H. Al-Ahmadie, W. L. Gerald, W. Bornmann and J. F. Bromberg, Stat3 is tyrosine-

phosphorylated through the interleukin-6/glycoprotein 130/Janus kinase pathway in breast cancer, *Breast Cancer Res.*, 2007, **9**(3), R32.

43. H. Hu, H. J. Lee, C. Jiang, J. Zhang, L. Wang, Y. Zhao, Q. Xiang, E. O. Lee, S. H. Kim and J. Lu, Penta-1,2,3,4,6-O-galloyl-beta-D-glucose induces p53 and inhibits STAT3 in prostate cancer cells in vitro and suppresses prostate xenograft tumor growth *in vivo*, *Mol. Cancer. Ther.*, 2008, **7**(9), 2681–2691.

44. H. Song, X. Jin and J. Lin, Stat3 upregulates MEK5 expression in human breast cancer cells, *Oncogene*, 2004, **23**(50), 8301–8309.

45. D. Bullinger, H. Neubauer, T. Fehm, S. Laufer, C. H. Gleiter and B. Kammerer, Metabolic signature of breast cancer cell line MCF-7: profiling of modified nucleosides *via* LC-IT MS coupling, *BMC Biochem.*, 2007, **8**, 25.

46. N. R. Nielsen and M. Gronbaek, Stress and breast cancer: a systematic update on the current knowledge, *Nature Clin. Pract. Oncol.*, 2006, **3**(11), 612–620.

Discovery and Validation Case Studies, Recommendations

CHAPTER 10

Discovery and Validation Case Studies, Recommendations: A Pipeline that Integrates the Discovery and Verification Studies of Urinary Protein Biomarkers Reveals Candidate Markers for Bladder Cancer

YI-TING CHEN,[a] CAROL E. PARKER,[b]
HSIAO-WEI CHEN,[a,c] CHIEN-LUN CHEN,[d]
DOMINIK DOMANSKI,[b] DEREK S. SMITH,[b]
CHIH-CHING WU,[a,e] TING CHUNG,[a] KUNG-HAO LIANG,[f]
MIN-CHI CHEN,[c,g] YU-SUN CHANG,[a,c]
CHRISTOPH H. BORCHERS*[b,h] AND JAU-SONG YU*[a,c]

[a] Molecular Medicine Research Center, Chang Gung University, Taoyuan 333, Taiwan; [b] University of Victoria - Genome BC Proteomics Centre, University of Victoria, Victoria, British Columbia, Canada ; [c] Graduate Institute of Biomedical Sciences, College of Medicine, Chang Gung University, Taoyuan 333, Taiwan ; [d] Chang Gung Bioinformatics Center, Department of Urology, Chang Gung Memorial Hospital, Taoyuan 333, Taiwan; [e] Graduate Institute of Medical Biotechnology and Department of Medical Biotechnology and Laboratory Science, Chang Gung University, Taoyuan 333, Taiwan;

RSC Drug Discovery Series No. 33
Comprehensive Biomarker Discovery and Validation for Clinical Application
Edited by Péter Horvatovich and Rainer Bischoff
© The Royal Society of Chemistry 2013
Published by the Royal Society of Chemistry, www.rsc.org

[f] Liver Research Center, Chang Gung Memorial Hospital, Taoyuan 333, Taiwan; [g] Department of Public Health and Biostatistics Consulting Center, Chang Gung University, Taoyuan 333, Taiwan; [h] Department of Biochemistry and Microbiology, University of Victoria, Victoria, BC, Canada
*Email: yusong@mail.cgu.edu.tw; christoph@proteincentre.com

10.1   Introduction

10.1.1   Importance of Bladder Cancer

The function of the urinary bladder is to collect and store urine excreted by the kidneys and then expel it during urination. Urothelial carcinoma of the bladder is the fourth most common malignancy in men in the United States, accounting for more than 52 020 new cases according to the most recent estimates of the American Cancer Society for 2011.[1] 10 670 estimated deaths were caused by bladder cancer annually. The ratio of men to women that develop bladder cancer is approximately 3 : 1. Approximately 70% to 80% of patients diagnosed with bladder cancer currently present with hematuria. Approximately 75–85% of patients with bladder cancer present with a disease that is confined to the mucosa [stage Ta, carcinoma *in situ* (CIS)] or submucosa (stage T1).[2] Most of these patients can be managed with transurethral resection (TUR) and intra-vesical therapy. Muscle invasion in bladder is present in approximately 25% of patients at initial diagnosis. Approximately 50–70% of bladder cancers will recur and 10–30% of all bladder cancers will progress to muscle-invasive disease.[3,4] Radical cystectomy is the preferred and standard therapy for muscle-invasive bladder cancer. Overall survival rates for bladder cancer are stage dependent and 5-year survival for tumors confined to the mucosa (70%) are significantly higher than muscle-invasive (40%) or metastatic (less than 30%) cancers.[5,6] The earlier this cancer is found and treated, the better the outcome, which underscores the importance of finding biomarkers that can detect the early stages of this cancer.

10.1.2   Other Methods of Diagnosis

Hematuria is the most common sign of bladder cancer, and simply means that blood is found in the urine. Hematuria may be visible to the naked eye or visible only under a microscope. In both cases, it is usually painless, which might be neglected by patients. Incidence of bladder cancer in patients with microscopic hematuria, however, has a very low percentage. Patients with hematuria are commonly recommended to undergo further evaluation, which typically includes cystoscopy and cytology.[3] Diagnosis of bladder cancer has tradi-tionally relied on cystoscopy, *i.e.* endoscopy of the urinary bladder *via* the urethra. Cystoscopy is an invasive and costly procedure, and is usually used in conjunction with noninvasive urine cytology. Urinary cytology relies on the presence of shed cancer cells in voided urine. Normal bladders and bladders with small tumors or low-grade tumors are less likely to shed cells spontaneously into the urine because of stronger intercellular attachments. Cells from high-grade cancers are more likely to be shed into the urine than low-grade tumors.[7,8] As a result of this phenomenon, sensitivities range from 20–60% and 80% for detecting low-grade tumors and high-grade tumors in urine by cytology.[7,8] However, the low sensitivity of this technique, particularly for low-grade bladder tumors, has resulted in clinical limitations for treatment, progression outcome, and management of bladder cancer. Other bladder-cancer symptoms include frequent urination and pain upon urination which frequently happens in other noncancerous diseases as well. These characteristics have therefore prompted the search for more reliable noninvasive markers of bladder cancer and there is a compelling need to develop more reliable bladder-cancer biomarkers, particularly those which can be measured in body fluids.

10.1.3 Other Urinary Biomarker Studies for Bladder Cancer

Urine is an attractive biofluid for proteomic analysis in urological diseases,[9] in part because it is regarded as a proximal fluid for the bladder but also because urine can be obtained easily, noninvasively, and in large quantities. Several recent studies have been performed with the goal of discovering multiple biomarkers in the urinary proteome of bladder-cancer patients.[6,8,10] These newly discovered urine markers are gaining importance for diagnosis and follow up. Both nuclear matrix protein (NMP22) and bladder-tumor antigen (BTA) are US Food and Drug Administration-approved markers for surveillance and detection of bladder cancer in combination with cystoscopy.[6,10] Nuclear matrix proteins (NMPs) are parts of the internal structural framework of the cell nucleus. The average level of urine NMP22 in the transitional cell carcinoma (TCC) bladder-cancer patients was approximately 5-fold higher than that in individuals without bladder TCC.[11] The sensitivity of the NMP22 immunoassay has ranged from 47% to 100% and its specificity from 60% to 90% in several studies.[6] However, this test cannot be trusted for the detection of superficial bladder cancer, particularly stage Ta cancer (a very early stage).[11] Pyuria and hematuria may also significantly affect the detection of urinary NMP22.[12] The bladder-tumor antigen (BTA) assay is designed to detect a bladder-tumor-associated antigen, human complement factor H-related protein, in the urine of patients with bladder cancer.[6,7] The sensitivity and specificity of BTA ranges from 66 to 70% and 65 to 78%.[13] Because of the high false-positive rates, both NMP22 and BTA are not recommended for use without cystoscopy. Some preliminary studies[4,6,7] for discovery and verification of biomarkers such as hyaluronic acid and hyaluronidase, immunoCyt, BLCA-1 and BLCA-4, surviving, telomerase, microRNA,[14,15] and metabolites[16,17] are currently in progress. However, the properties of these candidate markers either are not superior to existing detection methods with respect to sensitivity or specificity, or their clinical utility has not been examined in large patient populations. Verification or validation of current proposed biomarkers using a high-throughput platform are required.

10.1.4   Protein-Based Biomarker Discovery

Proteomics-based biomarker discovery is based on the detection of up- or downregulated proteins in diseased *versus* healthy cells or organisms. Ideally, these cells or organisms would be identical except for the disease. In practice, this can be achieved only for cell culture, and can only be approximated, even for animal models. It is nearly impossible to achieve for humans, partly because of the large individual variation (which is why age- and sex-matched subjects are used), but even then, the diet still would have to be controlled. There are no laboratory inbred strains of identical humans, and there is no "Purina human chow".

Biomarkers for diseases such as cancer can be discovered by comparing tumor biopsies with normal tissue, but even this is impossible for other diseases such as Alzheimers where it is difficult to get brain biopsies from control individuals. Plasma is considered as the biofluid of choice for the discovery of biomarkers, because it is easily accessible and because tumors are in contact with blood and "shed" proteins into the blood stream. However, plasma has a 10^{10} dynamic range of protein concentrations, which makes it challenging for biomarker discovery.[18] Recently, urine is becoming more widely considered as an alternative biofluid, because it is easily available. Urine should be a particularly appropriate source of biomarkers for bladder cancer, because urine is the proximal fluid for bladder-cancer tumors, which are expected to shed proteins into the urine.

Biomarker discovery depends on protein quantitation, usually relative quantitation, looking at protein up- or downregulation as a function of disease. Mass-spectrometry-based relative quantitation is usually performed on peptides, rather than intact proteins, and peptide-based quantitation is most accurately performed when isotopically labeled internal standards are used, which can compensate for ion-suppression effects.[19,20] Probably the most accurate relative quantitation method would be metabolic labeling methods that results in stable-isotope-labeled proteins. This is because the labeled and unlabeled proteins could be combined before digestion, and the digestion of the labeled and unlabeled analogs would be identical. However, SILAC (Stable-Isotope Labeling by Amino Acids in Cell Culture)[21] can only be done in cell culture, by either feeding the cells ^{15}N- or ^{13}C-labeled buffers or amino acids (and it can only be used for those organisms that cannot make their own amino acids from environmentally available nitrogen). The recent publication by Yates's group of comparative proteomics using rats fed ^{15}N-labeled algae[22] and the recent introduction of ^{15}N-labeled mouse[23] means that metabolic labeling can now be done now more easily for mice and rats. However, this approach is out of the question for human subjects – other methods must be used for human samples.

Recently, label-free software-based methods have been introduced (*e.g.,* Ion Accounting[24] and Spectral Counting[25–27] approaches) that assume that the concentration of either a reference protein or a spiked-in reference standard is unchanged by the disease, or that the total ion current (TIC) from the sample is unchanged (see [28] and references contained therein for a review of these techniques). Methods that use isotopic labeling are more expensive, but do not require this assumption and the presence of the label can compensate for suppression effects from the sample matrix. These stable-label biomarker discovery methods including Applied Biosystem's iTRAQ and mTRAQ, Thermo's TMT tags, and acetylation/deuteroacetylation, involve a post-digestion labeling step. These peptide-based methods involve a different set of assumptions – that of complete or at least reproducible digestion of the protein into peptides as mentioned above, plus reproducible labeling of the peptides in the samples to be compared. Both mTRAQ and the acetylation method are chemical labeling methods, where nonisobaric reagents are used to modify the peptides in the protein digest, which have different molecular weights for the different tags. The iTRAQ method is different in that all of the labels result in the same molecular weight shift in all of the labeled peptides.

10.1.5   Introduction to iTRAQ for Biomarker Discovery

The iTRAQ method was developed in 2004 by Ross *et al.*,[29] and is now marketed by Applied Biosystems. It consists of a set of 4 or 8 tags that are used to derivatize digested proteins from up to 4 or 8 different treatments, after which the digests are combined and analyzed in a single LC/MS/MS run. This gives relative quantitation data from up to 4 or 8 treatments or samples from a single mass-spectrometric analysis. As mentioned above, the iTRAQ technique is an MS/MS-based technique that uses "isobaric labels", meaning that the total mass shift from all of the different forms of the labels is the same in the MS mode. Thus, one advantage of this technique is that only one precursor mass needs to be selected in order to compare all of these forms. Each tag consists of 3 parts: the amine-reactive portion that attaches to the digested peptides, the mass-balance region, and the reporter region. Because the tag is attached to the N-terminal amino group of the peptide, the y-ions (which can be used for the identification of the peptide and therefore the protein) have the same m/z values for all versions of the tagged peptides. This is important because the "protein identification" portion of the spectrum actually consists of the sum of all of the ions for the differentially labeled versions. Thus, the signal is not split between the differentially labeled peptides and the LOD is not compromised. In the MS/MS mode, the reporter portion of the tag appears at different masses due to the different compositions of the reporter region – this is the region that gives the information on the relative abundances of the peptide in the different treatments. There is some indication that the fold-changes from iTRAQ are

Multiple Reaction Monitoring

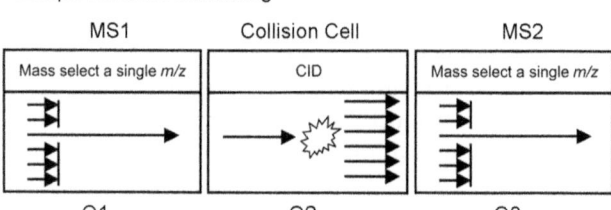

Figure 10.1 Diagram of an MRM experiment on a triple quadrupole instrument.

"compressed", but at least the indications are that the up- or downregulation determinations are still valid.[30–32]

Biomarker discovery using iTRAQ is usually performed with a "MudPIT" approach.[33,34] "MudPIT" (which stands for Multidimensional Protein identification Technology) is a way of fractionating the proteome (or, in this case, a digest of the proteome) in order to "dig deeper" and find lower-abundance differentially expressed proteins. In practice, this usually means SCX fractionation, with the collection of multiple LC fractions, each of which is subjected to a separate C18 reversed-phase LC/MS/MS analysis. Thus, an iTRAQ experiment is quite instrument-time intensive, with each sample turning into as many as 42 separate LC/MS/MS runs, as in the study described here.

Instrumentally, iTRAQ is performed on instruments that can generate fragments from the peptides, because the fragments of the reporter region only appear in the MS/MS spectra. Operationally, this means LC/MS/MS instruments with electrospray or MALDI sources can be used. Typically, these instruments are comprised of a source, the first mass analyzer (Q1, where the precursor m/z is selected), a collision cell to fragment this peptide (Q2), and a second mass analyzer (Q3) to separate the product ions by their m/z (Figure 10.1). Note that the HCD option (with higher collision energies) must be used on Orbitrap instruments in order to generate sufficient ion abundances of these reporter ions. For LC/MS/MS instruments, each SCX fraction is injected separately. For LC-MALDI-TOF/TOF instruments, normally each SCX fraction is separated offline using C18 reversed-phase HPLC, with the effluent being spotted directly onto MALDI target plates. These plates are then subsequently analyzed in a MALDI-TOF/TOF instrument.

10.1.6   Introduction to MRM for Biomarker Discovery and Verification

While iTRAQ is considered to be the "gold standard" for relative quantitation, it is too expensive in terms of reagent costs as well as instrument time for the large numbers of samples required for biomarker validation and verification. In addition, it only provides a relative up- or down- fold-change, not a concentration measurement (*e.g.*, concentration in ng/mL), and this actual protein concentration in urine or plasma is what is required for a clinical analysis. MRM is emerging as an alternative proteomics-based approach for biomarker validation. Like iTRAQ, it is a peptide-based technique, but the target peptides are not chemically labeled. For accuracy, quantitation is performed using known amounts of stable-isotope-labeled internal standard peptides (SIS peptides). This means that, unlike iTRAQ, actual plasma protein concentrations, can be determined, with the caveats that the proteolytic digestion must be complete (or at least repro-ducible), and that the measured peptides should originate only from the target protein.[35]

MRM is also a type of MS/MS experiment. In this case, because the peptide targets are known in advance, the instruments select a *narrow* Q1 window around each target mass of precursor ion instead of scanning the entire mass range. This results in improved sensitivity over full-scan instruments. Instead of a full scan for Q3 to detect all of the fragments from the targeted precursor ion, Q3 only scans a very narrow *m/z* window centered on each targeted fragment ion. Thus, a precursor/product ion pair is selected – which is normally termed a "transition". These transitions must be specific to each targeted protein (*i.e.* "proteotypic), and other selection criteria are also applied. For example, peptides with oxidizable residues such as methionine are usually avoided, as are cysteine-containing peptides, or peptides that may result in a missed cleavage with the protease used (usually trypsin). A detailed description of how these transitions are selected, and how their sensitivities are optimized is given in two publications by Kuzyk *et al.*[36,37]

In a large multilaboratory study, MRM with SIS peptides was found to be reproducible between analysis sites,[38] and MRM has been proposed as an ideal technique for the verification and validation of proposed biomarkers.[28,39,40] MRM panels with varying numbers of proteins that target specific diseases are already being developed (*e.g.*, 67 proteotypic proteins for the diagnosis of cardiovascular disease,[41] and peptides for the diagnosis of different forms of anemia[42]). Once these multiprotein panels have been developed, they can be turned around and used for biomarker discovery as well as verification. This involves looking at patterns of protein up- and downregulation involving multiple proteins in multiple pathways. This is the case for the bladder-cancer example described in this book chapter. Here, a panel of proteins originally developed for plasma analysis was used for the discovery of protein in urine whose expression patterns are characteristic of different forms of bladder cancer.

10.1.7 Biomarker Discovery Pipeline

The goal of clinical proteomics is to identify proteins showing different expression levels between different clinical statuses and to develop quantitative assays for their routine and reliable measurement that can be used in a clinical setting. These clinical assays would contribute to solutions for various clinical needs including diagnosis, prognosis, or treatment monitoring. A complete biomarker-discovery pipeline includes the generation of candidate markers using a limited number of well-defined samples, followed by multiplexed verification of the differential expression of these molecules using a targeted platform on a larger number of sample sets, and, eventually, unbiased validation of a protein signature or a marker panel using clinical assays in large retrospective and prospective collections of specimens from intended target population.[43]

The current trend is to use biomarker *panels* for diagnosis, prognosis, or therapeutic purposes. Panels of biomarkers have greater specificity and sensitivity than each of the individual biomarkers alone, as each biomarker contributes to the classification power of diagnostic test.[44,45] The use of a biomarker panel usually relies on the interpretation of the pattern of the different biomarkers in relationship to one another rather than relying on the absolute level of each biomarker. Combining multiple biomarkers into one biomarker panel for disease screening could produce a more cost-effective test with higher specificity and sensitivity. The panel might, for example, combine a biomarker with high sensitivity and low specificity (which would detect potentially cancers but would result in many false-positives) with a second biomarker having less sensitivity but higher specificity. It has recently been proposed that many newly discovered biomarkers should be measured simultaneously by multiplexed approaches and in combination with imaging or other routine clinical techniques.[46,47]

In this chapter, we describe the workflow that we used to develop a noninvasive assay for the detection of a panel of potential biomarkers of human bladder tumor in urine specimens. In this study, both the biomarker discovery and biomarker verification stages were done by mass spectrometry, but using different mass-spectrometric techniques.

10.2   Experimental

10.2.1   Materials

10.2.1.1   Samples

All urine samples were collected at Chang Gung Memorial Hospital, Taoyuan, Taiwan. The study protocol was approved by the Medical Ethics and Human Clinical Trial Committee at Chang Gung Memorial Hospital. In this study, hernia patients were defined as the control group. Urological disease groups included UTI, HU, UTI plus HU, low-grade/early-stage bladder cancer (LgEs), high-grade/early-stage bladder cancer (HgEs), and high-grade/advanced-stage (HgAs) bladder cancer. Table 10.1 lists the clinical sample information for all the samples used in experiments of discovery phase (iTRAQ) and verification phase (MRM-MS).

Table 10.1 Summary of clinical sample information for all samples used in experiments of discovery phase (iTRAQ) and verification phase (LC-MRM-MS). Adapted from,[48,53] with permission.

Classification-iTRAQ tag	Diagnosis status	Discovery phase by iTRAQ (Pooled samples)						Validation phase by MRM-MS (Individual samples)		
		Clinical sample set-1			Clinical sample set-2					
		Sex	Number	Age	Sex	Number	Age	Male(#)	Female(#)	Age
Nontumor control (NT)	Hernia	male	7	63.4 ± 12.6	male	7	70.1 ± 15.4	56	1	63.9 ± 14.0
Low-grade/Early-stage (LgEs)	pTa	male	2	63.0 ± 12.7	male	4	62.8 ± 18.0	16	5	67.5 ± 12.5
High-grade/Early-stage (HgEs)	Tis,pTa,pT1	male	4	70.3 ± 5.7	male	6	68.7 ± 10.1	27	14	71.4 ± 9.8
High-grade/Advanced-stage (HgAs)	>pT2	female	2	78.0 ± 4.2	male	5	80.2 ± 6.1	8	6	70.5 ± 11.4
UTI/HU	UTI,HU,UTI+HU							17	6	57.3 ± 11.9

10.2.2   Sample Preparation for iTRAQ and MRM-MS

10.2.2.1   Concentration and Desalting of Urine Samples

The first morning-urine clinical specimens were collected as previously described, with the addition of a protease inhibitor.[48] The collected samples were centrifuged at $5000 \times g$ for 30 min at 4 °C within 5 h, to remove cells and debris, and the clarified supernatants were stored at –80 °C for further processing. Urinary proteins were enriched from 12.5 ml of urine using a 10-kDa centrifugal filter at $5000 \times g$ for 30 min at 4 °C, recentrifuged with 12.5 ml of 20% aceto-nitrile/H_2O, and finally solubilized with water for desalting. The final volume was approximately 500 µL. The amount of protein in each concentrated/desalted urine sample was measured using a DC™ (Detergent Compatible) protein assay kit (Bio-Rad, USA). Samples were then lyophilized and stored at –20 °C for subsequent processing. The total amount (in µg) of protein in each individual sample was then converted into a concentration (µg/mL), based on the initial raw urine volume prior to desalting and concentration.

10.2.2.2   Sample preparation for iTRAQ

10.2.2.2.1   iTRAQ Labeling and Fractionation by Strong Cationic exchange (SCX) and Basic Reversed-Phase Chromatography.   In the discovery stage of this study, age-matched hernia patients were chosen as the control subgroup. Pooled urinary protein samples (100 µg) from each subgroup were processed according to the manufacturer's protocol for 4-plexiTRAQ (Applied Biosystems, Foster City, CA) as previously described.[48] Briefly, pooled urinary proteins from each subgroup were reduced, the cysteines were blocked with methyl methanethiosulfonate, and the proteins were digested with trypsin (see[49] for detailed protocols for each of these steps). Tryptic peptides from the NT, LgEs, HgEs, and HgAs subgroups (see explanation of the subgroup abbreviations in Table 10.1) were labeled with 114, 115, 116, and 117 iTRAQ tags, respectively. The pooled mixtures of iTRAQ-labeled peptides from clinical sample set 1 were fractionated by SCX chroma-tography. The pooled mixtures of iTRAQ-labeled peptides from clinical sample set 2 were fractionated in parallel by offline SCX and reversed-phase (RP) chromatography, as described previously.[48] Briefly, for the SCX chro-matography, the iTRAQ-labeled peptide mixture was loaded onto a 2.1 × 150 mm BioBasic SCX column containing 5-µm particles and a 300-µm pore size for separation. For the RP chromatography under basic conditions, the iTRAQ-labeled peptide mixture was loaded onto a 4.6 × 150 mm Gemini C18 column (3-µm particles and a 160-µm pore size, HPLC Phenomenex, Torrance, CA, USA), and the peptides were eluted at a flow rate of 400 µL/min. This basic fractionation step is orthogonal with respect to the usual acidic elution conditions normally used for the online analysis step, which is generally performed using formic acid. Elution was monitored by measuring absorbance at 220 nm, and fractions were collected every 1 min. After pooling the eluate into 42 fractions from each experiment and vacuum drying, samples were ready for nano-ESI-LC-MS/MS analysis. Thus, there were a total of 42 × 3 fractions analyzed in the discovery phase.

10.2.3 LC-ESI MS/MS Analysis of iTRAQ-Labeled Peptides by LTQ-Orbitrap Pulsed-Q Dissociation

A 2-µg aliquot of the peptides from each fraction was loaded onto a trap column (Zorbax 300SB-C18, 0.3 × 5 mm, Agilent Technologies, Wilmington, DE, USA) separated on a resolving 10-cm analytical BioBasic C18 PicoFrit column (inner diameter, 75 µm). Peptides were eluted from the analytical column at a flow rate of 0.25 µL/min with a 51-min gradient. The LC was coupled online to a linear ion trap-orbitrap (LTQ-Orbitrap, Thermo Fisher, San Jose, CA, USA) operated using Xcalibur 2.0 software (Thermo Fisher). Intact peptides were detected in the Orbitrap at a resolution of 30 000. Peptides were selected for MS/MS using the pulsed-Q dissociation (PQD) operating mode. A data-dependent procedure that alternated between a single MS scan followed by three MS/MS scans was applied for the three most-abundant precursor ions in the MS survey scan. The m/z values selected for MS/MS were dynamically excluded for 180 s.

10.2.3.1 Data Analysis of iTRAQ Results for Candidate Selection Protein Identification and Quantitation by Sequence Database Searching

The resulting MS/MS spectra were searched against the European Bio-informatics Institute (http://www.ebi.ac.uk/) nonredundant International Protein Index (IPI) human sequence database (v3.27, March 2007) containing 67 528 sequences and 28 353 548 residues using the MASCOT engine (Matrix Science, London, UK; version 2.2.04) with the Mascot Daemon program (Matrix Science, version 2.2.0). For protein identification, a mass tolerance of 10 ppm was permitted for intact peptide masses and 0.5 Da for PQD fragmented ions, with allowance for two missed cleavages in the trypsin digests, oxidized methionine as a potential variable modification, and iTRAQ (N terminal), iTRAQ (K), and methyl methanethiosulfonate (C) as fixed modifications. The charge states of peptides for database searching were limited to +2 and +3. Protein identification and quantification were validated using the open-source Trans-Proteomic Pipeline (TPP) software (Version 4.0) that includes Peptide-Prophet and ProteinProphet scoring programs. In this study, we used Peptide-Prophet and ProteinProphet probability (48) scores ≥0.95 to ensure an overall false-positive rate of less than 0.7%. The ratio for each protein was quantified using the Libra program,[50] a module within the TPP software package that performs quantification on MS/MS spectra that have multiple forms of the labels, as is the case for iTRAQ. The minimum intensity threshold of a reporter ion in the spectrum was set at 20 for quantitation using the Libra software. Each quantified protein had to contain at least one Libra peptide.

10.2.3.2   Sample Preparation for MRM-MS

10.2.3.2.1   Tryptic Digestion of Urine Proteins for MRM-MS. In the optimized tryptic digestion procedure, 15 μg of urine protein was dissolved in 15 μL of 25 mM ammonium bicarbonate. Sequencing-grade-modified trypsin (Promega, Madison, WI) was added to the urine protein at a 1 : 50 enzyme:substrate ratio. The first digestion was carried out for 8 h at 37 °C. Then, disulfide bonds were reduced with 0.45 μL of 200 mM dithiothreitol by incubation at 37 °C for 30 min. The reduced urinary proteins were then alkylated with 1.35 μL of 200 mM iodoacetamide for 1 h at room temperature in the dark. After alkylation, 200 mM DTT was added to the samples to consume any leftover alkylating agent, and kept at 37 °C for 30 min. The samples were incubated with 1 μg/μL trypsin (at a 1 : 10 enzyme:substrate ratio) at 37 °C for 16 h to give as complete a proteolysis as possible. The tryptic peptides were then lyophilized in a 96-well plate and stored at −80 °C for subsequent processing.

10.2.3.2.2   Addition of Stable-Isotope-Labeled Standard (SIS) Peptides and Desalting by Solid-Phase Extraction (SPE). All sample manipulations, including the addition of the SIS peptides and the SPE desalting steps were performed using a Tecan Freedom Evo150 liquid-handling robot.[41,42,51] The [$^{13}C_6$]Lys and [$^{13}C_6$]Arg-coded SIS peptides were synthesized at the UVic-Genome BC Proteomics Centre, and were purified as previously described.[37] Briefly, the purity of the HPLC-purified peptides was then determined by capillary zone electrophoresis (CZE) and amino acid analysis. Urinary tryptic peptides in the 96-well plates were rehydrated in 0.7% formic acid and spiked with a constant level of a SIS peptide mixture containing 65 SIS peptides. The SIS peptide mixture was prepared as previously described,[52] and was concentration balanced for the protein composition of plasma, not for urine. Samples were then desalted and concentrated by SPE using a Waters Oasis 2-mg HLB Elution plate (Waters, MA), according to the manufacturer's protocol. The eluted samples were frozen and lyophilized to dryness. Prior to LC-MRM/MS analysis, samples were rehydrated with 0.1% v/v formic acid to produce a solution with a 1-μg/μL protein concentration.

10.2.4   AB 4000 Qtrap Mass Spectrometry for MRM-MS Samples

10.2.4.1   Selection of Protein Targets

The starting point for this assay was a previously determined procedure for the analysis of 45 plasma proteins from undepleted plasma.[37] An additional 18 proteins were added to this procedure, giving it the ability to quantitate a total of 63 proteins per analysis. This assay used optimized transitions (precursor/product ion pairs) that were specific for the particular target proteins. Scheduled MRM, meaning that particular transitions were monitored at specific times during the gradient elution was used. This procedure increased the number of peptides that could be monitored in each run – in other words, it increased the multiplexing capabilities of the assay compared to a method where all transitions were monitored during the entire chromatographic separation. Three transitions were used per target peptide (with 65 proteotypic peptides representing 63 proteins), and each was optimized for the highest possible sensitivity by tuning the declustering potential (DP) and the collision energy (CE) voltages.

10.2.4.2 MRM Analysis

An AB/MDS Sciex 4000 QTRAP with a nanoelectrospray ionization source controlled by Analyst 1.5 software (Sciex) was used for all LC-MRM/MS analyses. Isolation of Q1 and Q3 ions were set to unit resolution (0.6–0.8 Da full width at half-height). The scheduled MRM option was used for all data acquisition, with a target cycle time of 2 s and a 4-min MRM detection window. A Nanoflow Eksigent NanoLC-1Dplus HPLC was used for the separation of desalted urinary tryptic peptides. LC-MRM/MS analysis of each sample took 42 min, with 1 μg of protein from the original sample being injected onto the analytical column (a 75 μm × 150 mm Magic C18AQ column packed in house, with 5-μm particles, 100 Å pore size, (Michrom, Auburn, CA)),[49] and preconcentrated at 300 nL/min of 2% buffer (98% v/v acetonitrile, 0.1% v/v formic acid) for 6 min. The gradient was run at a flow rate of 300 nL/min using a series of linear gradients (from 2 to 12% B (in 2 min), from 12 to 18% B (in 15 min), from 18 to 25% B (in 6 min), from 25 to 40% B (in 3.9 min), and finally from 40 to 80% B (in 1.1 min), giving a total analysis time of 42 min. A blank solvent injection was run between all samples to prevent sample carryover on the HPLC column using a 20-min method (from 2 to 42% B in 9.5 min, from 42 to 80% B in 2.5 min, hold at 80% B for 0.5 min and re-equilibrate with 2% B for 7.5 min; flow rate 600 nL/min).

10.2.4.3 Data Analysis of MRM-MS for Candidate Verification

The MultiQuant 1.2 (Applied Biosystems) software with the MQL algorithm was used for peak integration. Quantitation was based on external calibration curves for each of the 65 peptides as described previously.[37] However, it is not possible to prepare a "blank" sample that contains the appropriate matrix but that contains no endogenous proteins. Therefore, a "standard-addition" type of calibration curve was generated for each target protein covering the range from 0.1 to 3 times the concentration of a reference sample. These calibration curves were prepared by using 6 different amounts of a tryptic digest from a standard urine of one patient sample from a bladder-cancer patient, spiked into a solution containing a constant level of SIS peptides (see Supplemental Table 4 in reference 53). A $1/x$ (x = concentration ratio) weighting was used to allow the coverage of a wide dynamic range.

Because the accurate concentration of each SIS peptide is known, the concentrations of the protein in the unknown samples can be determined from the observed peak-area ratios. The coefficient of variations (%CV) for technical replicates using MRM were $<10\%$,[51] so only one LC-MRM-MS run was performed for each clinical sample. All integrated peaks were manually inspected to ensure correct peak detection and accurate integration. Instead of performing additional technical replicates, additional transitions per peptide were added to the method, to give 3 transitions per target peptide. The Agilent software uses one transition per peptide as the "quantifier", with the additional transitions being used as "qualifiers" to verify the retention time and to reveal any potential signal interference. This interference determination step is important, particularly for nanoscale separations.[54]

The urinary concentration of each target *protein* is expressed in pmol/µg of urinary protein, which is derived from the determined molar level of each proteotypic *peptide*, assuming complete tryptic digestion and 100% peptide recovery. The LOD (based on 5 aliquots of a concentrated and desalted urine sample) was defined as the lowest level at which a signal was observed for the endogenous peptide with a signal-to-noise ratio $(S/N) > 20$ in all five replicates, while the LOQ was defined as the lowest concentration of endogenous peptide that can be measured with $<20\%$ CV $(n = 5)$. The concentration of each target protein was expressed in pmol/µg of urinary *protein*, which is derived from the determined molar level of each proteotypic *peptide*, assuming complete tryptic digestion and 100% peptide recovery

10.2.5   Verification of MRM-MS Results

10.2.5.1   Quantitative Measurement of APOA1 by ELISA

The levels of APOA1 in diluted urine or desalted/concentrated urinary protein samples were determined using a sandwich ELISA kit (Matritech, Newton, MA, USA) according to the instructions provided by the manufacturer (Mabtech, Cincinnati, OH, USA). Purified APOA1 (A0722, Sigma-Aldrich, USA) was used as the calibration standard. Urine samples (100 µl of NT urine at a 1:2 dilution, or 100 µl of bladder-cancer urine at a 1:100 dilution) were added and processed as previously described.[48] The APOA1 concentration in urine samples was calculated by reference to a standard curve prepared using purified APOA1. The reference level of APOA1 was set at 0.2–20 ng/mL by the manufacturer.

10.2.5.2   Immunobead-Based Suspension Array System for Detection of Urinary Apolipoproteins (APOA1, APOA2, APOB, APOC3, and APOE)

The urinary APOA1, APOA2, APOB, APOC3, and APOE levels were determined using the MILLIPLEX MAP™ Human Apolipoprotein Panel kit (Millipore, MA, USA). A 10-µL aliquot of each standard and urine sample (2-fold dilution) was incubated with 90 µL of immunobeads in filter-bottom microplates (Millipore) for 50 min. The plates were washed three times with wash buffer using a vacuum manifold (Millipore). The biotin-conjugated detection antibody was added and incubated for 30 min. After three washes, 50 μl of the phycoerythrin–streptavidin solution was added. After 30 min, the immunobeads were resuspended in sheath buffer (Bio-Rad Laboratories, CA, USA) and analyzed using the Bio-Plex 200 system (Bio-Rad Laboratories). The standard curves and analyte concentrations were obtained using the Bio-Plex Manager software version 4.2 (Bio-Rad Laboratories).

10.2.6 Statistical Analysis

10.2.6.1 Statistical Analysis for Verification of a Single Biomarker

The statistical package SPSS 13.0 (SPSS Inc., Chicago, IL, USA) was used for all analyses. Both Pearson's (γ) and Spearman's rank (ρ) correlation coefficients were used to evaluate correlations between MRM-MS, ELISA and Bio-plex systems. Differences in concentration levels of targeted urinary proteins between different clinical parameters measured by MRM-MS assay was analyzed using the nonparametric Mann–Whitney test. The Receiver Operator Characteristic (ROC) curve analysis and the Area-Under-the-Curve (AUC) were applied to determine the optimal cutoff point that yielded the highest total accuracy with respect to discriminating different clinical classifications. The optimal cutoff point was determined using Youden's index (J), calculated as $J = 1 - (\text{false-positive rate} + \text{false-negative rate}) = 1 - [(1 - \text{sensitivity}) + (1 - \text{specificity})] = \text{sensitivity} + \text{specificity} - 1$ as stated in ref. 55. All tests were two-sided and p-value < 0.05 was considered statistically significant.

10.2.6.2 Statistical Analysis for Development of a Marker Panel

Two groups of clinically distinct study subjects were recruited for an evaluation of the joint classification performance of multiple differentially expressed peptides as previously described.[52] The cancer group consisted of 76 bladder-cancer patients, and the control group consisted of 80 noncancer subjects (57 hernia and 23 UTI/HU specimens). A nonparametric Mann–Whitney test was first used to individually compare the peptide abundances between these clinically distinct groups. We then applied a hierarchical clustering method, using the Cluster 3.0 software,[56] which reduced the number from 65 peptides to 43 peptides, which showed significant differences (p-value < 0.0008) based on their Spearman's rank correlation coefficients. A heatmap was used for the visualization of peptide abundance, where red and green colors are used to show higher and lower abundance levels compared with the peptide-specific average abundance of the 156 samples. The hierarchical relationships of the peptides are also shown on the left-hand side of the heatmap. Using this approach, the list of statistically significant peptides was further shortened from 43 to 6, based on their representative peptides derived from the hierarchy tree, with the goal of reducing redundancy in the biomarker panel while preserving an adequate degree of classification accuracy. Peptides for the biomarker panel were selected so that the absolute values of the Spearman's rank correlation coefficients between them do not exceed 0.7, a threshold value for high correlation.[57] The logistic regression method was used to further evaluate the joint performance of the short list of peptides for the classification of cancerous and noncancerous subjects, as previously described.[52] An ROC curve was plotted for every 0.1 increment of the cut value on the p-value to demonstrate the classification performance of the model. The Hosmer and Lemeshow test with 8 degrees-of-freedom was also used to evaluate the goodness-of-fit of the constructed model, where a p-value < 0.05 shows that the model does not have an adequate goodness-of-fit.[57] SPSS (Statistical Package for the Social Sciences) software[58] was used for the above analyses unless otherwise stated.

10.2.7 Functional Annotation and Network Analysis of Differential Proteins

ProteinCenter bioinformatics software (version 2.8.1, Proxeon Biosystems, Odense, Denmark) was used to annotate the urine proteome with Gene Ontology terms (molecular function, cellular component, and biological process). A comparison of the two independent clinical sample data sets from the iTRAQ experiments was also performed using this same ProteinCenter bioinformatics software. To link these data sets, the IPI accession numbers for each of the two sets of data were loaded onto the ProteinCenter server. Individual data sets were then linked *via* their accession numbers in the ProteinCenter database.

A network analysis of differentially expressed proteins identified in each clinical set was performed using the MetaCore analytical suite version 5.4 (GeneGo, Inc., St. Joseph, MI, USA). Hypothetical networks of proteins from our experiments and proteins from the MetaCore database were built using the algorithms integrated into MetaCore. The statistical significance of the identified networks was based on p-values, which are defined as the probability that a given number of proteins from the input list will match a certain number of gene nodes in the network. The relevant pathway maps were then prioritized based on their statistical significance (p-value < 0.001).

10.3 Results and Discussion

10.3.1 iTRAQ for Candidate Selection

Our previous work had validated the performance and reproducibility of the iTRAQ workflow for urinary proteomics, including the processes for urine–protein desalting, concentration, proteolytic digestion, and iTRAQ labeling.[48] To select proteins as potential bladder-cancer biomarkers, two independent sets of biological samples (clinical sample set 1 and clinical sample set 2), pooled from different individuals, were used for the iTRAQ experiments.

A total of 352 nonredundant proteins (1074 distinct peptides) were identified by combining LC-MS/MS data files from the 42 SCX fractions from clinical sample set 1, and 552 nonredundant proteins (1589 distinct peptides) were identified by combining LC-MS/MS data files from 42 SCX and the 42 basic RP fractions from clinical sample set 2. Integration of the proteins identified from these two clinical sample sets yielded a total of 638 urine proteins (PeptideProphet and ProteinProphet probability scores ≥0.95, false-positive rate<0.7%).

The iTRAQ ratios of 274 proteins (based on 1500 peptides that met the Libra acceptance criteria) and 429 proteins (based on 3276 peptides that met the Libra acceptance criteria) in clinical sample sets 1 and 2, respectively. were quantified using TPP software;[59] 196 proteins (38.7% of the total 507 quantified proteins) were quantified in both sample sets. Among the quantified proteins, those displaying a fold-difference ≥2.0 or ≤0.5 between any tumor subgroup and the NT subgroup were defined as significantly up- or downregulated. Importantly, 22 proteins were upregulated (BC>NT) in both clinical sample sets and 33 proteins were downregulated (BC<NT) in both sets.[48] A brief summary of these 22 urinary proteins that showed increased levels in bladder cancer in the two iTRAQ experiments is presented in Table 10.2. These proteins represent potential urinary bladder-cancer biomarkers.

10.3.2   MetaCore™ Analysis of Biological Networks Associated with Differentially Expressed Proteins in Urine

We used MetaCore software to analyze the 196 proteins of clinical sample set 1 and the 338 proteins of clinical sample set 2 whose levels were significantly different between hernia controls and bladder-cancer patients in order to determine which were the biological networks with which they are associated. The analysis revealed 14 pathways that were significantly associated with these dysregulated urinary proteins (*p*-value <0.001), including blood coagulation, cell adhesion, immune response, and lipoprotein metabolism-related pathways (Table 10.3). Most of these pathways were identified in both of the independent clinical sample sets. As shown in Figure 10.2, the most significant biological network identified was the blood-coagulation pathway with a *p*-value of 3.23×10^{-18}. Of the 39 network objects (*i.e.* the molecules that comprise the network) for this pathway, 12 and 16 were quantified as differentially expressed proteins in clinical sample sets 1 and 2, respectively. In our iTRAQ results, we detected upregulation of plasma kallikrein; alpha-2-antiplasmin precursor; plasmin; alpha-2-macroglobulin precursor; alpha-1-antitrypsin precursor; antithrombin III variant; heparin cofactor 2 precursor; prothrombin precursor; and fibrinogen α-, β-, and γ-chains; in addition, kininogen-1 precursor was downregulated in the urine samples from bladder-cancer patients. Our findings are consistent with a previous study by Tsihlias and Grossman,[60] who reported that fibrin/fibrinogen degradation products were either absent or present at exceedingly low levels in the urine of

Table 10.2 iTRAQ ratios and verification results of the 22 proteins whose expression increased in urine specimens from patients with hernia and bladder cancer. Adapted from, [48,53] with permission.

| Accession Number | Gene Name | Protein Name | Discovery by iTRAQ (Cancer/Hernia ratio) | | | | | | Verification by MRM-MS | |
| | | | Clinical sample set 1 | | | Clinical sample set 2 | | | Average concentration in hernia urine (pmol/μg) | Average concentration in bladder-cancer urine (pmol/μg) |
			115/114 (LgEs)	116/114 (HgEs)	117/114 (HgAs)	115/114 (LgEs)	116/114 (HgEs)	117/114 (HgAs)		
IPI00164623.4	C3	187-kDa protein	1.8	19.8	27.8	10.3	3.4	13.8		
IPI00553177.1	SERPINA1	alpha-1-antitrypsin precursor	1.2	12.5	11.2	3.9	3.3	6	3.9E-03	2.0E-02
IPI00478003.1	A2M	alpha-2-macroglobulin precursor	1.3	14	20.9	10.2	4.8	16.6	2.2E-04	8.8E-03
IPI00019943.1	AFM	Afamin precursor	1.2	5.1	7.1	4	3.6	5.6	1.7E-03	2.5E-03
IPI00022434.2	ALB	ALB protein	1.4	11.8	15.9	5.1	3.7	7.5	1.5E+00	5.7E+00
IPI00022429.3	ORM1	Alpha-1-acid glycoprotein 1 precursor	2.5	4.8	2.2	1.5	3.2	1.5	1.3E-01	2.9E-01
IPI00032179.2	SERPINC1	Antithrombin III variant	1.1	7.3	4.3	3.3	2.6	4.4	4.2E-03	1.6E-02
IPI00021841.1	APOA1	Apolipoprotein A-I precursor	2	32.3	41.7	14.4	4.5	18.1	8.0E-04	5.9E-02
IPI00021854.1	APOA2	Apolipoprotein A-II precursor	2.4	71.9	124.9	2.2	1.3	2.6	1.3E-03	5.4E-02
IPI00304273.2	APOA4	Apolipoprotein A-IV precursor	1.1	5.2	10.2	3.6	3.6	3.5	4.4E-04	9.1E-03

IPI	Gene	Protein								
IPI00298828.3	APOH	b2-glycoprotein 1 precursor	1.2	6.7	13.7	3.1	1.9	3.6	7.2E−04	6.0E−03
IPI00017601.1	CP	Ceruloplasmin precursor	1	6.9	6.5	2.6	1.8	3	5.2E−03	1.3E−02
IPI00303963.1	C2	Complement C2 precursor (Fragment)	1.2	5.1	6.1	2.1	1.2	3.2		
IPI00298497.3	FGB	Fibrinogen b-chain precursor	5.2	8.6	45.4	15.1	3.2	54	1.4E−04	9.4E−03
IPI00745363.3	-	Immunoglobulin heavy chain variable region	1.2	4.5	7.1	2.4	1.5	3.6		
IPI00021885.1	FGA	Isoform 1 of Fibrinogen alpha chain precursor	2.9	16.9	35.7	2.8	2.4	12.9	1.1E−03	2.1E−02
IPI00021891.5	FGG	Isoform Gamma-B of Fibrinogen gamma chain precursor	1.6	11.1	29.1	9	2	30.9	1.3E−04	1.1E−02
IPI00017526.1	S100P	Protein S100-P	1.4	50.8	29.3	4.8	1.5	1.9		
IPI00022463.1	TF	Serotransferrin precursor	1.4	21	21.7	5.4	3.6	7.2	2.6E−02	7.2E−02
IPI00019399.1	SAA4	Serum amyloid A-4 protein precursor	1.4	15.8	15.4	3.9	2.8	5.3		
IPI00022432.1	TTR	Transthyretin precursor	1.3	3.3	4	4	2.3	6.3	2.1E−03	1.6E−02
IPI00555812.4	GC	Vitamin D-binding protein precursor	1.8	15.6	25.3	5.8	3.3	8.9	3.7E−03	2.0E−02

Table 10.3 Significant biological pathways (*p*-value <0.001) in which differentially expressed urine proteins participate.

Biological pathway	*p-value*	Network objects	Clinical sample set
Blood coagulation_Blood coagulation	5.40E–15	12/39	1
	3.23E–18	16/39	2
Cell adhesion_Cell-matrix glycoconjugates	8.05E–04	4/38	1
	1.33E–09	10/38	2
Cell adhesion_ECM remodeling	2.67E–04	5/52	1
	4.17E–04	6/52	2
Cell adhesion_Plasmin signaling	5.85E–04	4/35	1
	4.70E–04	5/35	2
Cell adhesion_PLAU signaling	8.89E–04	4/39	1
	7.85E–04	5/39	2
Cell adhesion_Role of tetraspanins in the integrin-mediated cell adhesion	6.12E–04	5/37	2
Fructose metabolism/ Rodent version	9.53E–04	7/83	2
Immune response_Alternative complement pathway	6.64E–05	5/39	1
	8.18E–05	6/39	2
Immune response_Classical complement pathway	1.95E–10	10/52	1
	5.15E–06	8/52	2
Immune response_Lectin-induced complement pathway	3.71E–12	11/49	1
	3.38E–05	7/49	2
LDL metabolism during development of fatty streak lesion – conventional	7.08E–06	7/39	2
Lipoprotein metabolism I. Chylomicron, VLDL and LDL metabolism – conventional	5.08E-07	6/28	1
	1.12E-05	6/28	2
Lipoprotein metabolism II. HDL metabolism – conventional	3.35E-05	5/34	1
Niacin-HDL metabolism	1.65E-04	5/47	1

healthy persons, but tended to be higher in patients with tumors of increasing grade and stage.

10.3.3 Verification of iTRAQ-Discovered Biomarkers in a Larger Number of Individual Samples Using Western Blot Analyses

Four proteins (APOA1, APOA2, PRDX2, and HCII), found in our earlier biomarker discovery work,[48] were successfully verified by Western blot analyses where they showed significant difference in urinary concentrations between bladder-cancer and hernia patients. In addition, our iTRAQ results showed that S100P was the only member of this family upregulated in both clinical sample sets. In the Western blot study, several additional members of the S100 protein family, including S100A2, S100A4, S100A6, S100A8, S100A9, S100A11 and S100P, were quantified and found to be differentially expressed proteins in urine from bladder-cancer patients. Both S100A8 and S100A9 were upregulated in one pooled clinical sample set, but were slightly downregulated in the second sample set, indicating individual variability in S100A8/A9

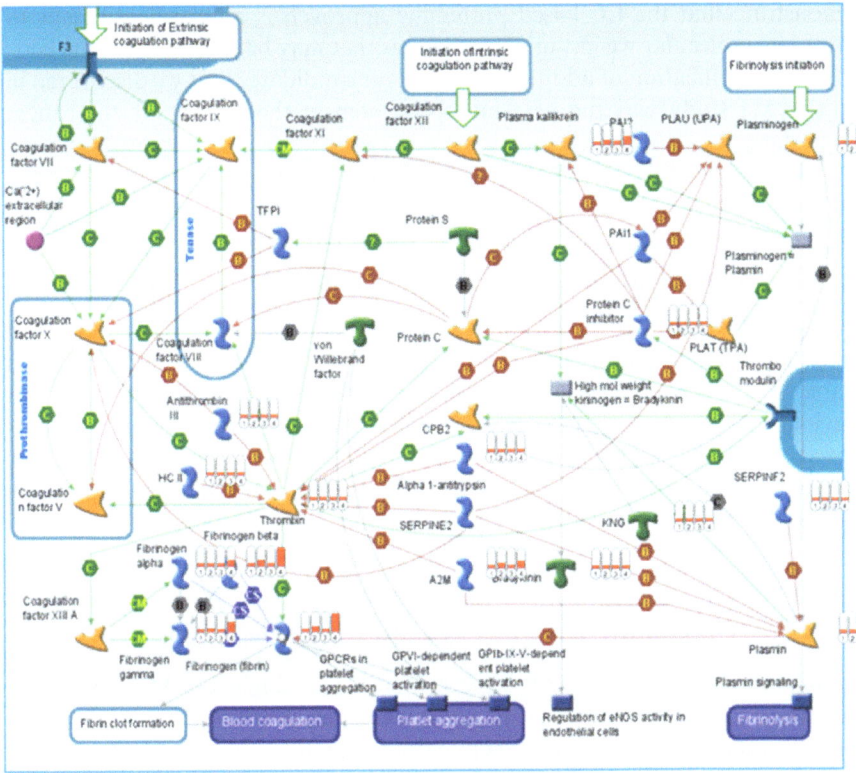

Figure 10.2 Biological networks associated with the differentially expressed proteins identified in the iTRAQ analysis, generated using MetaCore software.

expression in urine. S100A4, S100A6 and S100A11 were only able to be quantified by iTRAQ in one clinical sample set. S100 proteins are characterized by calcium-binding EF-hand motifs and act as multifunctional signaling factors that are involved in the regulation of diverse cellular process.[61] They have molecular weights of approximately 10 000 Da and are difficult to resolve by gel-based proteomic approaches. S100A8 and S100A9 are elevated early in tissue and serum in many pathological conditions associated with inflammation, such as inflammatory disease, arthritis, tumor growth, and early metastasis.[62–66] S100 family members are considered to be potential biomarkers of numerous types of cancer.[67–70] The expression of the S100 family of genes has been systematically evaluated in mouse and rat bladder-tumor models by microarray and RT-PCR, which showed that S100A8 and S100A9 are significantly overexpressed in bladder tumors; similar results have been obtained in humans.[71] However, using Western blot analyses of S100A6 and S100A8 in individual urine samples, we found that neither protein was able to differentiate hernia, bladder cancer, and UTI/hematuria patients with statistical confidence (data not shown). Although S100 family members are not likely to be disease-specific biomarkers of bladder cancer, our data still demonstrate that the LC-based proteomic approach is an excellent means to detect low molecular weight urinary proteins that may be potential biomarkers.

Further verification of additional biomarker candidates that we discovered in the iTRAQ study was the next important step in the search for the cancer biomarkers.

10.3.4   Verification of Additional Biomarkers in a Larger Number of Individual Urine Samples Using MRM-MS

The majority of the 22 proteins that showed differential changes in the iTRAQ study of bladder-cancer urine were common abundant plasma proteins (Table 10.2). While quantitative profiling of the plasma proteome has been previously explored,[37,38,51,72,73] we decided to use our high-throughput multiplexed MRM-based plasma-protein platform to accurately quantify these "plasma" proteins in urine and to evaluated their performance as biomarkers for bladder cancer.

10.3.4.1   *Quantitation of Urinary Proteins from Patients with Common Urological Diseases by MRM-MS*

To achieve quantitation of 63 urine proteins in a single LC-MS/MS run, a mixture of 65 SIS peptides was spiked into each urinary tryptic peptide digest. The selection of the appropriate peptide(s) for each target protein to generate SIS peptides was determined from our previous plasma studies.[53,74] The transition with highest S/N was used for quantitation because higher signals in general produce more accurate quantitative data.[37] Chromatograms for the multiplexed MRM quantitation of 63 proteins (with 65 peptides) in a single LC-MRM/MS analysis of the urine samples from a hernia patient and a bladder-cancer patient are shown in Figure 10.3A and Figure 10.3B, respectively. This multiplexed MRM analysis was performed on the urinary protein digests from each of the 156 patients in this study. The concentration distributions of the 65 target peptides in the hernia group ($n = 57$) and bladder cancer ($n = 76$) are summarized as Figures 10.4A and 4B, respectively. The average concentrations and standard derivations of these 65 peptides representing 63 proteins in the 156 clinical urine samples are summarized in Table 10.4. In plasma, the 30 most-abundant proteins comprise approximately 99% of the total protein mass.[75,76] To explore the quantitative profile of urinary proteome, we plotted the urinary proteome as a pie chart using the data obtained by MRM-MS. Of all 63 proteins, albumin was the most-abundant protein in hernia as well as bladder cancer in all groups. The average concentrations of albumin were 1.49 and 5.71 pmol/µg (*i.e.* 103.13 ng/µg and 396.20 ng/µg), which accounted for 10.3% and 39.6% of total urine protein in hernia and bladder cancer, respectively (Figure 10.3A and 3B). The percentage is much lower than the percentage of albumin in serum or plasma, where albumin comprises approximately 50% of the total serum protein content.[77]

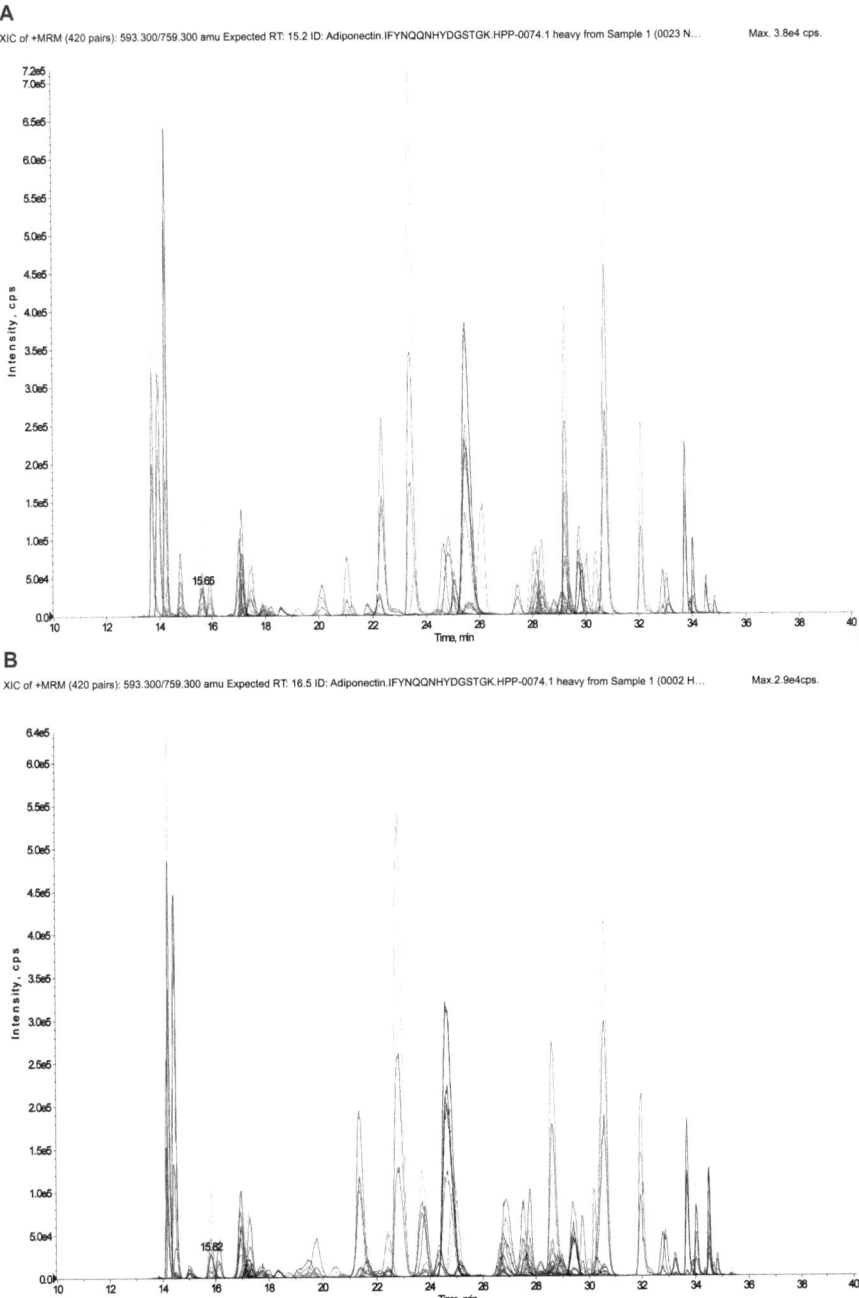

Figure 10.3   Multiplexed MRM quantitation of 63 proteins (with 65 peptides) in a single multiplexed LC-MRM/MS analysis of the urine samples from (A) a hernia patient and (B) a bladder-cancer patient.

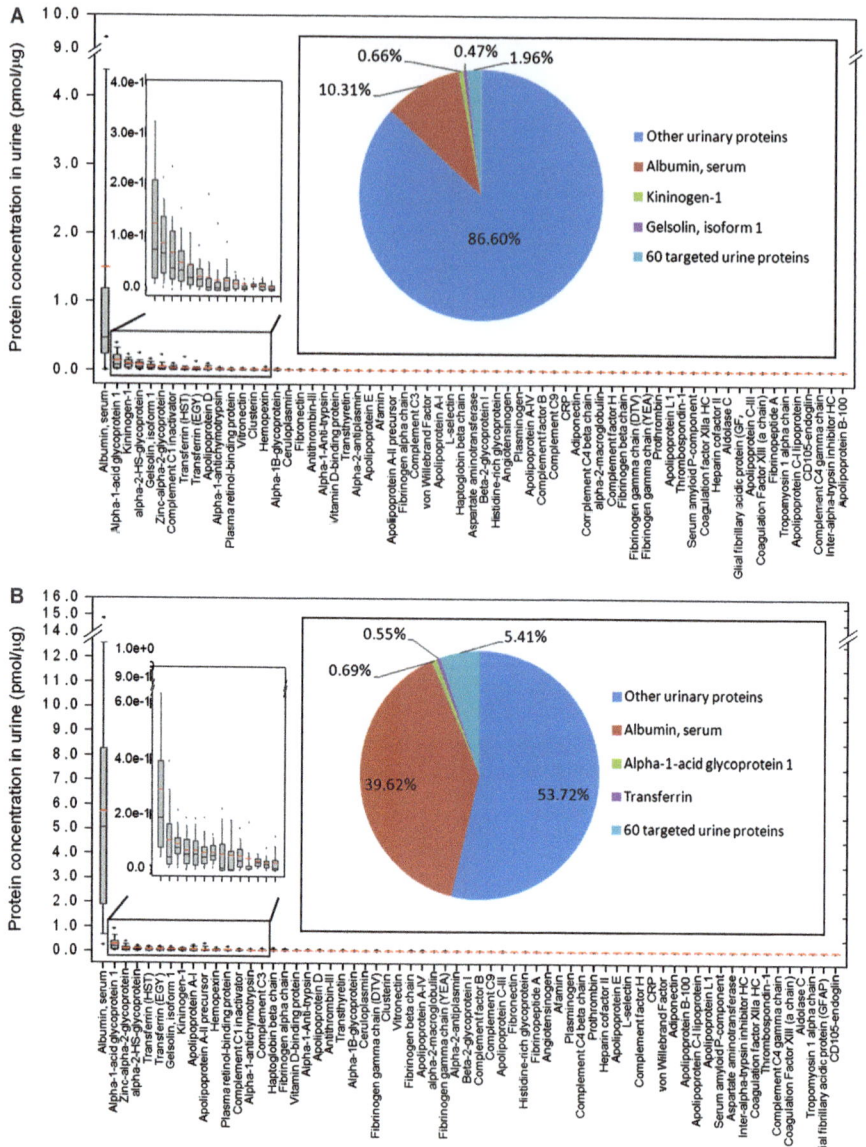

Figure 10.4 The concentration distributions of the 65 target peptides in the (A) hernia group ($n = 57$), and (B) bladder cancer ($n = 76$) determined using the MRM-MS approach.
Adapted from,[48] with permission.

The concentration of the remaining targeted 62 proteins accounted for 3.1–6.7% of total urine proteome. These results indicated that the remaining untargeted urine proteins comprised 86.6% of the total urinary proteins for hernia patients and 53.7% of the total urinary proteins for bladder-cancer patients.

Table 10.4 Urinary protein concentration of 156 individual samples as determined by LC-MRM-MS. Reprinted from[48] with permission. The protein concentration of each target is expressed in pmol/μg of urine protein, which is exactly the same as the determined level of the proteotypic peptide by assuming complete tryptic digestion (*i.e.* 100% recovery). Calibration curves could not be generated for ten of the 63 proteins (adiponectin, aldolase C, aspartate aminotransferase, CD105-endoglin, coagulation factor XIII (a chain), complement C4 gamma chain, glial fibrillary acidic protein, thrombospondin-1, tropomyosin 1 alpha chain, and von Willebrand factor) whose concentrations in the standard urine sample were below the LOQ. For these ten proteins, the concentrations were determined based on the known concentration of the corresponding SIS peptide, making the assumption that the calibration curve goes thru zero. Reprinted from,[53] with permission.

Protein Name	Hernia (n = 57)	LgEs (n = 21)	HgEs (n = 41) (pmol/μg)	HgAs (n = 14)	UTI/HU (n = 23)
Adiponectin	2.7E-4 ± 3.3E-4	5.1E-4 ± 4.2E-4	7.8E-4 ± 8.8E-4	5.6E-4 ± 4.5E-4	3.8E-4 ± 2.8E-4
Afamin	1.7E-3 ± 1.1E-3	2.4E-3 ± 1.7E-3	2.4E-3 ± 1.1E-3	3.2E-3 ± 1.1E-3	1.9E-3 ± 1.3E-3
Albumin, serum	1.5E+0 ± 2.8E+0	3.8E+0 ± 4.4E+0	5.6E+0 ± 3.7E+0	9.0E+0 ± 5.2E+0	3.5E+0 ± 2.0E+0
Aldolase C	6.2E-5 ± 1.1E-4	7.5E-5 ± 1.0E-4	4.9E-5 ± 7.3E-5	1.1E-4 ± 8.4E-5	1.1E-4 ± 2.1E-4
Alpha-1-acid glycoprotein 1	1.3E-1 ± 1.3E-1	3.0E-1 ± 3.5E-1	3.0E-1 ± 3.2E-1	2.8E-1 ± 2.9E-1	2.7E-1 ± 3.5E-1
Alpha-1-antichymotrypsin	1.3E-2 ± 1.1E-2	2.0E-2 ± 1.6E-2	2.9E-2 ± 2.4E-2	3.5E-2 ± 3.3E-2	2.2E-2 ± 2.5E-2
Alpha-1-Anti-trypsin	3.9E-3 ± 5.3E-3	1.3E-2 ± 1.7E-2	2.0E-2 ± 3.0E-2	3.2E-2 ± 2.5E-2	1.1E-2 ± 1.7E-2
Alpha-1B-glycoprotein	8.1E-3 ± 8.5E-3	1.4E-2 ± 1.4E-2	1.4E-2 ± 9.3E-3	2.1E-2 ± 1.4E-2	1.2E-2 ± 9.6E-3
Alpha-2-antiplasmin	2.1E-3 ± 2.1E-3	3.9E-3 ± 5.2E-3	6.6E-3 ± 5.8E-3	9.3E-3 ± 7.4E-3	4.7E-3 ± 3.6E-3
alpha-2-HS-glycoprotein	7.3E-2 ± 7.0E-2	7.2E-2 ± 6.2E-2	1.1E-1 ± 7.0E-2	7.9E-2 ± 3.5E-2	5.4E-2 ± 3.9E-2
alpha-2-macroglobulin	2.2E-4 ± 8.9E-4	3.8E-3 ± 7.6E-3	8.4E-3 ± 1.2E-2	1.7E-2 ± 1.3E-2	3.8E-3 ± 5.3E-3
Angiotensinogen	6.1E-4 ± 6.9E-4	2.1E-3 ± 3.0E-3	3.2E-3 ± 3.2E-3	2.6E-3 ± 2.1E-3	1.8E-3 ± 2.9E-3
Antithrombin-III	4.2E-3 ± 4.5E-3	1.1E-2 ± 1.0E-2	1.6E-2 ± 1.3E-2	2.2E-2 ± 1.5E-2	1.1E-2 ± 6.0E-3
Apolipoprotein A-I	8.0E-4 ± 6.1E-3	3.7E-2 ± 7.8E-2	4.8E-2 ± 6.7E-2	1.2E-1 ± 1.5E-1	2.7E-2 ± 3.3E-2
Apolipoprotein A-II precursor	1.3E-3 ± 7.6E-3	3.1E-2 ± 6.4E-2	5.0E-2 ± 8.7E-2	1.0E-1 ± 1.0E-1	1.1E-2 ± 2.5E-2
Apolipoprotein A-IV	4.4E-4 ± 8.5E-4	1.7E-2 ± 5.5E-2	6.9E-3 ± 1.9E-2	3.6E-3 ± 5.1E-3	6.0E-3 ± 1.6E-2

Table 10.4 *(Continued)*

Protein Name	Hernia (n = 57)	LgEs (n = 21)	HgEs (n = 41) (pmol/µg)	HgAs (n = 14)	UTI/HU (n = 23)
Apolipoprotein B-100	7.7E−6 ± 3.8E−5	1.7E−4 ± 3.5E−4	6.0E−4 ± 1.8E−3	1.4E−3 ± 1.9E−3	1.6E−4 ± 4.2E−4
Apolipoprotein C-I lipoprotein	3.3E−5 ± 3.7E−5	5.6E−4 ± 1.7E−3	4.5E−4 ± 1.4E−3	9.7E−4 ± 1.6E−3	3.4E−4 ± 1.1E−3
Apolipoprotein C-III	5.4E−5 ± 3.8E−4	6.2E−3 ± 2.3E−2	1.4E−3 ± 4.3E−3	5.1E−3 ± 9.9E−3	2.8E−3 ± 8.7E−3
Apolipoprotein D	1.9E−2 ± 3.0E−2	1.5E−2 ± 1.7E−2	1.7E−2 ± 2.3E−2	1.6E−2 ± 1.8E−2	1.3E−2 ± 2.0E−2
Apolipoprotein E	1.7E−3 ± 1.5E−3	1.4E−3 ± 1.5E−3	1.1E−3 ± 1.5E−3	2.1E−3 ± 1.4E−3	8.7E−4 ± 6.3E−4
Apolipoprotein L1	1.1E−4 ± 1.1E−4	3.5E−4 ± 4.5E−4	5.4E−4 ± 5.2E−4	8.4E−4 ± 1.0E−3	2.6E−4 ± 2.5E−4
Aspartate aminotransferase, mitochondrial (m-type)	7.4E−4 ± 1.4E−3	7.6E−4 ± 1.6E−3	4.9E−4 ± 9.2E−4	1.7E−4 ± 2.2E−4	4.3E−4 ± 8.6E−4
Beta-2-glycoprotein I	7.2E−4 ± 1.1E−3	4.5E−3 ± 8.5E−3	5.3E−3 ± 7.4E−3	1.0E−2 ± 1.0E−2	6.1E−3 ± 1.3E−2
CD105-endoglin	3.0E−5 ± 4.4E−5	3.4E−5 ± 6.8E−5	2.7E−5 ± 3.0E−5	3.7E−5 ± 3.5E−5	2.6E−5 ± 3.7E−5
Ceruloplasmin	5.2E−3 ± 7.1E−3	7.1E−3 ± 7.5E−3	1.3E−2 ± 1.3E−2	2.0E−2 ± 1.9E−2	8.1E−3 ± 8.4E−3
Clusterin	9.2E−3 ± 7.8E−3	1.1E−2 ± 8.6E−3	9.9E−3 ± 9.1E−3	9.6E−3 ± 5.7E−3	1.4E−2 ± 9.0E−3
Coagulation factor XIIa HC	7.8E−5 ± 1.3E−4	2.1E−4 ± 3.6E−4	3.1E−4 ± 3.6E−4	5.2E−4 ± 4.4E−4	2.7E−4 ± 5.7E−4
Coagulation Factor XIII (a chain)	4.7E−5 ± 8.8E−5	1.1E−4 ± 1.8E−4	7.1E−5 ± 1.2E−4	2.1E−4 ± 3.4E−4	1.0E−4 ± 1.4E−4
Complement C1 inactivator	2.7E−2 ± 2.5E−2	2.4E−2 ± 1.8E−2	3.4E−2 ± 2.4E−2	4.3E−2 ± 2.8E−2	2.7E−2 ± 2.0E−2
Complement C3	1.0E−3 ± 2.6E−3	1.9E−2 ± 3.6E−2	2.3E−2 ± 3.2E−2	4.8E−2 ± 5.0E−2	1.1E−2 ± 1.2E−2
Complement C4 beta chain	2.7E−4 ± 2.9E−4	1.4E−3 ± 2.2E−3	1.7E−3 ± 1.9E−3	3.7E−3 ± 4.4E−3	1.1E−3 ± 1.0E−3
Complement C4 gamma chain	2.4E−5 ± 3.0E−5	9.5E−5 ± 2.5E−4	1.1E−4 ± 1.7E−4	4.1E−4 ± 4.4E−4	2.3E−4 ± 7.1E−4
Complement factor B	4.2E−4 ± 7.1E−4	2.5E−3 ± 4.2E−3	4.0E−3 ± 5.1E−3	6.8E−3 ± 6.7E−3	1.8E−3 ± 1.9E−3
Complement factor H	1.4E−4 ± 3.2E−4	9.5E−4 ± 1.9E−3	8.4E−4 ± 1.2E−3	1.5E−3 ± 2.1E−3	3.7E−4 ± 8.6E−4
Complement C9	3.1E−4 ± 8.4E−4	1.7E−3 ± 2.8E−3	3.6E−3 ± 6.8E−3	7.0E−3 ± 8.4E−3	2.3E−3 ± 5.8E−3
CRP	3.0E−4 ± 3.5E−4	3.7E−4 ± 5.0E−4	1.3E−3 ± 3.9E−3	9.8E−4 ± 1.3E−3	1.0E−3 ± 2.1E−3
Fibrinogen alpha chain	1.1E−3 ± 3.9E−3	1.9E−2 ± 5.5E−2	1.7E−2 ± 2.9E−2	3.5E−2 ± 3.7E−2	5.4E−2 ± 1.6E−1
Fibrinogen beta chain	1.4E−4 ± 9.8E−4	3.2E−3 ± 6.4E−3	6.9E−3 ± 1.4E−2	2.6E−2 ± 2.7E−2	1.2E−2 ± 3.0E−2
Fibrinogen gamma chain (DTV)	1.3E−4 ± 9.2E−4	5.2E−3 ± 9.6E−3	7.1E−3 ± 1.4E−2	3.1E−2 ± 3.4E−2	1.9E−2 ± 5.1E−2

Fibrinogen gamma chain (YEA)	$1.3E-4 \pm 9.4E-4$	$2.9E-3 \pm 6.0E-3$	$6.1E-3 \pm 1.3E-2$	$2.4E-2 \pm 2.5E-2$	$9.9E-3 \pm 2.4E-2$
Fibrinopeptide A	$3.7E-5 \pm 1.6E-4$	$9.9E-4 \pm 2.3E-3$	$2.0E-3 \pm 4.9E-3$	$8.8E-3 \pm 1.5E-2$	$1.1E-3 \pm 1.9E-3$
Fibronectin	$4.5E-3 \pm 4.4E-3$	$4.1E-3 \pm 4.0E-3$	$2.7E-3 \pm 2.5E-3$	$4.1E-3 \pm 3.2E-3$	$2.3E-3 \pm 3.0E-3$
Gelsolin, isoform 1	$5.4E-2 \pm 5.1E-2$	$6.2E-2 \pm 6.6E-2$	$7.1E-2 \pm 5.2E-2$	$4.7E-2 \pm 5.4E-2$	$5.2E-2 \pm 3.6E-2$
Glial fibrillary acidic protein (GFAP)	$5.7E-5 \pm 8.7E-5$	$6.1E-5 \pm 8.3E-5$	$4.4E-5 \pm 6.0E-5$	$3.8E-5 \pm 4.1E-5$	$4.9E-5 \pm 6.0E-5$
Haptoglobin beta chain	$7.5E-4 \pm 2.7E-3$	$4.6E-3 \pm 1.2E-2$	$2.2E-2 \pm 3.5E-2$	$4.8E-2 \pm 5.3E-2$	$2.4E-2 \pm 9.0E-2$
Hemopexin	$8.2E-3 \pm 1.3E-2$	$2.5E-2 \pm 2.7E-2$	$5.2E-2 \pm 5.0E-2$	$8.2E-2 \pm 5.8E-2$	$2.8E-2 \pm 2.3E-2$
Heparin cofactor II	$7.8E-5 \pm 2.2E-4$	$1.0E-3 \pm 1.6E-3$	$1.5E-3 \pm 2.0E-3$	$2.0E-3 \pm 1.6E-3$	$6.5E-4 \pm 6.5E-4$
Histidine-rich glycoprotein	$6.3E-4 \pm 9.3E-4$	$1.7E-3 \pm 2.7E-3$	$3.0E-3 \pm 4.3E-3$	$5.9E-3 \pm 5.8E-3$	$2.0E-3 \pm 2.1E-3$
Inter-alpha-trypsin inhibitor HC	$1.3E-5 \pm 5.0E-5$	$2.2E-4 \pm 4.6E-4$	$3.4E-4 \pm 7.9E-4$	$9.9E-4 \pm 1.0E-3$	$4.9E-5 \pm 9.3E-5$
Kininogen-1	$9.1E-2 \pm 7.9E-2$	$5.7E-2 \pm 4.6E-2$	$6.4E-2 \pm 3.3E-2$	$7.1E-2 \pm 4.4E-2$	$6.7E-2 \pm 4.7E-2$
L-selectin	$7.6E-4 \pm 6.8E-4$	$8.6E-4 \pm 6.1E-4$	$1.1E-3 \pm 6.2E-4$	$1.3E-3 \pm 7.6E-4$	$9.8E-4 \pm 6.0E-4$
Plasma retinol-binding protein	$1.3E-2 \pm 4.8E-2$	$5.0E-2 \pm 1.4E-1$	$5.2E-2 \pm 1.9E-1$	$9.6E-3 \pm 7.7E-3$	$5.6E-2 \pm 1.5E-1$
Plasminogen	$5.0E-4 \pm 7.8E-4$	$1.8E-3 \pm 2.8E-3$	$2.2E-3 \pm 2.7E-3$	$4.4E-3 \pm 3.8E-3$	$1.1E-3 \pm 1.4E-3$
Prothrombin	$1.3E-4 \pm 2.6E-4$	$1.1E-3 \pm 1.7E-3$	$1.6E-3 \pm 1.8E-3$	$3.2E-3 \pm 3.3E-3$	$5.2E-4 \pm 5.4E-4$
Serum amyloid P-component	$8.1E-5 \pm 1.2E-4$	$3.3E-4 \pm 6.2E-4$	$4.3E-4 \pm 4.2E-4$	$1.1E-3 \pm 8.1E-4$	$2.6E-4 \pm 2.3E-4$
Thrombospondin-1	$9.5E-5 \pm 1.1E-4$	$1.0E-4 \pm 1.3E-4$	$1.8E-4 \pm 2.1E-4$	$3.7E-4 \pm 4.0E-4$	$8.8E-5 \pm 1.5E-4$
Transferrin (EGY)	$2.0E-2 \pm 4.4E-2$	$3.3E-2 \pm 3.2E-2$	$7.6E-2 \pm 5.9E-2$	$1.1E-1 \pm 6.6E-2$	$3.8E-2 \pm 3.7E-2$
Transferrin (HST)	$2.6E-2 \pm 5.4E-2$	$3.9E-2 \pm 3.6E-2$	$7.7E-2 \pm 5.5E-2$	$1.1E-1 \pm 6.6E-2$	$4.5E-2 \pm 3.4E-2$
Transthyretin	$2.1E-3 \pm 4.2E-3$	$7.5E-3 \pm 1.1E-2$	$1.7E-2 \pm 1.8E-2$	$2.4E-2 \pm 2.0E-2$	$1.0E-2 \pm 1.3E-2$
Tropomyosin 1 alpha chain	$3.5E-5 \pm 4.1E-5$	$7.4E-5 \pm 1.8E-4$	$4.6E-5 \pm 4.8E-5$	$5.6E-5 \pm 5.6E-5$	$4.0E-5 \pm 5.8E-5$
Vitamin D-binding protein	$3.7E-3 \pm 4.4E-3$	$1.3E-2 \pm 1.9E-2$	$2.1E-2 \pm 1.8E-2$	$3.1E-2 \pm 2.0E-2$	$1.2E-2 \pm 7.6E-3$
Vitronectin	$1.1E-2 \pm 6.4E-3$	$1.1E-2 \pm 5.9E-3$	$9.5E-3 \pm 4.4E-3$	$1.1E-2 \pm 4.0E-3$	$1.1E-2 \pm 5.6E-3$
von Willebrand Factor	$8.2E-4 \pm 6.7E-4$	$6.8E-4 \pm 5.0E-4$	$6.4E-4 \pm 4.9E-4$	$8.1E-4 \pm 5.0E-4$	$5.0E-4 \pm 4.7E-4$
Zinc-alpha-2-glycoprotein	$4.8E-2 \pm 6.7E-2$	$9.2E-2 \pm 9.8E-2$	$1.3E-1 \pm 1.6E-1$	$7.2E-2 \pm 1.1E-1$	$6.7E-2 \pm 7.7E-2$

Thus, a large protein of the proteome still remains unstudied and is worthy of further investigation, particularly in statuses of renal and urological diseases.

10.3.4.2   Assessment of Technical Variation, LOD, and LOQ in the MRM Assay

Since the protein concentration of urine ($\sim$ 100 µg/mL) is much lower than that of plasma (50–80 mg/mL)[78] in order to load 1 µg of digested protein on the column a larger volume must be injected, but this is not possible (particularly with nano-LC) because of urine's high salt content. Our previous iTRAQ-based quantitative study of human urine samples[48] indicated that the %CV resulting from sample preparation is <15%). The same sample preparation procedure for urine sample concentration and desalting was used in the MRM study, and 5 aliquots of a standard urine sample (HgAs_06) were used to determine the technical variation, LOD, and LOQ of the LC-MRM/MS. The results are shown in Table 10.1. Ten of the 65 failed to generate a standard curve, probably because their presence was below the LOQ. If these ten peptides are ignored, 44 peptides (80% of 55 peptides) and 51 peptides (93% of 55 peptides) showed CVs of <20% and <30%, respectively. These results are similar to the reproducibilities reported for MRM-based quantitation of plasma, where 27 of the 45 proteins showed CVs of <20%.[37] This indicates that MRM-MS can provide reproducible urinary peptide quantitation. Four of the MRM-MS assays (alpha-2-HS-glycoprotein, apolipoprotein C-I lipoprotein, complement factor H, and fibronectin) had CVs of >30%. This high CV probably resulted from the use of a concentration-balanced SIS peptide mixture that was based on plasma protein concentrations,[37] instead of urinary protein compositions. In addition, the dynamic range of urinary proteins is quite different from that of plasma. The use of a concentration-balanced SIS-peptide mixture based on urinary protein concentrations would probably result in lower %CVs for these four proteins.

The LOQ was defined as the lowest concentration of endogenous peptide that can be measured with a <20% CV, while the LOD was defined as the lowest level at which a signal is observed for the endogenous target peptide with an S/N > 20 in all of the replicates. The LOD and LOQ data for the detection of urinary peptides in the standard urine sample are summarized in Table 10.3. The LOQ ranges from 0.0143 fmol/µg (hemopexin) to 314.00 fmol/µg (angiotensinogen) for endogenous urinary proteins, while the LOD values ranged from 0.0023 (endogenous beta-2-glycoprotein I) to 39.6 fmol/µg (antithrombin-III).

10.3.4.3   Correlation of MRM-MS Data with Bio-Plex and ELISA Data for APOA1, APOA2, APOB, APOC3, and APOE

The members of apolipoprotein family have shown altered levels of protein expression in body fluids of numerous diseases including cancer,[79–83] gastric adenocarcinoma,[84] and atypical endometrial hyperplasia.[85] Apolipoprotein is also a well-known plasma biomarker for the risk of developing cardiovascular heart disease.[86,87] Therefore, APOA proteins were selected as a model protein family to determine the correlation between the urinary protein concentration as determined by MS-based (MRM-MS) and by antibody-based (Bio-Plex) assays (Table 10.5A and Figures 10.5A–5E). Additionally, APOA1 was measured using three quantitative assays including MRM-MS, ELISA, and Bio-plex systems (Table 10.5B and Figures 10.5A, 5F, and 5G). The correlation results of these assays are summarized in Table 10.5 and Figure 10.5, which show that any two of the assays are moderately correlated ($\gamma = 0.607$ (APOA1)-0.868 (APOE)) with high statistical significance (p-value < 0.001) for measurement of the five apolipoproteins. Using APOA1 as a model protein, our results indicate that the quantitative correlation between MRM-MS (peptide-level quantitation) and the two immunoassays (protein-level quantitation) is moderate to strong ($\gamma = 0.607$ for MRM-MS compared with ELISA, and $\gamma = 0.760$ for MRM-MS compared to Bio-Plex). The correlations between MRM-MS and the immunoassays were found to be similar to that between the two immunoassays themselves (*i.e.* $\gamma = 0.750$ for ELISA compared with Bio-Plex).[52] The observed discrepancies between the ELISA and the Bio-Plex assays might be due to differences in specificities of the antibodies used in the two immunoassays.

Table 10.5 (A) Correlation between MRM-MS and Bioplex assays for APOA1, APOA2, APOB, APOC3, and APOE quantification in urine samples (B) Correlation between ELISA, MRM-MS and Bioplex assays for APOA1 quantification in urine samples ($n = 47$). Reprinted from,[53] with permission. Part B was reprinted from,[48] with permission.

(A)

	MRM (ng/ml) vs. Bio-plex (ng/ml)		
Protein Name	*Correlation (r)*	*Significance (p-value)*	*n*
APOA1	0.760	<0.001 [p]	47
APOA2	0.741	<0.001 [p]	39
APOB	0.761	<0.001 [s]	24
APOC3	0.693	<0.001 [s]	29
APOE	0.868	<0.001 [p]	85

[p]The correlation was calculated based on Person's correlation for $n > 30$.
[s]The correlation was calculated based on Spearman's correlation for $n < 30$.

(B)

	Pearson	
APOA1	*Correlation (r)*	*Significance (p-value)*
MRM (ng/mL) *vs.* ELISA (ng/mL)	0.607	<0.001
MRM (ng/mL) *vs.* Bio-plex (ng/mL)	0.760	<0.001
ELISA (ng/mL) *vs.* Bio-plex (ng/mL)	0.750	<0.001

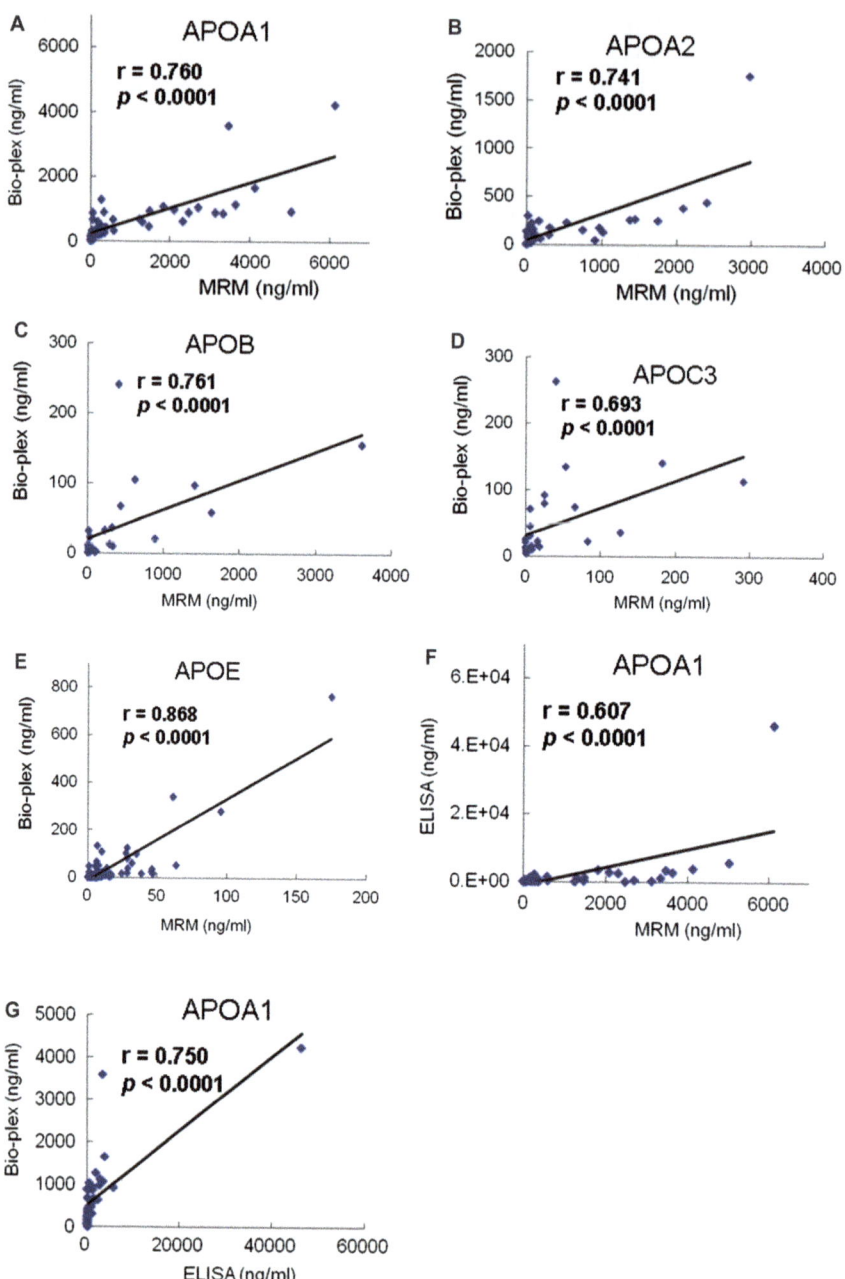

Figure 10.5   The correlations of concentrations determined by MS-based (MRM-MS)
and by antibody-based assays. (A) Bioplex *versus* MRM for APOA1.
(B) Bioplex *versus* MRM for APOA2. (C) Bioplex *versus* MRM for
APOB. (D) Bioplex *versus* MRM for APOC3. (E) Bioplex *versus* MRM
for APOE. (F) ELISA *versus* MRM for APOA1. (G) Bioplex *versus*
ELISA for APOA1.
Parts A, F and G were reprinted from,[48] with permission.

10.3.4.4   Assessment of the Diagnostic Efficacy of Individual Urinary Proteins as Bladder-Cancer Biomarkers

The data on the levels of the 63 urine proteins quantified by MRM-MS in all 156 specimens were used to evaluate the diagnostic efficacy of these proteins as biomarkers. The mean and range of the concentrations of these 63 urinary proteins in these 156 clinical specimens from the different clinical groups have been shown in our previous study.[52] Forty-seven out of 63 proteins showed significant differences between the hernia and bladder-cancer groups with *p*-values < 0.05 and AUC values ranging from 0.613 to 0.896. Thirty-three individual proteins were able to differentiate hernia from LgEs bladder cancer with AUC values from 0.580 to 0.833 ($n = 78$). As shown in Table 10.6, thirty-one individual urinary proteins could differentiate early-stage bladder cancer from late-stage bladder cancer ($n = 76$), 30 individual urinary proteins could differentiate high-grade bladder cancer from low-grade bladder cancer ($n = 76$); and 12 proteins showed higher concentrations in the bladder-cancer subgroup than in either the hernia urine or the UTI/HU subgroups (*p*-value < 0.05). These 12 proteins (afamin, alpha-1-anti-trypsin, alpha-2-HS-glycoprotein, angiotensinogen, apolipoprotein A-II precursor, apolipoprotein L1, complement C9, inter-alpha-trypsin inhibitor HC, plasminogen, prothrombin, thrombospondin-1 and transferrin) are potential urinary biomarker candidates for detection of bladder cancer. The AUC values of the ROC curves for these 12 proteins used to differentiate bladder cancer from the noncancerous groups range from 0.625 to 0.796. Prothrombin is the best single biomarker with the highest AUC value of 0.796 (which means 71.1% sensitivity and 75.0% specificity with 0.268 fmol/μg as the cutoff value), for differentiating bladder cancer ($n = 76$) from patients without bladder cancer ($n = 80$).

10.3.4.5   Assessment of the Diagnostic Efficacy of a 6-Protein Biomarker Panel in Detecting Bladder Cancer

To improve the accuracy of diagnosis, biomarker panels can be used.[44,45] To select proteins for inclusion in this panel, we used the MRM results from the bladder-cancer group ($n = 76$) and the control group (hernia and UTI/HU, $n = 80$). A hierarchical clustering analysis was performed on the abundance levels of these 43 proteotypic peptides that showed significant differences (*p*-value < 0.0008) between these groups, based on the Mann–Whitney U statistics. These peptides were grouped based on their pair-wise Spearman's rank correlation coefficients (ρ), and then visualized as a heat map (Figure 10.6).[52] The bladder-cancer disease group is shown on right side of the heat map, while the left side shows the hernia groups. We then used the hierarchical clustering to organize the peptides into 6 groups, with peptides within each group having similar color profiles. One representative peptide/protein was selected for each group. The shortlist of target peptides consisted of afamin, adiponectin, complement C4 gamma chain, apolipoprotein A-II precursor, ceruloplasmin, and prothrombin. As predicted, the AUC value for this 6-peptide

Table 10.6 List of proteins showing significant differences between different groups (p-values <0.05) and their respective AUC values. BC, bladder-cancer group; Lg, low-grade BC group; Es, early-stage BC group; Hg, high-grade BC group; As, advanced-stage BC group; UTI, urinary tract infection; HU, hematuria. Reprinted from[53] with permission.
*: p-value <0.05

Protein	AUC value					
	Hernia vs. LgEs (n=78)	Hernia vs. Es (n=119)	Hernia vs. BC (n=133)	Es vs. As (n=76)	Lg vs. Hg (n=76)	UTI/HU vs. BC (n=99)
Adiponectin	*, 0.728	*, 0.777	*, 0.779			
Afamin		*, 0.687	*, 0.716	*, 0.704		*, 0.674
Albumin, serum	*, 0.725	*, 0.820	*, 0.839	*, 0.734	*, 0.696	
Aldolase C				*, 0.698		
Alpha-1-acid glycoprotein 1	*, 0.693	*, 0.721	*, 0.715			
Alpha-1-antichymotrypsin		*, 0.680	*, 0.692			
Alpha-1-Anti-trypsin	*, 0.736	*, 0.783	*, 0.807	*, 0.714		*, 0.652
Alpha-1B-glycoprotein		*, 0.670	*, 0.697			
Alpha-2-antiplasmin		*, 0.712	*, 0.739		*, 0.705	
alpha-2-HS-glycoprotein	*, 0.783	*, 0.643	*, 0.637		*, 0.694	*, 0.721
alpha-2-macroglobulin	*, 0.668	*, 0.867	*, 0.884	*, 0.768	*, 0.716	
Angiotensinogen	*, 0.787	*, 0.768	*, 0.780			*, 0.650
Antithrombin-III	*, 0.682	*, 0.835	*, 0.850		*, 0.681	
Apolipoprotein A-I	*, 0.622	*, 0.774	*, 0.801	*, 0.719	*, 0.667	
Apolipoprotein A-II		*, 0.722	*, 0.745	*, 0.685	*, 0.657	*, 0.634
Apolipoprotein A-IV precursor	*, 0.681	*, 0.790	*, 0.789		*, 0.662	
Apolipoprotein B-100	*, 0.609	*, 0.637	*, 0.668	*, 0.707		
Apolipoprotein C-I lipoprotein		*, 0.671	*, 0.667			
Apolipoprotein C-III	*, 0.580	*, 0.661	*, 0.676			
Apolipoprotein D						
Apolipoprotein E		*, 0.618		*, 0.714		
Apolipoprotein L1	*, 0.651	*, 0.772	*, 0.803		*, 0.686	*, 0.648

Protein						
Aspartate aminotransferase, mitochondrial (m-type)						
Beta-2-glycoprotein I	*, 0.664	*, 0.781	*, 0.811	*, 0.732	*, 0.683	
CD105-endoglin		*, 0.684	*, 0.713	*, 0.682	*, 0.660	
Ceruloplasmin						
Clusterin						**, 0.648
Coagulation factor XIIa HC						
Coagulation Factor XIII (a chain)		*, 0.725	*, 0.759	*, 0.709	*, 0.717	
Complement C1 inactivator	*, 0.739					
Complement C3	*, 0.760	*, 0.834	*, 0.850	*, 0.688	*, 0.663	
Complement C4 beta chain		*, 0.841	*, 0.855	*, 0.679	*, 0.676	
Complement C4 gamma chain		*, 0.662	*, 0.703	*, 0.770	*, 0.695	
Complement factor B	*, 0.712	*, 0.796	*, 0.808	*, 0.669	*, 0.663	
Complement factor H		*, 0.667	*, 0.694			
Complement C9	*, 0.647	*, 0.735	*, 0.758	*, 0.676		**, 0.655
CRP						
Fibrinogen alpha chain	*, 0.723	*, 0.814	*, 0.840	*, 0.725	*, 0.677	
Fibrinogen beta chain	*, 0.736	*, 0.808	*, 0.836	*, 0.806	*, 0.684	
Fibrinogen gamma chain (DTV)	*, 0.815	*, 0.830	*, 0.854	*, 0.786		
Fibrinogen gamma chain (YEA)	*, 0.699	*, 0.768	*, 0.804	*, 0.804	*, 0.678	
Fibrinopeptide A	*, 0.696	*, 0.771	*, 0.805	*, 0.798	*, 0.656	
Fibronectin						*, 0.650
Gelsolin, isoform 1						
Glial fibrillary acidic protein (GFAP)						
Haptoglobin beta chain	*, 0.647	*, 0.776	*, 0.807	*, 0.776	*, 0.760	
Hemopexin	*, 0.717	*, 0.828	*, 0.851	*, 0.727	*, 0.731	
Heparin cofactor II	*, 0.833	*, 0.886	*, 0.896		*, 0.657	
Histidine-rich glycoprotein		*, 0.769	*, 0.798	*, 0.743	*, 0.722	

Table 10.6 (Continued)

Protein	AUC value					
	Hernia vs. LgEs (n=78)	Hernia vs. Es (n=119)	Hernia vs. BC (n=133)	Es vs. As (n=76)	Lg vs. Hg (n=76)	UTI/HU vs. BC (n=99)
Inter-alpha-trypsin inhibitor HC	*, 0.769	*, 0.815	*, 0.831	*, 0.726		*, 0.696
Kininogen-1						
L-selectin		*, 0.642	*, 0.658			
Plasma retinol-binding protein		*, 0.633	*, 0.631			
Plasminogen	*, 0.656	*, 0.710	*, 0.748	*, 0.720		*, 0.638
Prothrombin	*, 0.792	*, 0.846	*, 0.852			*, 0.660
Serum amyloid P-component	*, 0.647	*, 0.775	*, 0.800	*, 0.774	*, 0.649	
Thrombospondin-1			*, 0.613	*, 0.684	*, 0.713	
Transferrin (EGY)	*, 0.737	*, 0.817	*, 0.836	*, 0.724	*, 0.648	*, 0.657
Transferrin (HST)	*, 0.721	*, 0.790	*, 0.808	*, 0.698	*, 0.765	*, 0.666
Transthyretin	*, 0.653	*, 0.815	*, 0.829	*, 0.675	*, 0.733	
Tropomyosin 1 alpha chain					*, 0.745	
Vitamin D-binding protein	*, 0.688	*, 0.793	*, 0.821	*, 0.711	*, 0.704	
Vitronectin						
von Willebrand Factor						
Zinc-alpha-2-glycoprotein	*, 0.671	*, 0.709	*, 0.684			

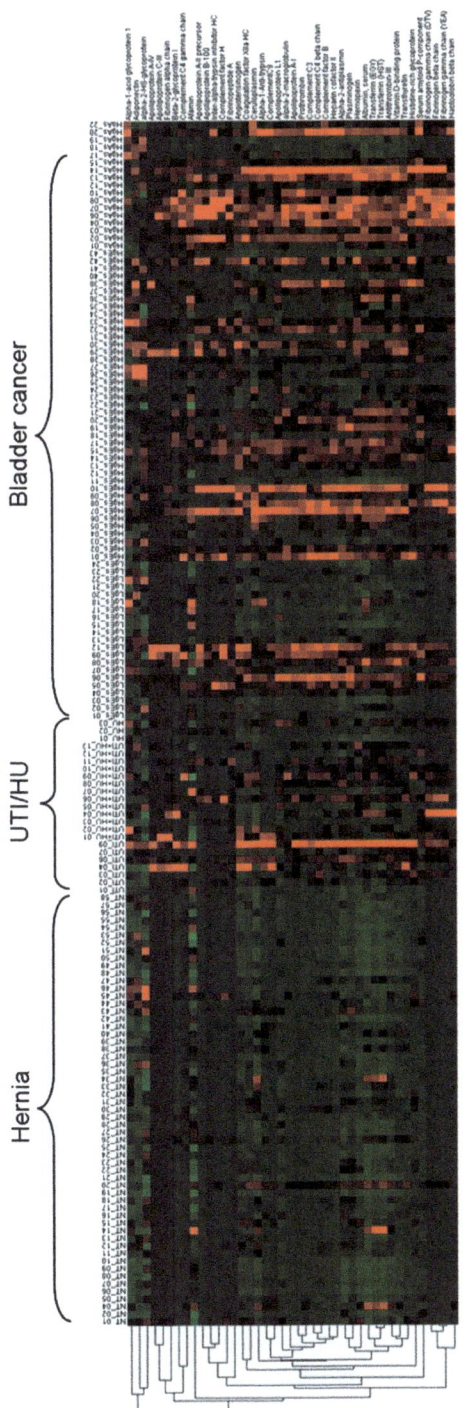

Figure 10.6 The heat map of 43 peptides whose abundance levels are significantly different (p-value < 0.0008) between the bladder-cancer and the nonbladder-cancer (hernia and UTI/HU) groups. The color represents a standardized value of peptide concentration where the mean peptide concentration for each row is 0. Red represents high abundance, while green represents low abundance. The color brightness indicates the degree of standard deviations from the average concentration of a specific peptide. Reprinted from,[48] with permission.

panel is 0.814, which is better than the AUC for the best single-protein biomarker alone (prothrombin, 0.796). This 6-peptide panel has a classification performance of 76.9%, with sensitivity = 76.3%, specificity = 77.5%, positive predictive value = 76.3% and negative predictive value = 77.5%, when the cutoff value of the probability of occurrence (*p*-value) is 0.4.

10.3.4.6   Biological Significance of the Members of 6-Peptides Biomarker Panel in Disease

The Afamin superfamily. Afamin, a protein in our final 6-protein panel, is a member of the albumin superfamily, which includes albumin, vitamin D-binding protein, α-fetoprotein, and afamin.[88,89] The concentration of afamin has been reported as being significantly higher in breast cancer serum samples than control serum sample by SELDI-TOF MS.[90] Plasma proteomic analysis revealed that afamin was overexpressed in patients with squamous cell carcinoma of the uterine cervix.[91] However, afamin concentrations decreased in ovarian-cancer patients compared to healthy controls and patients with benign disease.[92,93]

Adiponectin. Adiponectin is an adipocyte-specific peptide hormone, which increases sensitivity to insulin and has anti-inflammatory properties in nonadipose tissues,[52] and exerts antioxidative effects in prostate carcinoma cells.[94] Many other studies have also attempted to clarify the link between adiponectin and different stages of liver diseases and obesity-related diseases. Levels of serum adiponectin were found to be significantly higher in the cirrhosis and cirrhotic hepatocellular carcinoma groups than in the normal subjects.[95] Increased leptin and decreased adiponectin levels were found to disrupt the homeostatic signaling pathways involved in cell proliferation, survival, cell-cycle regulation, and angiogenesis with the implication of insulin-leptin-adiponectin in triple-negative breast cancers.[96]

Complement C4. A derivative of complement C4 was reported as a serum biomarker that could distinguish prostate cancer samples from age-matched controls.[97] The serum level of complement C4 was also significantly higher in hepatocellular carcinoma group than in controls. Detection of hepatocellular carcinoma in HCV-related liver cirrhosis patients was improved by using alpha-fetoprotein and complement C4 as a marker panel.[98]

Apolipoprotein A-II. Apolipoprotein A-II, one of the abundant apolipoproteins, has been verified as an upregulated urinary protein in bladder cancer in previous studies using Western blot and MRM-MS, with statistically significant *p*-values.[16,52] A diagnostic panel using CA19-9, apolipoprotein C-I, and apolipoprotein A-II has been shown to improve the diagnostic capability for pancreatic adenocarcinoma.[99] Apolipoprotein A-II is highly overexpressed in the CSF of pediatric brain-tumor patients, which most likely is due to a disrupted blood/brain barrier.[100] Ongoing studies are expected to confirm the significance of apolipoprotein A-II as a disease biomarker for numerous clinical applications.

Ceruloplasmin. Ceruloplasmin, a copper-containing acute-phase protein, is mainly synthesized in the liver as a secretory form. Ceruloplasmin regulates body iron homeostasis thought its ferroxidase capabilitity.[101,102] A decrease in ceruloplasmin ferroxidase activity results in accumulation of intracellular iron and related cellular injury that is associated with free-radical injury in neuro-degenerative diseases.[101] Ceruloplasmin have been previously reported to be present at elevated levels in blood samples of patients with hepatocellular carcinoma.[103]

Prothrombin. Prothrombin is the urinary protein with the best differentiation performance as a single-protein biomarker in this study and was also selected as a member of the 6-peptide panel. Differentiation of hernia *vs.* LgEs bladder cancer and differentiation of hernia *vs.* all bladder cancers are both significant with *p*-values of less than 0.001, which indicates its potential as an early diagnosis biomarker. Previous studies have revealed that des-v-carboxy prothrombin is an early predictor of microinvasion of small hepatocellular carcinoma ($\leqq 2$ cm).[104]

All six of the selected peptides showed significant differences in concentration levels between hernia and bladder cancer. However, only adiponectin, apolipoprotein A-II and prothrombin are capable of use for the early detection of bladder cancer. Our new biomarker panel will be evaluated repeatedly in the upcoming verification and validation phases to determine its performance metrics, including sensitivity, specificity, clinical convenience, and cost.

10.3.5   Comparison of iTRAQ and MRM Results and the Biomarker Discovery Pipeline

Seventeen proteins of the 22 proteins that were found to be overexpressed in BC patients in the iTRAQ study were included in the MRM-MS study. These proteins are afamin, albumin, alpha-1-acid glycoprotein 1 precursor, alpha-1-anti-trypsin, alpha-2-macroglobulin, antithrombin-III, apolipoprotein A-I precursor, apolipoprotein A-II precursor, apolipoprotein A-IV, beta-2-glycoprotein I, ceruloplasmin, fibrinogen alpha chain, fibrinogen gamma chain transferrin, transthyretin, fibrinogen beta chain, and vitamin D-binding protein. Three of these proteins (afamin, apolipoprotein A-II precursor, and ceruloplasmin,) made it onto the "shortlist" of 6 biomarkers for the biomarker paned. Thus, this MRM study has confirmed the potential clinical utility of several proteins that were discovered in the iTRAQ discovery.[48] Several new potential biomarkers were also discovered, demonstrating that multiplexed LC-MRM/MS analysis is useful for both verifying previously discovered biomarker candidates and for discovering new potential biomarkers.

10.4   Conclusions

In this work, we have demonstrated the combination of untargeted mass-spectrometry-based biomarker discovery using iTRAQ, with biomarker verification using a different, targeted, mass-spectrometry-based technique such as MRM-MS. We also demonstrated that multiplexed MRM-MS can be used as a biomarker discovery tool. In this study, several new potential biomarkers were discovered and verified by MRM-MS. This workflow indicates the important roles of quantitative proteomics and targeted proteomics in the biomarker discovery and verification pipeline. To further validate the outcome of this preclinical study and to proceed towards translation of these results to the clinic, additional studies using the six-protein biomarker panel must now be examined in a larger clinical cohort of patients.

Acknowledgements

This research project was supported by grants to Chang Gung University from the Ministry of Education (EMRPD190041) of Taiwan, Republic of China (YTC, HWC, KHL, CCW, CLC, TC, MCC, YSC, and JSY). In addition, this research was supported by grants to YTC, CLC, and JSY from the Chang Gung Memorial Hospital (CMRPG371252; CMRPG371253, CMRPD160099, CMRPD180032, CMRPD190601), National Health Research Institutes (NHRI-EX100-10015BI and HMRPD1A0091) and the National Science Council of Taiwan, Republic of China (NSC99-2320-B-182-017-MY3, NSC99-2923-B-182-002-MY2). The UVic-Genome BC Proteomics Center is partially supported by grants from Genome Canada and Genome BC (DD, DSS, CEP, CHB).

References

1. R. Siegel, E. Ward, O. Brawley and A. Jemal, *CA Cancer J. Clin.*, 2011, **61**, 212–236.
2. A. Anastasiadis and T. M. de Reijke, *Ther. Adv. Urol.*, 2012, **4**, 13–32.
3. D. S. Kaufman, W. U. Shipley and A. S. Feldman, *Lancet*, 2009, **374**, 239–249.
4. B. L. Jacobs, C. T. Lee and J. E. Montie, *CA Cancer J. Clin.*, 2010, **60**, 244–272.
5. Y. Lotan, K. Elias, R. S. Svatek, A. Bagrodia, G. Nuss, B. Moran and A. I. Sagalowsky, *J. Urol.*, 2009, **182**, 52–57, discussion 58.
6. D. Tilki, M. Burger, G. Dalbagni, H. B. Grossman, O. W. Hakenberg, J. Palou, O. Reich, M. Roupret, S. F. Shariat and A. R. Zlotta, *Eur. Urol.*, 2011, **60**, 484–492.
7. J. Parker and P. E. Spiess, *Scientific World Journal*, 2011, **11**, 1103–1112.
8. P. Villicana, B. Whiting, S. Goodison and C. J. Rosser, *Biomark Med.*, 2009, **3**, 265.
9. P. Zurbig, H. Dihazi, J. Metzger, V. Thongboonkerd and A. Vlahou, *Proteomics Clin. Appl.*, 2011, **5**, 256–268.
10. F. Abogunrin, H. F. O'Kane, M. W. Ruddock, M. Stevenson, C. N. Reid, J. M. O'Sullivan, N. H. Anderson, D. O'Rourke, B. Duggan, J. V.

Lamont, R. E. Boyd, P. Hamilton, T. Nambirajan and K. E. Williamson, *Cancer*, 2011.

11. H. Jamshidian, K. Kor and M. Djalali, *J. Urol.*, 2008, **5**, 243–247.
12. N. Atsu, S. Ekici, O. O. Oge, A. Ergen, G. Hascelik and H. Ozen, *J. Urol.*, 2002, **167**, 555–558.
13. M. L. Quek, K. Sanderson, S. Daneshmand and J. P. Stein, *Curr. Opin. Urol.*, 2004, **14**, 259–264.
14. J. W. Catto, A. Alcaraz, A. S. Bjartell, R. De Vere White, C. P. Evans, S. Fussel, F. C. Hamdy, O. Kallioniemi, L. Mengual, T. Schlomm and T. Visakorpi, *Eur. Urol.*, 2011, **59**, 671–681.
15. M. Hanke, K. Hoefig, H. Merz, A. C. Feller, I. Kausch, D. Jocham, J. M. Warnecke and G. Sczakiel, *Urol. Oncol.*, 2010, **28**, 655–661.
16. Z. Huang, L. Lin, Y. Gao, Y. Chen, X. Yan, J. Xing and W. Hang, *Mol. Cell Proteomics*, 2010, **10**, M111 007922.
17. N. Putluri, A. Shojaie, V. T. Vasu, S. K. Vareed, S. Nalluri, V. Putluri, G. S. Thangjam, K. Panzitt, C. T. Tallman, C. Butler, T. R. Sana, S. M. Fischer, G. Sica, D. J. Brat, H. Shi, G. S. Palapattu, Y. Lotan, A. Z. Weizer, M. K. Terris, S. F. Shariat, G. Michailidis and A. Sreekumar, *Cancer Res.*, 2011, **71**, 7376–7386.
18. N. L. Anderson and N. G. Anderson, *Molecular & Cellular Proteomics*, 2002, **1**, 845–867.
19. D. R. Goodlett, A. Keller, J. D. Watts, R. Newitt, E. C. Yi, S. Purvine, J. K. Eng, P. von Haller, R. Aebersold and E. Kolker, *Rapid Commun. Mass Spectrom.*, 2001, **15**, 1214–1221.
20. S. A. Gerber, J. Rush, O. Stemman, M. W. Kirschner and S. P. Gygi, *Proceeding of the National Academy of Sciences U S A*, 2003, **100**, 6940–6945.
21. S.-E. Ong, B. Blagoev, I. Kratchmarova, D. B. Kristensen, H. Steen, P. Akhilesh and M. Mann, *Molecular & Cellular Proteomics*, 2002, **1**, 376–386.
22. D. B. McClatchy, M.-Q. Dong, C. C. Wu, J. D. Venable and J. R. Yates, *J. Proteome Research*, 2007, **6**, 2005–2010.
23. http://www.silantes.com/silanmouse.htm, 2012.
24. J. C. Silva, M. V. Gorenstein, G.-Z. Li, J. P. C. Vissers and S. J. Geromanos, *Molecular & Cellular Proteomics*, 2006, **5**, 144–156.
25. H. Liu, R. G. Sadygov and J. R. Yates, III, *Anal. Chem.*, 2004, **76**, 4193–4201.
26. K. L. Simpson, A. D. Whetton and C. Dive, *Journal of Chromatography, B: Analytical Technologies in the Biomedical and Life Sciences*, 2009, **877**, 1240–1249.
27. J. M. Asara, H. R. Christofk, L. M. Freimark and L. C. Cantley, *Proteomics*, 2008, **8**, 994–999.
28. M. Elliott, D. Smith, M. Kuzyk, C. E. Parker and C. H. Borchers, *Journal of Mass Spectrometry*, 2009, **44**, 1637–1660.
29. P. L. Ross, Y. N. Huang, J. N. Marchese, B. Williamson, K. Parker, S. Hattan, N. Khainovski, S. Pillai, S. Dey, S. Daniels, S. Purkayastha,

P. Juhasz, S. Martin, M. Bartlet-Jones, F. He, A. Jacobson and D. J. Pappin, *Molecular & Cellular Proteomics*, 2004, **3**, 1154–1169.

30. L. V. DeSouza, A. D. Romaschin, T. J. Colgan and K. W. M. Siu, *Anal. Chem.*, 2009, **81**, 3462–3470.

31. L. V. DeSouza, A. M. Taylor, W. Li, M. S. Minkoff, A. D. Romaschin, T. J. Colgan and K. W. M. Siu, *J. Proteome Research*, 2008, **7**, 3525–3534.

32. L. V. DeSouza, J. Grigull, S. Ghanny, V. Dube, A. D. Romaschin, T. J. Colgan and K. W. M. Siu, *Molecular & Cellular Proteomics*, 2007, **6**, 1170–1182.

33. M. P. Washburn, D. Wolters and J. R. Yates, 3rd, *Nature Biotech.*, 2001, **19**, 242–247.

34. D. A. Wolters, M. P. Washburn and J. R. Yates, III, *Analytical Chemistry*, 2001, **73**, 5683–5690.

35. J. L. Proc, M. A. Kuzyk, D. B. Hardie, J. Yang, D. S. Smith, A. M. Jackson, C. E. Parker and C. H. Borchers, *J. Proteome Research*, 2010, **9**, 5422–5437.

36. M. A. Kuzyk, C. E. Parker and C. H. Borchers, in *Methods in Molecular Biology*, ed. H. Backvall, Humana Press, 2012, in press.

37. M. A. Kuzyk, D. Smith, J. Yang, T. J. Cross, A. M. Jackson, D. B. Hardie, N. L. Anderson and C. H. Borchers, *Mol. Cell. Proteomics*, 2009, **8**, 1860–1877.

38. T. A. Addona, S. E. Abbatiello, B. Schilling, S. J. Skates, D. R. Mani, D. M. Bunk, C. H. Spiegelman, L. J. Zimmerman, A.-J. L. Ham, H. Keshishian, S. C. Hall, S. Allen, R. K. Blackman, C. H. Borchers, C. Buck, H. L. Cardasis, M. P. Cusack, N. G. Dodder, B. W. Gibson, J. M. Held, T. Hiltke, A. Jackson, E. B. Johansen, C. R. Kinsinger, J. Li, M. Mesri, T. A. Neubert, R. K. Niles, T. C. Pulsipher, D. Ransohoff, H. Rodriguez, P. A. Rudnick, D. Smith, D. L. Tabb, T. J. Tegeler, A. M. Variyath, L. J. Vega-Montoto, A. Wahlander, S. Waldemarson, M. Wang, J. R. Whiteaker, L. Zhao, N. L. Anderson, S. J. Fisher, D. C. Liebler, A. G. Paulovich, F. E. Regnier, P. Tempst and S. A. Carr, *Nature Biotechnology*, 2009, **27**, 633–641.

39. M. Bantscheff, M. Schirle, G. Sweetman, J. Rick and B. Kuster, *Analytical and Bioanalytical Chemistry*, 2007, **389**, 1017–1031.

40. C. E. Parker, T. W. Pearson, N. L. Anderson and C. H. Borchers, *Analyst*, 2010, **135**, 1830–1838.

41. D. Domanski, A. J. Percy, J. Yang, A. G. Chambers, J. S. Hill, G. V. Cohen Freue and C. H. Borchers, *Proteomics*, 2012, **12**, 1222–1243.

42. D. Domanski, G. Cohen Freue, L. Sojo, M. A. Kuzyk, C. E. Parker, Y. P. Goldberg and C. H. Borchers, *J. Proteomics*, 2011, **75**, 3514–3528.

43. S. Surinova, R. Schiess, R. Huttenhain, F. Cerciello, B. Wollscheid and R. Aebersold, *J. Proteome Res.*, 2010, **10**, 5–16.

44. D. L. Clarke-Pearson, *N. Engl. J. Med.*, 2009, **361**, 170–177.

45. B. M. Nolen, C. J. Langmead, S. Choi, A. Lomakin, A. Marrangoni, W. L. Bigbee, J. L. Weissfeld, D. O. Wilson, S. Dacic, J. M. Siegfried and A. E. Lokshin, *Cancer Biomark*, 2011, **10**, 3–12.

46. Q. Fu, F. S. Schoenhoff, W. J. Savage, P. Zhang and J. E. Van Eyk, *Proteomics Clin Appl*, 2010, **4**, 271–284.

47. G. Bertino, S. Neri, C. M. Bruno, A. M. Ardiri, G. S. Calvagno, M. Malaguarnera, A. Toro, S. Clementi, N. Bertino and I. Di Carlo, *Minerva. Med.*, 2011, **102**, 363–371.

48. Y. T. Chen, C. L. Chen, H. W. Chen, T. Chung, C. C. Wu, C. D. Chen, C. W. Hsu, M. C. Chen, K. H. Tsui, P. L. Chang, Y. S. Chang and J. S. Yu, *J. Proteome Res.*, 2010, **9**, 5803–5815.

49. L. B. Ohlund, D. B. Hardie, M. H. Elliott, D. S. Smith, J. D. Reid, G. V. Cohen-Freue, A. P. Bergman, M. Sasaki, L. Robertson, R. F. Balshaw, R. T. Ng, A. Mui, B. M. McManus, P. A. Keown, W. R. McMaster, C. E. Parker and C. H. Borchers, in *Sample Preparation in Biological Mass Spectrometry*, eds. A. Ivanov and A. Lazarev, Springer, New York, 2011, pp. 575–624.

50. http://tools.proteomecenter.org/wiki/index.php?title = Software:Libra, 2012.

51. D. Domanski, D. S. Smith, C. A. Miller, Y. Yang, A. M. Jackson, G. Cohen Freue, J. S. Hill, C. E. Parker and C. H. Borchers, *Clinics in Laboratory Medicine*, 2011, **31**, 371–384.

52. M. J. Chen, Y. T. Yeh, K. T. Lee, C. J. Tsai, H. H. Lee and S. N. Wang, *J. Surg. Oncol.*, 2012.

53. Y.-T. Chen, H.-W. Chen, D. Domanski, D. S. Smith, K.-H. Liang, C.-C. Wu, C.-L. Chen, T. Chung, M.-C. Chen, Y.-S. Chang, C. E. Parker, C. H. Borchers and J.-S. Yu, *J. Proteomics*, 2012, **75**, 3529–3545.

54. A. J. Percy, A. G. Chambers, J. Yang, D. Domanski and C. H. Borchers, *Analytical and Bioanalytical Chemistry*, 2012, **404**, 1089–1101.

55. W. J. Youden, *Cancer*, 1950, **3**, 32–35.

56. M. B. Eisen, P. T. Spellman, P. O. Brown and D. Botstein, *Proc. Natl. Acad. Sci. U S A*, 1998, **95**, 14863–14868.

57. D. E. Hinkle, W. Wiersma and S. G. Jurs, *Boston: Houghton Mifflin Co.*, 1988.

58. International_Business_Machines, in *Statistical Package for the Social Sciences*, http://www-01.ibm.com/software/analytics/spss/.

59. http://tools.proteomecenter.org/wiki/index.php?title = Software:TPP, 2012.

60. J. Tsihlias and H. B. Grossman, *Urol. Clin. North Am.*, 2000, **27**, 39–46.

61. D. B. Zimmer, E. H. Cornwall, A. Landar and W. Song, *Brain Res. Bull.*, 1995, **37**, 417–429.

62. J. Roth, T. Vogl, C. Sorg and C. Sunderkotter, *Trends Immunol.*, 2003, **24**, 155–158.

63. C. Ryckman, K. Vandal, P. Rouleau, M. Talbot and P. A. Tessier, *J. Immunol.*, 2003, **170**, 3233–3242.

64. G. Srikrishna and H. H. Freeze, *Neoplasia*, 2009, **11**, 615–628.

65. K. Sunahori, M. Yamamura, J. Yamana, K. Takasugi, M. Kawashima, H. Yamamoto, W. J. Chazin, Y. Nakatani, S. Yui and H. Makino, *Arthritis Res. Ther.*, 2006, **8**, R69.

66. P. L. van Lent, L. Grevers, A. B. Blom, A. Sloetjes, J. S. Mort, T. Vogl, W. Nacken, W. B. van den Berg and J. Roth, *Ann. Rheum. Dis.*, 2008, **67**, 1750–1758.

67. S. Diederichs, E. Bulk, B. Steffen, P. Ji, L. Tickenbrock, K. Lang, K. S. Zanker, R. Metzger, P. M. Schneider, V. Gerke, M. Thomas, W. E. Berdel, H. Serve and C. Muller-Tidow, *Cancer Res.*, 2004, **64**, 5564–5569.

68. R. Harpio and R. Einarsson, *Clin. Biochem.*, 2004, **37**, 512–518.

69. R. Yao, A. Lopez-Beltran, G. T. Maclennan, R. Montironi, J. N. Eble and L. Cheng, *Anticancer Res.*, 2007, **27**, 3051–3058.

70. I. Salama, P. S. Malone, F. Mihaimeed and J. L. Jones, *Eur. J. Surg. Oncol.*, 2008, **34**, 357–364.

71. R. Yao, D. D. Davidson, A. Lopez-Beltran, G. T. MacLennan, R. Montironi and L. Cheng, *Histol. Histopathol*, 2007, **22**, 1025–1032.

72. L. Anderson and C. L. Hunter, *Mol. Cell Proteomics*, 2006, **5**, 573–588.

73. J. R. Whiteaker, L. Zhao, C. Lin, P. Yan, P. Wang and A. G. Paulovich, *Mol. Cell Proteomics*, 2011.

74. M. A. Kuzyk, D. Smith, J. Yang, T. J. Cross, A. M. Jackson, D. B. Hardie, N. L. Anderson and C. H. Borchers, *Molecular & Cellular Proteomics*, 2009, **8**, 1860–1877.

75. N. L. Anderson and N. G. Anderson, *Mol. Cell Proteomics*, 2002, **1**, 845–867.

76. P. Kaur, M. D. Reis, G. R. Couchman, S. N. Forjuoh, J. F. Greene and A. Asea, *J. Proteomics Bioinform*, 2010, **3**, 191–199.

77. R. S. Tirumalai, K. C. Chan, D. A. Prieto, H. J. Issaq, T. P. Conrads and T. D. Veenstra, *Molecuar and Cellular Proteomics*, 2003, **2**, 1096–1103.

78. M. F. Carroll and J. L. Temte, *Am. Fam. Physician*, 2000, **62**, 1333–1340.

79. J. S. Vermaat, I. van der Tweel, N. Mehra, S. Sleijfer, J. B. Haanen, J. M. Roodhart, J. Y. Engwegen, C. M. Korse, M. H. Langenberg, W. Kruit, G. Groenewegen, R. H. Giles, J. H. Schellens, J. H. Beijnen and E. E. Voest, *Ann. Oncol.*, 2010, **21**, 1472–1481.

80. M. Cohen, R. Yossef, T. Erez, A. Kugel, M. Welt, M. M. Karpasas, J. Bones, P. M. Rudd, J. Taieb, H. Boissin, D. Harats, K. Noy, Y. Tekoah, R. G. Lichtenstein, E. Rubin and A. Porgador, *PLoS One*, 2011, **6**, e14540.

81. Y. Murakoshi, K. Honda, S. Sasazuki, M. Ono, A. Negishi, J. Matsubara, T. Sakuma, H. Kuwabara, S. Nakamori, N. Sata, H. Nagai, T. Ioka, T. Okusaka, T. Kosuge, M. Shimahara, Y. Yasunami, Y. Ino, A. Tsuchida, T. Aoki, S. Tsugane and T. Yamada, *Cancer Sci.*, 2011, **102**, 630–638.

82. F. J. van Duijnhoven, H. B. Bueno-De-Mesquita, M. Calligaro, M. Jenab, T. Pischon, E. H. Jansen, J. Frohlich, A. Ayyobi, K. Overvad, A. P. Toft-Petersen, A. Tjonneland, L. Hansen, M. C. Boutron-Ruault, F. Clavel-Chapelon, V. Cottet, D. Palli, G. Tagliabue, S. Panico, R. Tumino, P. Vineis, R. Kaaks, B. Teucher, H. Boeing, D. Drogan, A. Trichopoulou, P. Lagiou, V. Dilis, P. H. Peeters, P. D. Siersema,

L. Rodriguez, C. A. Gonzalez, E. Molina-Montes, M. Dorronsoro, M. J. Tormo, A. Barricarte, R. Palmqvist, G. Hallmans, K. T. Khaw, K. K. Tsilidis, F. L. Crowe, V. Chajes, V. Fedirko, S. Rinaldi, T. Norat and E. Riboli, *Gut.*, 2011, **60**, 1094–1102.

83. M. Linden, S. B. Lind, C. Mayrhofer, U. Segersten, K. Wester, Y. Lyutvinskiy, R. Zubarev, P. U. Malmstrom and U. Pettersson, *Proteomics*, 2012, **12**, 135–144.

84. H. S. Ahn, Y. S. Shin, P. J. Park, K. N. Kang, Y. Kim, H. J. Lee, H. K. Yang and C. W. Kim, *Br. J. Cancer*, 2012, **106**, 733–739.

85. Y. S. Wang, R. Cao, H. Jin, Y. P. Huang, X. Y. Zhang, Q. Cong, Y. F. He and C. J. Xu, *J. Hematol. Oncol.*, 2012, **4**, 15.

86. B. G. Nordestgaard, M. J. Chapman, K. Ray, J. Boren, F. Andreotti, G. F. Watts, H. Ginsberg, P. Amarenco, A. Catapano, O. S. Descamps, E. Fisher, P. T. Kovanen, J. A. Kuivenhoven, P. Lesnik, L. Masana, Z. Reiner, M. R. Taskinen, L. Tokgozoglu and A. Tybjaerg-Hansen, *Eur. Heart J.*, 2010, **31**, 2844–2853.

87. S. Mora, J. E. Buring, P. M. Ridker and Y. Cui, *Ann. Intern. Med.*, 2011, **155**, 742–750.

88. H. S. Lichenstein, D. E. Lyons, M. M. Wurfel, D. A. Johnson, M. D. McGinley, J. C. Leidli, D. B. Trollinger, J. P. Mayer, S. D. Wright and M. M. Zukowski, *J. Biol. Chem.*, 1994, **269**, 18149–18154.

89. I. Kratzer, E. Bernhart, A. Wintersperger, A. Hammer, S. Waltl, E. Malle, G. Sperk, G. Wietzorrek, H. Dieplinger and W. Sattler, *J. Neurochem.*, 2009, **108**, 707–718.

90. A. W. Opstal-van Winden, E. J. Krop, M. H. Karedal, M. C. Gast, C. H. Lindh, M. C. Jeppsson, B. A. Jonsson, D. E. Grobbee, P. H. Peeters, J. H. Beijnen, C. H. van Gils and R. C. Vermeulen, *BMC Cancer*, **11**, 381.

91. D. H. Jeong, H. K. Kim, A. E. Prince, D. S. Lee, Y. N. Kim, J. Han and K. T. Kim, *J. Gynecol Oncol.*, 2008, **19**, 173–180.

92. D. Jackson, R. A. Craven, R. C. Hutson, I. Graze, P. Lueth, R. P. Tonge, J. L. Hartley, J. A. Nickson, S. J. Rayner, C. Johnston, B. Dieplinger, M. Hubalek, N. Wilkinson, T. J. Perren, S. Kehoe, G. D. Hall, G. Daxenbichler, H. Dieplinger, P. J. Selby and R. E. Banks, *Clin. Cancer Res.*, 2007, **13**, 7370–7379.

93. H. Dieplinger, D. P. Ankerst, A. Burges, M. Lenhard, A. Lingenhel, L. Fineder, H. Buchner and P. Stieber, *Cancer Epidemiol. Biomarkers Prev.*, 2009, **18**, 1127–1133.

94. J. P. Lu, Z. F. Hou, W. C. Duivenvoorden, K. Whelan, A. Honig and J. H. Pinthus, *Prostate Cancer Prostatic Dis.*, 2012, **15**, 28–35.

95. N. A. Sadik, A. Ahmed and S. Ahmed, *Hum Exp Toxicol*, 2012.

96. A. A. Davis and V. G. Kaklamani, *Int. J. Breast Cancer*, 2012, **2012**, 809291.

97. M. T. Davis, P. Auger, C. Spahr and S. D. Patterson, *Proteomics Clin. Appl.*, 2007, **1**, 1545–1558.

98. O. S. Ali, M. A. Abo-Shadi and L. N. Hammad, *Egypt J. Immunol*, 2005, **12**, 91–99.

99. A. Xue, C. J. Scarlett, L. Chung, G. Butturini, A. Scarpa, R. Gandy, S. R. Wilson, R. C. Baxter and R. C. Smith, *Br. J. Cancer*, 2010, **103**, 391–400.

100. J. M. de Bont, M. L. den Boer, R. E. Reddingius, J. Jansen, M. Passier, R. H. van Schaik, J. M. Kros, P. A. Sillevis Smitt, T. H. Luider and R. Pieters, *Clin. Chem.*, 2006, **52**, 1501–1509.

101. N. Tapryal, C. Mukhopadhyay, D. Das, P. L. Fox and C. K. Mukhopadhyay, *J. Biol. Chem.*, 2009, **284**, 1873–1883.

102. V. Vassiliev, Z. L. Harris and P. Zatta, *Brain Res. Brain Res. Rev.*, 2005, **49**, 633–640.

103. T. Ishihara, I. Fukuda, A. Morita, Y. Takinami, H. Okamoto, S. Nishimura and Y. Numata, *J. Proteomics*, 2012, **74**, 2159–2168.

104. Y. I. Yamashita, E. Tsuijita, K. Takeishi, M. Fujiwara, S. Kira, M. Mori, S. Aishima, A. Taketomi, K. Shirabe, T. Ishida and Y. Maehara, *Ann Surg Oncol.*, 2012.

Discovery and Validation Case Studies, Recommendations: Discovery and Development of Multimarker Panels for Improved Prediction of Near-Term Myocardial Infarction

PETER JUHASZ,*[a] MOIRA LYNCH,[a]
MANUEL PANIAGUA,[a] JENNIFER CAMPBELL,[a]
ARAM ADOURIAN,[a] YU GUO,[a] XIAOHONG LI,[a]
BØRGE G. NORDESTGAARD[b] AND NEAL F. GORDON[a]

[a] BG Medicine, 610N Lincoln St., Waltham, MA 02451, USA; [b] Department of Clinical Biochemistry and the Copenhagen General Population Study, Herlev Hospital, Copenhagen University Hospital, Copenhagen, Denmark
*Email: pjuhasz@verizon.net

11.1 Introduction

Personalized management of cardiovascular disease requires physicians to have means of accurate risk assessment. As cardiovascular disease presents itself with an extremely complex phenomenology involving a multitude of biological phenomena, plausibly, it requires multiple measures to assess cardiovascular

RSC Drug Discovery Series No. 33
Comprehensive Biomarker Discovery and Validation for Clinical Application
Edited by Péter Horvatovich and Rainer Bischoff
© The Royal Society of Chemistry 2013
Published by the Royal Society of Chemistry, www.rsc.org

disease risk. The best-known examples of such multifactorial risk score are the Framingham Risk Score[1] and the European SCORE chart[2] that can be viewed as the current standard of cardiovascular risk assessment. The performance of these risk scores is limited, however, with the majority of severe cardiovascular events (such as heart attack and stroke) occurring in individuals assigned a low to medium risk. With the majority of the population being considered low risk, it is this group that is considered to be the most burdensome to public health.

A variety of imaging technologies have been developed to further aid the assessment of cardiovascular risk including coronary calcium imaging by X-ray tomography, carotid ultrasound, and MRI.[3–5] However powerful these technologies are, their use in the context of regular screening is not practical considering costs and health risks of radiation exposure. The measurement of soluble circulating biomarkers could provide physicians an attractive option to collect information on biological processes that contributes to the increase of disease risk. Newly developed "omics" technologies are well suited to the *de novo* discovery of circulating markers potentially associated with disease risk. The importance of the systematic analysis of the new generations of circulating markers has been emphasized recently in a Scientific Statement of the American Heart Association.[6]

Determination of circulating biomarkers that inform on the risk of future cardiovascular events is not trivial and presents serious challenges to current bioanalytical approaches. Differences in circulating concentrations of biomolecules that can distinguish individuals at different risk level are typically small (occasionally less than 10%) and influenced by broad ranges of anthropomorphic variables (age, gender, weight, life style, coexisting medical conditions, *etc.*) that cause substantial overlap between the concentration distributions across different risk categories. Consequently, discovery studies need to be adequately powered requiring large numbers of samples (hundreds to thousands) along with highly reproducible analytical measurements. These requirements constrain the measurement technologies that can be applied as well as the specific implementation strategy. Separately, under these circumstances it is unlikely to find single circulating markers suitable to identify risk of future cardiovascular events. This situation is not surprising given the multiple biological pathways involved in development of cardiovascular disease. Its pathogenesis is associated with developing arteriosclerosis, oxidative damage to lipoproteins, chronic inflammation in the vasculature, and the presence of thrombogenetic components in atherosclerotic plaques.[7–9] It is anticipated that a multimarker signature, representing multiple biological pathways will provide a greater improvement in cardiovascular event risk prediction above conventional risk factors alone.

Mass spectrometry (MS)-based proteomics technologies have proven to be the method of choice for conducting "unbiased" discovery of novel circulating protein markers.[10,11] These MS-based approaches are capable of identifying and quantifying hundreds to thousands of proteins in complex biological samples. Their sensitivity to detect lower abundance proteins in circulation is limited, however, without first applying extensive protein and peptide fractionation.[12] As such fractionation steps inevitably lead to decreased sample throughput, high-resolution proteomics workflows are not well suited to profile large sample collections. One workaround is to resort to pooled sample analysis as explored by multiple research groups.[13–16] To further improve analyte coverage MS-based marker discovery can be complemented with multiplexed immunoassay measurements to sample many important circulating proteins inaccessible to mass-spectrometric detection.[12,17,18] Independent of the discovery methodology employed, candidate markers that arise from initial discovery require subsequent verification requiring measurements in a large number of additional samples. Multiple Reaction Monitoring (MRM) mass spectrometry can be particularly useful for this end as a method capable of accurate, higher throughput quantification facilitating easily multiplexed protein assays.[19,20]

The application of these methodologies described above is illustrated on an example comprising the generation of a multimarker signature for better prediction of near-term risk of heart attack (MI).

11.2 Methods

11.2.1 Study Design

11.2.1.1 Sample Set from the Copenhagen General Population Study

Banked EDTA plasma samples were selected from the participants of the 2001–2003 examination of the Copenhagen City Heart Study and from the 2003–2007 examination of the Copenhagen General Population Study. Details describing the selection of the specific subset of patient samples used for this discovery study are described elsewhere.[21] Briefly, participants were identified with incident fatal or nonfatal MI within four years of blood collection. Controls with no history of cardiovascular disease were randomly selected from the same study as the cases, and matched for age (within 1 year), gender, year of examination/blood draw (within 1 year), and statin use. Since statins are widely prescribed for cholesterol reduction it would be impractical to exclude all subjects on statin therapy and hence matching cases and controls for statin use is an effective means of eliminating statin use as a variable. A total of 252 cases and 499 controls (751 samples) were thus available for analysis. Anthropomorphic data on this study is compiled in Table 11.1.

11.2.1.2 Pooling Strategies

Sample pooling has been utilized in "population proteomics" in order to effectively sample the full set of primary samples while reducing the number of analytical measurements to keep the time frame and cost of the analysis manageable. Early implementation of simple case-control pooling was employed by the Donahue *et al.*[15] More recent work has been reported on the

Table 11.1   Anthropomorphic characteristics of the sample set selected from the Copenhagen General Population Study; *p*-values were calculated by taking matching criteria into account (see Ref. 21).

Variable	Cases (n = 252)	Controls (n = 499)	p-value
Age, years	68.7 ± 10.9	68.6 ± 10.9	matched[a]
Gender, male %	62.3	62.3	matched[a]
Body mass index, kg/m²	27.7 ± 4.9	26.5 ± 4.0	<0.01
Smoking status, %	42.6	47.5	<0.01
Diabetes, %	12.4	6.2	<0.01
Family history of premature MI, %	5.9	3.9	0.24
Blood pressure, mmHg			
Diastolic	152.3 ± 22.2	147.4 ± 19.7	0.09
Systolic	85.2 ± 13.0	83.7 ± 11.2	<0.01
Statin use, %	14.3	13.6	0.81
Total cholesterol (mmol/L)	5.82	5.70	0.16
LDL cholesterol (mmol/L)	3.54	3.37	0.03
HDL cholesterol (mmol/L)	1.43	1.61	<0.01
Antihypertensive therapy, %	39.3	29.1	<0.01
Diuretic therapy, %	24.6	13.4	<0.01

[a]Cases and Controls were matched on this variable and hence deemed to be similar; no *p*-value is reported.

Women's Health Initiative Study where 16 pools, each combining aliquots from 100 patient samples, were generated from coronary heart disease and stroke cases (eight per disease cohort) and compared with 16 control pools.[16] A more controlled pooling approach was implemented here where 25 case and 25 control pools were generated, where in each pool, sample from five (5) individuals with similar age, gender, time frame for MI event (case only), smoking status and diabetes status, and heart therapy and statin usage was combined. Randomly generated case and control pools (eight apiece) were also compared where less-conclusive results were anticipated.

11.2.2   Discovery Proteomics – 8-plex iTRAQ 2D-LC MS/MS

Proteomic profiling for the 25 pairs (and eight pairs in a separate study) of case:control pools was completed using a plasma proteomics discovery platform described in Ref. 22. In brief, plasma samples were depleted in two stages using IgY antibody columns.[23] The flow-through protein fractions were reduced, alkylated, trypsin digested, and labeled with the 8-plex iTRAQ reagents.[24] Eight samples were combined into an iTRAQ mix that was analyzed by 2D-LC MALDI MS/MS.[22] The 50 plasma pools were organized into nine iTRAQ mixes where each iTRAQ experiments involved six samples (pools) and two replicates of a reference pool. The reference pool was created by combining an aliquot from each of the 751 primary samples and reflected the average molecular composition of the study samples. The remaining four extra slots in the iTRAQ experiments not used for individual pools were filled with extra replicates of the reference pool.

Peptide identifications were performed using Mascot, protein assignments and quantification *via* an inhouse-developed processing pipeline also described in Ref. 22. Quantitative results are reported as an array of log-transformed iTRAQ ratios containing the individual proteins as columns and the pool samples in the rows.

11.2.3   Multiplex Immunoassays (Luminex 200)

Multiplex immunoassays (MIA) were performed on all 751 primary study samples in a 96-well plate format at Myriad Genetics/Rules Based Medicine using an assay panel described as HumanMAP v1.6.[25] In each 96-well plate, 24 of the 96 wells were reserved for calibrator samples/controls and 72 wells contained study samples. Of the 72 sample wells, eight were reserved for replicates of the reference pool combined from small aliquots of each study sample. Reference samples were distributed across the 96-well plate to account for any positional bias if present and were used to correct for interplate variation. Primary samples were randomized and distributed over 12, 96-well plates placing cases and their matching controls on the same plate as much as possible.

These immunoassays report protein concentrations. If the reported concentrations were below the value of the limit of detection (LOD) for that protein, these measurements were treated as missing. Although imputed values could have been used here, as analytes with substantial percentage of below-LOD measurements were excluded from statistical analysis, this choice had negligible impact on the results. One protein (C-reactive protein, CRP) featured multiple occurrences of "above-detection-maximum" readings. In these cases, the stated maximum value of the reported concentration of $50\,\mu g/mL$ was imputed.

11.2.4   Multiple-Reaction Monitoring (MRM) Analysis

MRM analysis was performed on 731 out of the 751 primary study samples as 20 samples had insufficient volume for all the analyses. The analysis was performed on original samples without depletion. In this implementation primary samples were interleaved with reference sample replicates at regular intervals in the data acquisition queue for the purpose of signal normalization and trend correction. Advantages of such a workflow design include higher multiplexing capacity and quick assay turnaround by virtue of eliminating the need for using a large number of isotope labeled synthetic peptide standards, without sacrificing analytical performance.[26] The scheme of the acquisition batch is shown in Figure 11.1. Each batch consisted of 36 primary samples and 12 reference (QC) samples. Primary samples were randomized and distributed over 21 batches with cases and their matching controls analyzed in the same batch as much as possible. "Bulk" samples in Figure 11.1 refer to a stock plasma digest that was used to carry the LC-MS instrumentation through the first couple of analyses in the batch that had tendency to produce outlying

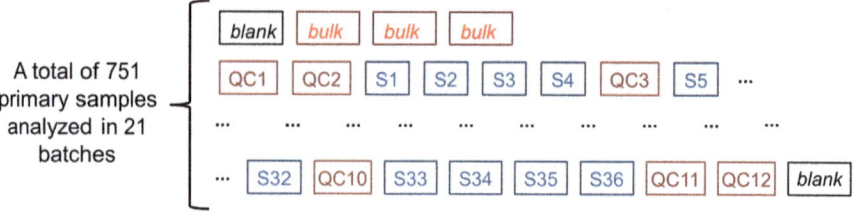

Figure 11.1 Scheme of the MRM analysis of 751 study samples; data processing (normalization, trend correction) was performed on a batch-by-batch basis.

measurements. Normalization was performed by dividing the peak areas of individual transitions by the median of the same transition in the QC samples. In this manner, peptide quantities were reported as ratios facilitating the conversion of peptide measurements into protein measurements through simple averaging.

Five (5) µL of plasma aliquots were processed in 96-well U-bottom plates. Ninety-five (95) µL of 100 mM triethylammonium bicarbonate/2M urea/10% acetonitrile/1% octyl β-D-glucopyranoside/10 mM TCEP was added to the plasma samples in each well. The plate was sealed and incubated for one hour at room temperature. Five (5) µL of 0.21 M iodoacetamide in 10% acetonitrile/2 M urea was added to each well and thoroughly pipette mixed. The plate was sealed again and incubated for 30 min at room temperature in the dark. A twenty (20)-µL aliquot of the reduced and alkylated plasma samples was transferred to a new well of a 96-well U-bottom plate for further processing. Six (6) µg sequencing grade trypsin (Promega, Madison, WI) was added to the 20 µL reduced and alkylated plasma samples and thoroughly pipette mixed. The plate was sealed and allowed to incubate overnight at room temperature. Digestion was stopped the next day by adding 26 µL of 2M urea/10 µg/mL fibrinopeptide A /1% formic acid to each well. Fifty (50) µL of each digested sample was transferred to a glass autosampler vials for LC-MRM analysis.

LC-MRM analysis was performed on a 4000QTrap linear ion trap instrument (AB/SCIEX, Concord, ON) interfaced with a U3000 HPLC system (Dionex, Sunnyvale, CA). Peptides were separated on a Targa C18 (5 µm) 150 × 1.0 mm column (Higgins Analytical) utilizing a 200-µL/min flow rate. Peptide elution was carried out over a 21-min gradient from 2% B to 32% B (A: 5% acetonitrile, 0.1% formic acid, B: 95% acetonitrile, 0.1% formic acid). The HPLC column compartment was kept at 50 °C during analysis.

For each target protein, two peptides were selected with two transitions (fragments) per peptide. Selection of these preceded by screening larger numbers of candidate peptides and transitions (typically five apiece). The set of candidate peptides were selected from those consistently observed in the pooled discovery experiments. Selection criteria for the final peptides and corresponding transition set were aimed to minimize background interference as well as distribute the peptides across the HPLC retention time space.

11.2.5 Data Analysis

Several different statistical analysis methodologies have been employed across the various data sets and are summarized in Table 11.2. For evaluating the mass-spectrometry-based discovery proteomics results for the pooled samples an univariate statistical analysis was performed using paired *t*-tests and Wilcoxon rank-sum tests that allows for non-normal distribution of protein ratios. Correction for multiple comparisons was accommodated.[27,28]

Targeted protein measurements were evaluated by both univariate and adjusted analyses in order to assess the performance of single-analyte predictors. In adjusted analyses conditional logistic regression was used to examine association between baseline variable levels and incidence of MI in multivariable models that included the following baseline covariates: total cholesterol, LDL and HDL cholesterol, smoking, hypertension, diabetes, family history of cardiovascular disease, and physical inactivity. The likelihood-ratio test was used to determine whether a logistic regression model that included the variable of interest provided a better fit than did a logistic regression model without the variable. In order to preserve statistical power, we considered men and women together in the primary analysis. To test for linear trend across categories, median levels within quintiles (based on control subjects) were used as a continuous measure.

Separately, a parsimonious set of biomarkers was selected using stepwise variable selection with retention *p*-value threshold of less than 0.05 per variable in iterative conditional logistic regression models in which the baseline covariates of total cholesterol, LDL and HDL cholesterol, smoking, hypertension, diabetes, family history of cardiovascular disease, and physical inactivity were forced to be retained. The estimated coefficients were subsequently used to weight each biomarker value in a multivariate score.

All *p*-values were two-tailed. Adjustment for multiple comparisons was performed using the false discovery rate (FDR) method of Benjamini and Hochberg.[28] All confidence intervals were calculated at the 95 percent level.

Table 11.2 Statistical analysis methodologies employed across the various data sets.

iTRAQ Px 50-pool	*MRM*	*MIA*
• Paired t-test • Wilcoxon rank sum	• Conditional logistic regression with and without adjustment for clinical covariates	• Conditional logistic regression with and without adjustment for clinical covariates • Stepwise variable selection with retention *p*-value threshold of less than 0.05 for multianalyte model construction

Abbreviations: Px = 8-plex iTRAQ 2D-LC MS/MS; MRM = Multiple-Reaction Monitoring; MIA = Multiplex Immunoassays.

11.3 Results

Discovery iTRAQ proteomics detected over 800 proteins[22] with about 600 of them regularly observed in a sufficient number of samples to perform a sufficiently powered statistical analysis. Using both paired *t*-tests and Wilcoxon ranked-sum tests ~40 markers were detected with a significance threshold of *p*-value < 0.05 and FDR corrected *p*-value < 0.2 (in other words, one out of five discoveries would be false at worst). Different statistical tests yielded slightly different numbers and identities of such markers. The top ten markers are listed in Table 11.3.

The 16-pool experiment did not yield any marker candidates with these significance thresholds. This result is not unexpected given the 10–20% reproducibility of iTRAQ protein measurements. In order to detect statistically significant mean fold changes below 20% requires measurements across a larger numbers of samples.

One of the limitations of the mass-spectrometry-based proteomics approach is detecting circulating proteins at low- and sub-ng/mL concentrations even when multiple stages of abundant protein depletion are used. Inflammation, immune response and tissue damage and repair all represent biological processes that are likely implicated in the lead up to a heart attack and involve proteins present in circulation at or below 10 ng/mL.[18] To sample some of these lower abundance proteins, 91 proteins were measured on the 751 primary samples using multiplex immunoassays.

Thirteen proteins with more than 40% of samples with missing measurements were excluded from subsequent data analysis. Two conditional logistic regression univariate models were used for analysis: Model 1 considered the analyte concentrations as the sole univariate predictor of outcome; Model 2 was a conditional logistic regression model with adjustment for the following baseline risk factors: total cholesterol, HDL cholesterol, BMI, smoking status,

Table 11.3 Top 10 markers of MI risk based on the iTRAQ analysis of 50 plasma pools (25 case *vs.* 25 control); *p*-values were derived from the Wilcoxon rank-sum test and *q*-values reflect the false discovery rate adjustment.[29]

Analyte name	Gene symbol	Median fold-change	p-value	q-value
Cholesterol ester transfer protein	CETP	0.69	<0.001	0.018
N-acetyllactosaminide beta-1,3-N-acetylglucosaminyltransferase 2	B3GNT2	0.84	<0.001	0.014
Matrix metalloproteinase 2	MMP2	0.88	<0.001	0.045
Insulin-like growth factor I	IGF1	0.94	<0.001	0.045
Thioredoxin	TXN	1.09	0.002	0.095
Reticulon-4 receptor 2-like	RTN4RL2	1.09	0.003	0.095
Orosomucoid 1	ORM1	1.24	0.003	0.095
Matrix Gla protein	MGP	0.91	0.003	0.095
Beta-2-glycoprotein	APOH	1.08	0.002	0.095
Transthyretin	TTR	0.76	0.003	0.095

hypertension, diabetes, family history of premature CV events, and physical inactivity. The results of the analysis can be represented in terms of p-values of significant differences across the quintiles of analyte levels, or p-values of a trend from the lowest to the highest quintile, or in terms of hazard ratios. p-values were corrected for multiple comparisons. With Model 1, 33 markers satisfied the (arbitrary) significance cutoff of <0.05 and false discovery rate less than 0.2. Only 15 of these remained significant after the correction for multiple risk factors in model 2. For brevity, p-values for trend are reported here and these are shown in Table 11.4 for ten of the strongest markers for predicting the risk of near-term MI.

Verification of discoveries from the pooled discovery study was performed using MRM-based targeted analysis. As abundant protein depletion has been shown to modulate the distribution of analyte concentrations[22] that skews the observed effect sizes, only proteins at sufficient abundance were considered that could be detected without depletion. A total of 15 proteins were selected for MRM measurement consisting of: i) ten (10) proteins from the pooled discovery study with a p-value <0.05, regardless of the false discovery rate (ORM1, APOH, TTR, PON1, APOA4, APOC3, APOE, FGA, FGB, and FGG), ii) three (3) candidate proteins emerging from the multiplexed immunoassay measurements that were also measured in the pooled discovery study but not identified as candidate markers (SERPINA1, C3, and APOA1) and iii) APOB and transferrin (TF) based on prior evidence of the potential involvement of these two proteins (see Table 11.1 and potential importance of iron-transporting proteins[21]).

MRM measurement results were evaluated with the same statistical procedures used for the analysis of multiplex immunoassay data. A compilation of the results is shown in Table 11.5.

Table 11.4 Selected markers of MI risk detected by multiplex immunoassays; p-values were determined from a quintiled model for trends; in Model 1 only analyte concentrations were considered as risk predictors; in Model 2 adjustments for traditional risk factors were incorporated.

Analyte name	Gene symbol	Median fold-change	p-value for trend – Model 1	p-value for trend – Model 2
Alpha-1 antitrypsin	SERPINA1	1.10	<0.001	<0.001
C reactive protein	CRP	1.58	<0.001	<0.001
Tumor necrosis factor receptor-like 2	TNFRSF14	1.14	<0.001	<0.001
Metalloproteinase inhibitor 1	TIMP1	1.13	<0.001	<0.001
Fibrinogen	FGA	1.12	<0.001	0.005
Interleukin 18	IL18	1.15	<0.001	0.009
Aspartate aminotransferase	GOT1	1.03	0.024	0.012
Alpha-Fetoprotein	AFP	0.92	0.029	0.018
Macrophage inflammatory protein 1α (MIP-1α)	CCL3	1.04	0.004	0.024
CD40	CD40	1.1	0.004	0.08

Table 11.5 Comparison of statistical analysis results for three analytical platforms; *p*-values for the discovery Px platform are derived from Wilcoxon's rank-sum test, Model 1 and Model 2 refer to Likelihood-ratio tests without or with adjustment for clinical covariates (see Section 11.2.5 on data analysis).

Protein	iTRAQ Px 50-pool MFC	p-value	%CV	MRM MFC	p-value (Model 1)	p-value (Model 2)	%CV	MIA MFC	p-value (Model 1)	p-value (Model 2)	%CV
α1-antitrypsin	1.065	0.264	12.2	1.066	<0.001	<0.001	4.9	1.103	<0.001	<0.001	6.2
Apolipoprotein A-I	0.981	0.937	27.7	0.955	0.011	0.999	3.4	0.858	<0.001	0.545	9.2
Apolipoprotein A-IV	1.204	0.003	5.5	1.043	0.100	0.071	3.4		Not measured		
Apolipoprotein B-100	0.966	0.230	10.9	1.030	0.895	0.795	12.7		Not measured		
Apolipoprotein C-III	1.095	0.020	26.6	1.045	0.79	0.998	5.3	1.075	0.077	0.829	20.0
Apolipoprotein E	1.056	0.042	28.0	1.038	0.060	0.525	3.4		Not measured		
Apolipoprotein H	1.084	0.002	11.7	1.012	0.407	0.271	4.5	1.031	0.044	0.366	6.9
Complement C3	0.998	0.052	23.8	1.062	0.003	0.484	3.8	1.102	<0.001	0.046	6.9
C-reactive protein	1.220	0.037	11.3		Not measured			1.58	<0.001	<0.001	8.0
Fibrinogen alpha	1.230	0.011	10.9	1.093	<0.001	0.002	5.1	1.123	<0.001	<0.001	11.0
Fibrinogen beta	1.070	0.230	22.2	1.040	<0.001	0.025	4.6				
Fibrinogen gamma	1.057	0.075	13.2	1.080	<0.001	0.002	6.2				
Orosomucoid	1.238	0.003	13.7	1.095	<0.001	0.003	9.6		Not measured		
Paraoxonase 1	0.923	0.007	9.2	0.945	0.063	0.437	4.8		Not measured		
Transferrin	1.074	0.063	22.9	1.01	0.505	0.674	5.4		Not measured		
Transthyretin	0.762	0.003	11.5	0.924	0.195	0.299	9.7		Not measured		

Abbreviations: Px = 8-plex iTRAQ 2D-LC MS/MS; CV = coefficient of variation; MRM = Multiple-Reaction Monitoring; MIA = Multiplex Immunoassays; MFC = mean fold change.

11.4 Discussion

Mass-spectrometry-based proteomics is ideally suited for unbiased discovery of protein markers. This technology is capable of measuring expression differences of hundreds to thousands of proteins in complex samples (such as blood) with minimal consumption of sample (for example, less than 100 μL of blood) using a variety of quantification methods ranging from simple direct peptide quantification to more precise stable isotope labeling.[26] Two key challenges to overcome, however, are the relatively low sample throughput associated with the required extensive upfront sample fraction/mass spectrometry and relatively poor sampling of the lower abundance proteins that may be relevant to the underlying disease mechanisms.

To overcome low sample throughput, pooling of individual samples to decrease the number of samples to be analyzed offers a powerful approach, but not without tradeoffs. The smaller number of samples decreases statistical power and prevents adjustment for sample specific covariates such as age, cholesterol level, *etc.* One can assess the importance of such adjustments by comparing *p*-values from Model 1 and Model 2 in Table 11.4. Consequently, sample pooling can be efficient to identify candidate markers early but one should be careful with the interpretation of *p*-values and projection of these to the measurements on larger individual sample sets. Care must also be taken in the pooling strategy selected. For example, we created a random pooling of patient samples to create eight (8) cases and eight (8) controls. Due to the small number of measurements, we were not able to show statistically significant differences between protein levels from the two groups (data not shown). A more elaborate pooling design that created a larger number of sample pools (25 cases and 25 controls) and attempted to match patient attributes such as age, gender and smoking status within each pool was used that successfully identified statistically significant protein differences between cases and controls. In spite of employing a two-stage protein depletion and two-dimensional peptide fractionation scheme, the mass-spectrometry-based discovery approach still falls short of detecting circulating proteins at the circulation levels of low ng/mL or below (leptin and myoglobin at around ~ 10 ng/mL levels represent the lowest concentration proteins measured as directly indicated by the Luminex measurements).

Multiplex immunoassays represent a useful complementary platform to detect low-abundance proteins, with a reasonable low sample volume needed for the analysis. Out of the 78 analytes profiled on the Luminex platform 53 were not detected (at least not in a sufficient number of pools) by iTRAQ proteomics. Not surprisingly, the lists of top candidate markers presented in Tables 11.3 and 11.4, based on the statistical significance thresholds that were established, reveal a good deal of complementarity. Where measured proteins are in common, the trends in mean fold change between cases and controls are largely in agreement (see Table 11.5) but statistical parameters differ consistently with lower *p*-values for the multiplexed immunoassay platform. The lower *p*-values for the results obtained with the multiplexed immunoassay platform likely results from one or more of the following two factors: i) greater statistical power due to greater than tenfold more samples analyzed and ii) generally higher reproducibility (lower coefficient of variation [%CV]) of the measurements. Analytical reproducibility is particularly important for detection of small effect sizes. For example, the three chains of fibrinogen have higher levels in cases than controls but only the results for the alpha chain qualify as statistically significant (*e.g.* $p < 0.05$). While the measurement results for beta and gamma chains show a similar mean fold change between cases and controls, due to higher variability in the measurements, the mean fold change does not appear statistically significant. Higher variability of measurements on the iTRAQ platform are generally seen with the more abundant proteins that are depleted to a larger degree. In particular the measurements of lipoproteins are impacted as observed previously[22] and here as well. Discordant results between mass spectrometric and immunoassay methodologies can also occur due to biases introduced by the assay procedure, however, this possibility is not discussed here any further.

Relative to the iTRAQ discovery platform, higher sample throughput platforms are required to accommodate larger numbers of samples used to confirm initial discoveries. Immunoassays, when available, are commonly employed for candidate marker qualification. However, in the absence of pre-existing antibodies and assays of verified specificity and reliability, targeted mass spectrometric multiple reaction monitoring (MRM) assays can be utilized[19] In our implementation of the MRM measurement platform, we were seeking to exploit the following features: i) rapid assay development, ii) ease of multiplexing with minimal sample volume, iii) high sample throughput and iv) generation of relative quantitative data avoiding the expense of isotope-coded peptide standards. Protein candidates for analysis by MRM were selected based on several criteria, starting with the verification of marker candidates from the pooled sample discovery study. As a first pass, all proteins measured in the discovery study that showed separation between cases and controls with a *p*-value of less than 0.05, regardless of the false discovery rate (*q*-value) were selected. Such relaxation of the selection criteria is easily justified as long as the multiplexing capacity of the MRM method can accommodate the additional protein targets without additional sample volume and acquisition time. In order to maximize sample throughput and accuracy of the measurement, it was also required that these proteins are present in circulation at an abundance level permitting their measurement without the need for upfront abundant protein depletion. Consequently, a substantial fraction of candidate markers had to be excluded from the selected MRM targets. A few additional proteins were added to the MRM panel that showed discordance between the iTRAQ discovery and multiplexed immunoassay measurements in order to understand the sources of these discrepancies (antitrypsin, apolipoprotein A-I, and complement C3).

As shown in Table 11.5, MRM measurements confirmed the association of $\alpha - 1$-antitrypsin, apolipoprotein A-I, complement C3, and all three fibrinogen chains with near-term risk of MI. There is generally good correlation between MRM and MIA measurement results, for those proteins measured on both

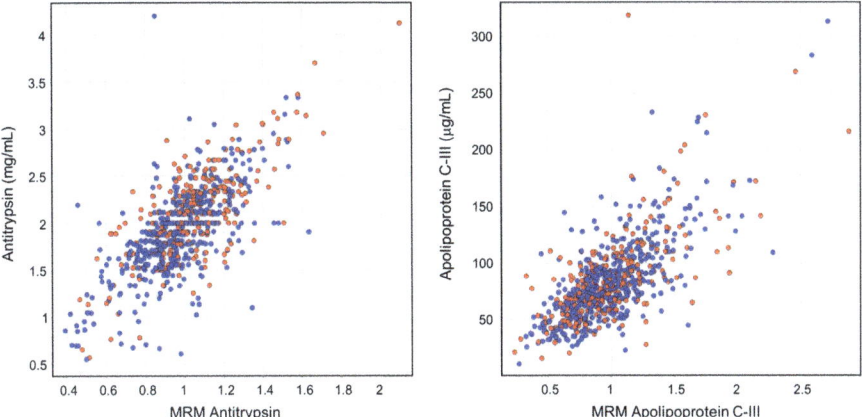

Figure 11.2   Typical consistency of MIA and MRM protein assays; MRM assay results are reported as ratio measurements relative to the reference (QC) sample; immunoassays are reported in concentration units; Red circles – cases, blue circles – controls.

platforms, which serves as a crossvalidation of both platforms. Two examples of strong correlation are shown in Figure 11.2. While immunoassays are more familiar to regulatory agencies and are preferred for clinical applications, MRM assays facilitate the rapid qualification of the biological signal on the path towards a clinical product.

For candidate protein markers where immunoassays measurement results are not available, MRM-based measurements across all the individual samples served to either confirm or refute initial discovery results. Specifically, orosomucoid (ORM1) was confirmed as a candidate risk marker and remained significant after adjusting for traditional risk factors. Paraoxonase 1 (PON1), a protein associated with cardiovascular disease risk in other studies,[30,31] remained unchanged with borderline significance. Apolipoprotein H (APOH) and transthyretin (TTR) could not be confirmed by MRM and are likely false discoveries. These validation results by MRM highlight the importance of careful interpretation of results obtained from pooled discovery studies in terms of marker *p*-values and false discovery rates. Actual values of these metrics of significance are not overly reliable indicators of the true rates of false and missed discoveries (false-negatives) when applied to a small number of biological samples (pools).

Several candidate markers identified in the discovery study have yet to be confirmed or discarded. For some, such as cholesterol ester transfer protein (CETP), insulin-like growth factor I (IGF1), S100 calcium-binding protein A9 (S100A9), and matrix Gla protein (MGP), evidence for association with cardiovascular disease (CVD) risk has been reported[32–37] that can be further evaluated using existing research-use-only (RUO) ELISA kits. Others, where no immunoassays are available, would have to be verified through MRM experiments. In its current implementation the dynamic range of MRM analysis did not allow the measurement of proteins such as N-acetyllactosaminide beta-1,3-N-acetylglucosaminyltransferase (B3GNT2), or thioredoxin (TXN) without similarly extensive abundant protein depletion as was utilized in the iTRAQ discovery proteomics workflow. Depletion – due to "off-target" retention – was shown to interfere with the linearity of quantification for many plasma proteins[22,38]. While linearity of quantification would not prevent the detection of significant differences between different clinical cohorts, at this stage of marker validation such an interference is highly undesirable. More extensive peptide fractionation has also been shown to improve the dynamic range of MRM-based peptide quantification.[39] As the throughput of the analysis would be negatively impacted by multidimensional peptide fractionation, the viability of such as workflow for large-scale marker validation is yet to be demonstrated.

One of the significant challenges to develop methods of risk prediction based on molecular markers is the very small effect sizes that separate cases and controls. From amongst the proteins listed in Tables 11.2–11.4 only CRP seems to exhibit a large enough size effect (and significance) to identify individuals at elevated risk as a univariate marker. However, CRP is known as a general inflammation marker[40] that is elevated in a variety of physiological conditions not necessarily related to cardiovascular disease. This makes CRP in itself unsuitable for the specific purpose of quantifying risk of near-term MI. As the development of cardiovascular disease is a process of multiple phases and phenomenologies it is anticipated that a multimarker signature, representing multiple biological pathways will be required to effectively predict risk of near-term MI.

11.5   Performance of Multimarker Panels

Developing panels comprising multiple biomarkers for more accurate risk prediction – especially for complex systemic diseases – has been transitioning into the mainstream of diagnostics development.[41,46] In constructing these panels one of the more important considerations is selecting analytes that are representative of multiple traits of disease pathology. For instance, in the case of cardiovascular risk prediction, one might want to select markers that represent progressing arteriosclerosis, presence of chronic inflammation, and proteins from pathways playing a role in hemostasis such as the panel derived by Wang et al.[41]

Based on the previously described results, stepwise selection of proteins to identify a parsimonious set of analytes that add predictive capacity to traditional risk factors yielded a panel of six proteins. This six-analyte panel comprised five proteins measured by MRM (alpha-1-antitrypsin, apolipoprotein A-I, apolipoprotein A-IV, apolipoprotein E, and fibrinogen gamma chain), and one protein measured by multiplexed immunoassay (TIMP1). The form of the six-analyte panel is a linear combination of log-transformed analyte values, with each term weighted by an estimated regression coefficient;

specifically, for the present panel, the formula for the six-analyte panel would be ("ln" denotes natural logarithmic transformation):

(1.31)*(ln(alpha-1-antitrypsin)) + (− 1.95)*(ln(apolipoprotein A-I))
+ (1.04)*(ln(apolipoprotein A-IV)) + (0.61)*(ln(apolipoprotein E))
+ (0.90)*(ln(fibrinogen gamma chain)) + (0.96)*(ln(TIMP1)).

It is noticeable that in this selection of analytes a set of distinct molecular pathways are represented. Without a detailed analysis of molecular pathways, this type of functional independence can also be recognized from the absence of significant correlation among the analytes making up the panel. As shown in Figure 11.3, the addition of this particular panel to a model comprising traditional risk factors improved the area under the receiver operating characteristic curve (ROC) from 0.65 (95% CI: 0.60–0.69) to 0.70 (95% CI: 0.66–0.74), $p = 0.005$ (comparison of areas takes into account the correlation of the ROC curves that arises when two distinct tests are performed on the same

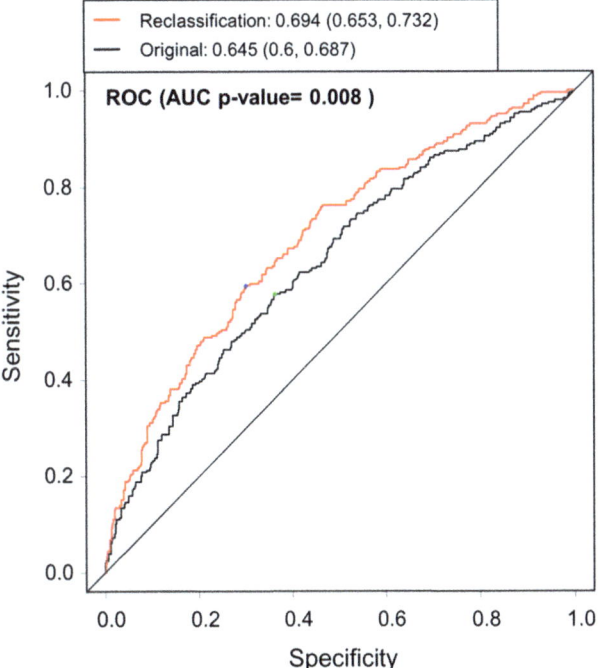

Figure 11.3   Receiver-operating characteristic curves for prediction of incident MI within 4 years, for logistic regression model comprising traditional risk factors only (black curve), and together with traditional risk factors augmented by a six-analyte panel (red curve). The six-analyte panel comprises five proteins measured by MRM (alpha-1-antitrypsin, apolipoprotein A-I, apolipoprotein A-IV, apolipoprotein.E, and fibrinogen. gamma.chain), and one protein measured by multiplexed immunoassay (TIMP1).

observed individuals). Applying more recent metrics of clinical utility,[42] the extent of risk reclassification provided by the derived exemplary multimarker panel can be summarized by category-free net reclassification improvement (NRI)of +36%. This NRI calculation represents the percent improvement in the correct prediction of heart attack (both positive and negative prediction) using the multimarker panel relative to the prediction using traditional risk factors. It is, however, essential to temper consideration of such performance metrics when reporting results solely on the derivation set, as is done here; all such multimarker panels, and indeed any novel biomarker, must be validated on independent test sets of samples, ideally from multiple independent studies, in order to assess potential clinical utility.[6]

11.6 Prepare FDA Filing for Multimarker Panel

For registration of multianalyte panels that are used to derive a single numerical score, the FDA has created an *in vitro* diagnostic multivariate index assay framework (IVDMIA).[43] The first devices to be approved under this relatively new framework were gene expression-based signatures[44] and more recently, two protein-based index assays to determine the likelihood that a pelvic mass is malignant; ROMA, a two protein-based index[45] and OVA1, based on five proteins.[46] The tests are a tool to help gynecologists appropriately refer women with masses that are likely to be malignant to a gynecologic oncologist for surgery. In both cases, already FDA-cleared protein assays were utilized. While not a requirement, the use of cleared assays facilitates the submission process in that the analytical performance characterization of the individual assays has already been reviewed and cleared and the submission can focus on the performance of the index assay. The use of noncleared assays requires the separate submission of their analytical performance data for review as components of the IVDMIA device.

Similar to the criteria that determine the clearance of individual assays, the submission for the IVDMIA must demonstrate clinical efficacy and safety. The algorithm or formula that converts the results of the individual protein tests into a single numerical score is first derived or demonstrated using a set of clinical samples where the IVDMIA test Score is correlated to clinical outcome. Very often, existing sample collections, such as the case-control samples from the Copenhagen General Population Study described in this chapter, can be applied. Once the formula is locked down, its clinical efficacy must be validated using a second independent sample set. The clinical performance criteria must be defined prospectively and care must be taken to power the study appropriately (*e.g.* use a sufficient number of samples) to demonstrate statistical significance. It is important to utilize a sample set that represents the clinical application that the device plans to address. In the cardiovascular risk application described here, while derivation of the IVDMIA can utilize a case:control design, the prevalence of cases is exaggerated relative to incidence rates in the general population for whom the test will be prescribed. Consequently, for the validation study a general population cohort that aligns with the intended population is required, and, preferably, reflects the distribution of ethnicities of the country/continent where the regulatory submission is planned. Samples from existing sample banks can be used if available. The alternative is to perform a prospective clinical sample collection, at considerable cost and time, however.

In addition to demonstrating clinical efficacy, there are two other major areas that are addressed in the regulatory submission. The first addresses the analytical performance of the IVDMIA test and includes an evaluation of measurement precision and specimen stability. As mentioned previously, if noncleared assays are incorporated in the device, a much more extensive analytical performance section is required for each of the noncleared assays. The second area covers specifics of reagent supply. A tolerance specification appropriate for the IVDMIA test is proposed. Specifications are then established for the individual reagent lot-to-lot tolerances that align with the allowable tolerance for the IVDMIA test. If the individual tests can be held within these required lot-to-lot tolerances, no further testing of reagent lots is required. However, if these tolerances are not achievable or are impractical, the assay reagents can be released as a specific combination that is tested together and verified to deliver IVDMIA test results that meet the device specification.

References

1. P. W. F. Wilson, R. B. D'Agostino, D. Levy, A. M. Belanger, H. Silbershatz and W. B. Kannel, *Circulation*, 1996, **97**, 1837.
2. Ž. Reiner, A. L. Catapano, G. De Backer, I. Graham, M.-R. Taskinen, O. Wiklund, S. Agewall, E. Alegria, M. J. Chapman, P. Durrington, S. Erdine, J. Halcox, R. Hobbs, J. Kjekshus, P. P. Filardi, G. Riccardi, R. F. Storey and D. Wood, *Eur. Heart J.*, 2011, **32**, 1769.
3. G. Mautner, S. Mautner, J. Froehlich, I. M. Feuerstein, M. A. Proschan, W. C. Roberts and J. L. Doppman, *Radiology*, 1994, **3**, 619.
4. Y. S. Chatzizisis, M. Jonas, A. U. Coskun, R. Beigel, B. V. Stone, C. Maynard, R. S. Gerrity, W. Daley, C. Rogers, E. R. Edelman, C. I. Feldman and P. H. Stone, *Circulation*, 2008, **117**, 993.
5. J. Sanz and Z. A. Fayad, *Nature*, 2008, **451**, 953.
6. AHA Scientific Statement, *Circulation*, 2009, **119**, 2408.
7. P. Libby, *Nature*, 2002, **420**, 868.
8. R. Virmani, F. Kolodgie and A. Burke, *Arterioscler Thromb Vasc Biol.*, 2005, **10**, 2054.
9. G. Stoll and M. Bendzus, *Stroke*, 2006, **37**, 1923.
10. R. Aebersold and M. Mann, *Nature*, 2003, **422**, 198.
11. X. Han, A. Aslanian and J. R. Yates III, *Curr. Opin. Chem. Biol.*, 2008, **12**, 483.
12. L. N. Anderson and N. G. Anderson, *Mol. Cell. Proteomics*, 2002, **1**, 845.
13. A. G. Paulovich, J. R. Whittaker, A. N. Hoofnagle and P. Wang, *Proteomics Clin. Appl.*, 2007, **2**, 1386.

14. N. A. Karp and K. S. Lilley, *Proteomics*, 2009, **9**, 388.
15. M. P. Donahue, K. Rose, D. Hochstrasser, J. Vonderscher, P. Grass, S. D. Chibout, C. L. Nelson, P. Sinnaeve, P. J. Goldschmidt-Clermont and C. B. Granger, *Am. Heart J.*, 2006, **152**, 478.
16. R. L. Prentice, S. Paczesny, A. Aragaki, L. M. Amon, L. Chen, S. J. Pitteri, M. McIntosh, P. Wang, T. B. Busald, J. Hsia, R. D. Jackson, J. E. Rossouw, J-A. E. Manson, K. Johnson and C. E., S. M. Hanash, *Genome Med.*, 2010, **2**, 48.
17. A. A. Ellington, I. J. Kullo, K. R. Bailey and G. G. Klee, *Clin. Chem.*, 2010, **56**, 186.
18. L. Anderson, *J. Physiol.*, 2005, **563**, 23.
19. L. Anderson and C. L. Hunter, *Mol. Cell. Proteomics*, 2006, **5**, 573.
20. T. A. Addona, X. Shi, H. Keshishian, D. R. Mani, M. Burgess, M. A. Gillette, K. R. Clauser, D. Shen, G. D. Lewis, L. A. Farrell, M. A. Fifer, M. S. Sabatine, R. E. Gerszten and S. A. Carr, *Nature Biotechnol.*, 2011, **29**, 635.
21. B. G. Nordestgaard, A. S. Adourian, J. J. Freiberg, Y. Guo, P. Muntendam and E. Falk, *Clin. Chem.*, 2010, **56**, 559.
22. P. Juhasz, M. Lynch, M. Sethuraman, J. Campbell, W. Hines, M. Paniagua, L. Song, M. Kulkarni, A. Adourian, Y. Guo, X. Li, S. Martin and N. Gordon, *J. Proteome Res.*, **10**, 34.
23. W. J. Qian, D. T. Kaleta, B. O. Petritis, H. Jiang, T. Liu, X. Zhang, H. M. Mottaz, S. M. Varnum, D. G. Camp, 2nd, L. Huang, X. Fang, W. W. Zhang and R. D. Smith, *Mol. Cell. Proteomics*, 2008, **7**, 1963.
24. P. L. Ross, Y. Huang, J. Marchese, B. Williamson, K. Parker, S. Hattan, N. Khainovski, S. Pillai, S. Dey, S. Daniels, B. Purkayastha, P. Juhasz, S. Martin, M. Bartlet-Jones, F. He, A. Jacobson and D. J. Pappin, *Mol. Cell. Proteomics*, 2004, **3**, 1154.
25. http://www.myriadrbm.com/products-services/humanmap-services/ humanmap/.
26. H. Zhang, Q. Liu, L. J. Zimmermann, A. L. Ham, R. J. C. Slebos, J. Rahman, T. Kikuchi, P. P. Masson, D. P. Carbone, D. Billheimer and D. C. Liebler, *Mol. Cell. Proteomics – Technological Innovation and Resources*, 10:10.1074/mcp.M110.006593, 2011, 1.
27. J. D. Storey and R. Tibshirani., *Proc. Natl. Acad. Sci. USA*, 2003, **100**, 9440.
28. Y. Benjamini and Y. Hochberg, Controlling the fase discovery rate: a practical and powerful approach to multiple testing, *J. R. Stat. Soc. B*, **57**, 289–300.
29. Z. Li, R. M. Adams, K. Chourey, G. B. Hurst‡, R. L. Hettich and C. Pan, *J. Proteome Res.*, 2010, **11**, 1582; A. Ritsch, H. Scharnagl, P. Eller, I. Tancevski, K. Duwensee, E. Demetz, A. Sandhofer, B. O. Boehm, B. R. Winkelmann, J. R. Patsch, W. März, *Circulation* 2010, **21**, 366.
30. N. Martinelli, D. Girelli, O. Olivieri, P. Guarini, A. Bassi, E. Trabetti, S. Friso, F. Pizzolo, C. Bozzini, I. Tenuti, L. Annarumma, R. Schiavon, P. Franco Pignatti and R. Corrocher, *Clin. Chem. Lab. Med.*, 2009, **47**, 432.

31. C. J. O'Donnell, M. K. Shea, P. A. Price, D. R. Gagnon, P. W. Wilson, M. G. Larson, D. P. Kiel, U. Hoffmann, M. Ferencik, M. E. Clouse, M. K. Williamson, I. A. Cupples, B. Dawson-Hughes and S. L. Booth, *Arterioscler Thromb Vasc Biol.*, 2006, **26**, 2769.

32. G. Engström, B. Hedblad, M. Rosvall, L. Janzon and F. Lindgärde, *Arterioscler. Thromb. Vasc. Biol.*, 2006, **26**, 643.

33. H. Soran, N. N. Younis, V. Charlton-Menys and P. Durrington, *Current Opin. Lipidol.*, 2009, **20**, 265.

34. A. M. Healy, M. D. Pickard, A. D. Pradhan, Y. Wang, Z. Chen, K. Croce, M. Sakuma, C. Shi, A. C. Zago, J. Garasic, A. I. Damokosch, T. L. Dowie, L. Poisson, J. Lillie, P. Libby, P. M. Ridker and D. I. Simon, *Circulation*, 2006, **113**, 2278.

35. A. Ritsch, H. Scharnagl, P. Eller, I. Tancevski, K. Duwensee, E. Demetz, A. Sandhofer, B. O. Boehm, B. R. Winkelmann, J. R. Patsch and W. März, *Circulation*, 2010, **121**, 366.

36. E. C. Cranenburg, R. Koos, L. J. Schurgers, E. J. Magdeleyns, T. H. Schoonbrood, R. B. Landewé, V. M. Brandenburg, O. Bekers and C. Vermeer, *Thromb Haemost.*, 2010, **104**, 811.

37. J. H. Page, J. Ma, M. Pollak, J. E. Manson and S. E. Hankinson, *Clin Chem.*, 2008, **54**, 1682.

38. S. Lin, T. A. Shaler and C. H. Becker, *Anal. Chem.*, 2006, **78**, 5762.

39. H. Keshishian, T. Addona, M. Burgess, E. Kuhn and S. A. Carr, *Mol. Cell. Proteomics*, 2007, **6**, 2212.

40. P. M. Ridker, *J. Am. Coll. Cardiol.*, 2007, **49**, 2129.

41. T. J. Wang, P. Gona, M. G. Larson, G. H. Tofler, D. Levy, C. Newton-Cheh, P. F. Jaques, N. Rifai, J. Selhub, S. J. Robins, E. J. Benjamin, R. D'Agostino and R. S. Vasan, *N. Engl. J. Med.*, 2006, **355**, 2631.

42. N. R. Cook, *Circulation*, 2007, **115**, 928.

43. http://www.fda.gov/MedicalDevices/DeviceRegulationandGuidance/GuidanceDocuments/ucm079148.htm.

44. M. J. van de Vijver, D. Yudong, Y. D. He, L. J. van't Veer, H. Dai, A. M. Hart, D. W. Voskuil, G. J. Schreiber, J. L. Peterse, C. Roberts and M. J. Marton, *N. Engl. J. Med.*, 2002, **347**, 1999.

45. R. G. Moore, D. S. McMeekin, A. K. Brown, P. DiSilvestro, M. C. Miller, W. J. Allard, W. Gajewski, R. Curman, R. C. Bast Jr. and S. J. Skates, *Gynecol Oncol.*, 2009, **112**, 40.

46. E. T. Fung, *Clin. Chem.*, 2010, **56**, 327.

CHAPTER 12

Discovery and Validation Case Studies, Recommendations: Bottlenecks in Biomarker Discovery and Validation by Using Proteomic Technologies

MARIA P. PAVLOU,[a,b] IVAN M. BLASUTIG[b,c] AND
ELEFTHERIOS P. DIAMANDIS*[a,b,c,d]

[a] Department of Pathology and Laboratory Medicine, Mount Sinai Hospital,
Toronto, ON, Canada; [b] Department of Laboratory Medicine and
Pathobiology, University of Toronto, Toronto, ON, Canada; [c] Department
of Clinical Biochemistry, University Health Network, Toronto, ON, Canada;
[d] Mount Sinai Hospital, Joseph & Wolf Lebovic Ctr., 60 Murray St [Box 32],
Flr 6 - Rm L6-201, Toronto, ON, M5T 3L9, Canada
*Email: ediamandis@mtsinai.on.ca

12.1 Introduction

With personalized medicine sailing away from empirical observation, the biomarkers of tomorrow hold promise for propelling decision making in a timely and reliably manner.Over the last few decades, there has been an escalated emphasis given to biomarker development that is highlighted by the number of research publications related to biomarker discovery and validation, as well as the amount of biomarker-related research grants awarded during the last 10 years.[1]

RSC Drug Discovery Series No. 33
Comprehensive Biomarker Discovery and Validation for Clinical Application
Edited by Péter Horvatovich and Rainer Bischoff
© The Royal Society of Chemistry 2013
Published by the Royal Society of Chemistry, www.rsc.org

Developing a biological marker entails the collaboration of academia, industry and government and the journey of a biomarker from bench to bedside is long and arduous and accompanied with a high risk of failure. This is reflected by the significant decline of biomarkers acquiring FDA approval over the last 10 years to less than one protein biomarker per year in the case of cancer biomarkers.[2,3]

Advancements in technology has given rise to the -omics fields, allowing researchers to perform large-scale analysis of the genome, transcriptome, proteome, peptidome and metabolome. However, exploiting the massive cloud of information generated by these analyses for the identification of biomarkers has proven to be challenging. The challenges of biomarker development are multidimensional but could be summarized in three key points:

- identifying the appropriate biomarker;
- delivering high-quality biomarker assays;
- assessing clinical performance of the marker.

These points also provide a snapshot of the biomarker development pipeline. Prior to initiating any biomarker-related study researchers should have a clear understanding of the unmet clinical need. This step is necessary to define the parameters of the study required to identify the type of biomarker that would address the clinical need. The current clinical practices should be extensively reviewed and major limitations should be noted so that the potential benefit such a biomarker would bring to the healthcare system can be weighed against the cost of developing and implementing the new biomarker. The value of a new marker can be translated in multiple ways. In the case of a predictive biomarker, the value will be two-fold: application of treatment with fewer side effects and optimal use of healthcare resources. Implementing a screening biomarker in the clinic would result in improved healthcare outcomes.[4] Once the desirable clinical use of the future marker is established, researchers should meticulously plan the next steps of the study. At that point, it should be noted that biomarker development requires a wide range of expertise and capabilities and carries an enormous financial burden. Suboptimal study design or execution could result in invalid findings that drain valuable clinical and financial resources.

To develop a high-quality study design, researchers need to be aware of the bottlenecks they may encounter during biomarker development, and plan ahead so as to bypass them or minimize their effects. Additionally, recognizing the limitations of a study will allow researchers to critically interpret the results and identify potential pitfalls. The present chapter provides an insight into the barriers that a potential biomarker has to overcome to reach clinical practice. A schematic representation of the considerations during biomarker development is depicted in Figure 12.1 (panels a–d).

12.2   Considerations during Biomarker Development

By definition, a biomarker must to be dynamic to recapitulate disease progress and/or pharmacological intervention. However, fluctuation of a biomarker

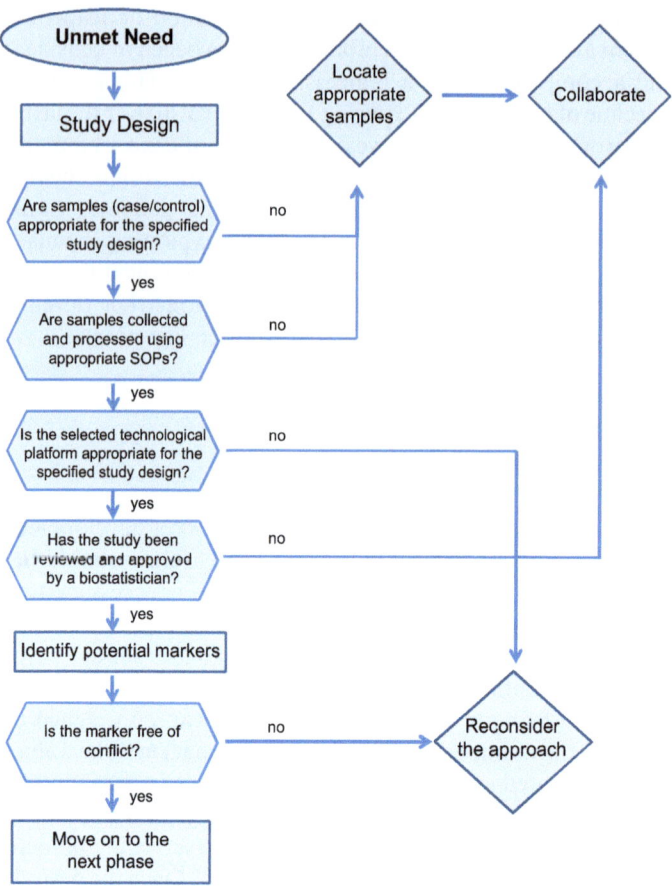

Figure 12.1a

between a healthy individual and a patient may be attributed to countless other biological or methodological factors that are irrelevant to the disease of interest.

12.2.1 Preanalytical Considerations

One of the largest sources of biomarker variations, and false-discoveries, are preanalytical factors. The impact of preanalytical variables on laboratory results has been addressed extensively.[5] Preanalytical sources of variation cover a wide array of factors but essentially fall under one of the following three categories: physiologic, specimen collection/handling and interferences[5] also depicted in Table 12.1. During biomarker development these factors should be carefully controlled either by using prespecified inclusion/exclusion criteria or by performing pilot studies assessing their effect on biomarker variation.[6]

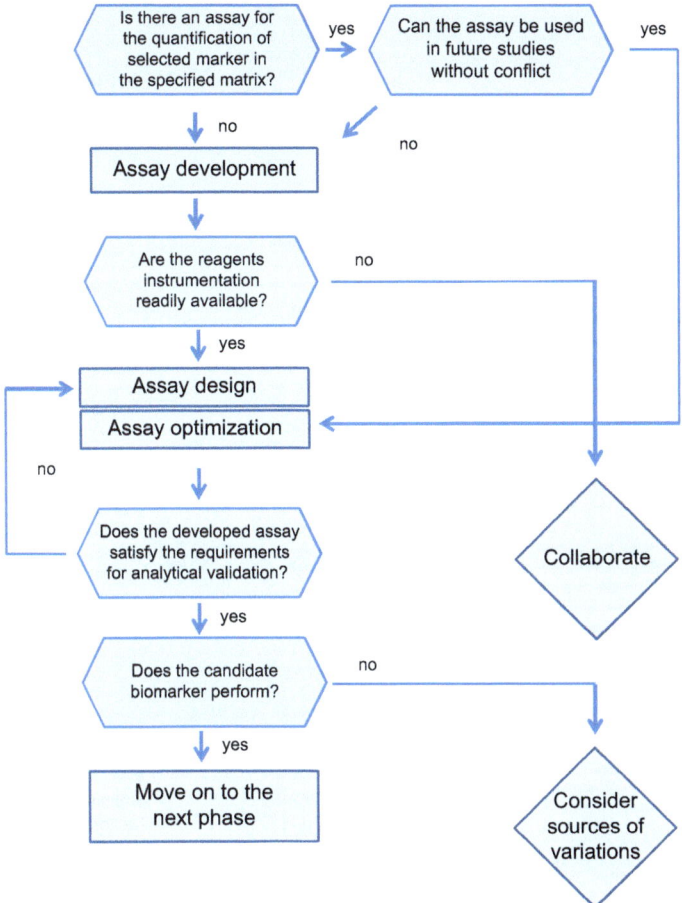

Figure 12.1b

Physiologic factors such as age, sex and lifestyle should be documented and carefully investigated to explore their relation to the analyte of interest. Control subjects should be matched as best as possible to case subjects so that confounding factors are controlled.[7] Age and sex are the most commonly matched factors but are not necessarily the most important. When samples cannot be appropriately matched, multivariate statistical analysis may be utilized to investigate and control for confounding factors.[8]

In any study, the collection, handling, processing and storage of specimens should be strictly controlled by using only predefined standard operating procedures (SOPs) to collect study samples. SOPs must remain consistent throughout the study. Differential collection or handling of "control" and "case" samples prior to analysis, such as unequal freezing–thaw cycles, may lead to altered levels of the analyte of interest between samples in a manner that is irrelevant to the disease state. Small differences in sample processing (such as

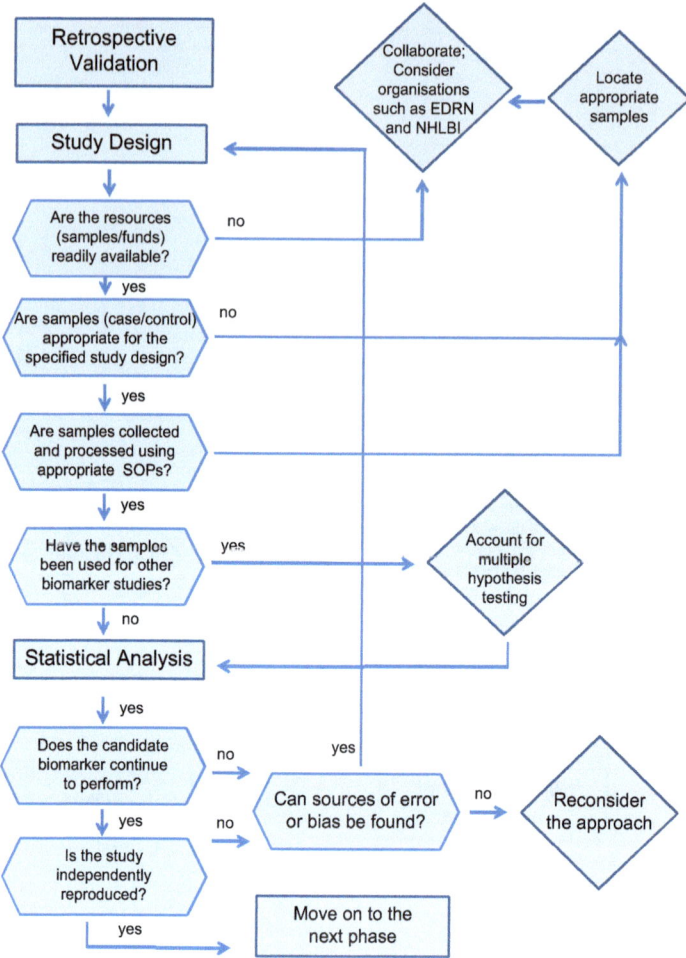

Figure 12.1c

the time and force of centrifugation) may lead to release of analytes that may affect biomarker measurement.[9] Contamination of samples by blood components and differences in storage conditions (duration, temperature) may affect significantly the level of a biomarker[10]

Special consideration should be given in tissue specimens, a popular material for development of tissue-based biomarkers, as tissue remains alive and reactive even after excision.[11] Between the time of excision and the point when the specimen is stabilized, such as fixed or snap-frozen in liquid nitrogen, the tissue will undergo reactive changes due to oxidative, hypoxic and metabolomic stress that will affect potential biomarker levels in an unpredictable manner. Careful consideration should also be taken during the fixation procedure given that the type of fixative, duration of fixation, size of tissue specimen, postfixation

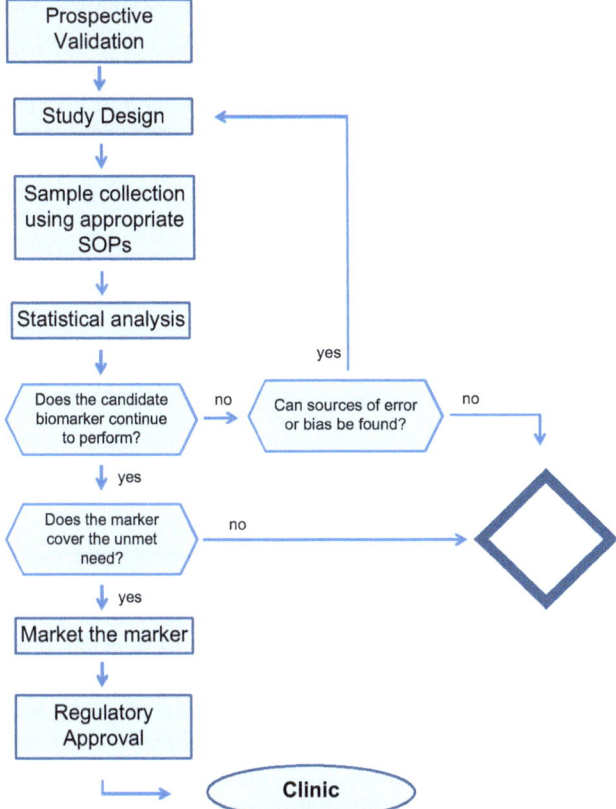

Figure 12.1d   Schematic representation of the considerations that researcher should take into account during biomarker development.

treatments and sectioning of the sample can also affect the levels or measurement of tissue biomarkers.[12] As with any sample collection, predefined SOPs should be generated and followed to ensure that all samples are collected and processed in a similar manner to avoid any unforeseen biases in the analyte of interest between samples. Analysis of surrogate stability markers is an attractive means to assess sample integrity and ensure that samples have been properly handled.[11]

12.2.2   Study Design

Every phase of the biomarker development pipeline has unique requirements in terms of the study population and evaluation of results that must be carefully considered.[7] Deficiencies in study design contribute to difficulties in data interpretation during biomarker development studies.[13]

In general, the study population should be selected based on the clinical context and must have clearly defined inclusion/exclusion criteria. The sample

Table 12.1 Preanalytical sources of variation in biomarker measurement. A wide range of factors can introduce variations in biomarker measurement and can be grouped into three categories: physiologic, specimen collection/handling and interferences.

Physiologic	Specimen Collection/Handling	Interference
Age	Anticoagulants	Circulating antibodies
Gender	Stabilizing additives	Drugs
Time	Duration of Tourniquet Application	Endogenous compounds
Menstruation	Size of needle gauge	Hemolysis
Pregnancy	Time /Temperature prior to centrifugation	Icterus
Lifestyle	Centrifugation speed, duration, temperature	Lipemia
Diurnal/seasonal variation	Storage/Transport temperature	
Exercise		
Fasting status		
Posture		
Infections		

size largely depends on the phase of development and typically increases as the biomarker moves towards the clinic[14] During verification and validation studies, sample size requirements should be calculated so as to ensure that adequate statistical power is achieved. In the case of exploratory studies, on the other hand, no simple rules exist for recommending sample sizes.[7] Inadequate sample size is a common factor contributing to inflated or false-positive results during biomarker development.[15]

In addition to the numeric size of the study population, the composition of the cohort is also very important. The study population should be appropriate for the intended clinical use of the biomarker and should match, as best as possible, the population for which the biomarker will be used in the clinic. Control subjects should be matched for age, gender and any presenting signs and symptoms that would warrant further testing for the disease in question. Case subjects should also be representative of the population that will be tested for the biomarker, this includes appropriate signs and symptoms as well as the disease stage. Additionally, case subjects should be proportionally represented within the study population to recreate the disease prevalence and allow for correct result evaluation (*i.e.* calculation of true positive rate; TPR, or false-positive rate; FPR). An example of poor study design can be seen in the seminal study of the diagnostic utility of carcinoembryonic antigen (CEA) in colorectal cancer in which CEA was found to be almost 100% sensitive and specific for colorectal cancer screening.[16] This performance failed to be reproduced in subsequent studies. The initial impressive performance of CEA was largely attributed to the fact that patients of the initial study had extensive cancer, whereas subjects participating in following studies had less-extensive asymptomatic cancer.[17,18] In this case, the inappropriate representation of study subjects with respect to population cohort led to overestimation of biomarker performance.

The study of tissue biomarkers offers an indispensable advantage in comparison to blood-based markers given that nondiseased specimens can also be collected from the patient and serve as controls, hence minimizing biological variability and seemingly serving as the perfect controls. However, depending on the disease of study, it may be difficult, or even impossible, to obtain control tissue from the same individual and even if adjacent tissue appears healthy, it may be predisposed to the disease and thus provide biased results.[19]

12.2.3 Analytical Considerations

Selecting and implementing the proper technological platform in each step of the biomarker development requires careful planning and validation. Every platform has technical limitations that researchers need to be aware of so as to critically interpret the results and recognize methodological artefacts. Below, we will focus on the main proteomic technologies utilized during the different phases of biomarker development.

12.2.3.1 Protein Biomarker Discovery Technologies

Mass spectrometry (MS) has risen as a powerful tool for protein biomarker discovery during the last decade. A variety of mass spectrometry-based approaches have been utilized in the search for disease biomarkers such as two-dimensional (2D) electrophoresis followed by tandem MS, gel-free liquid chromatography coupled with tandem MS, protein profiling using matrix-assisted laser desorption ionization (MALDI) ionization with time-of-flight (TOF) MS and quantitative MS-based proteomics. Despite its widespread use in biomarker discovery, MS-based analysis of complex biospecimens has a limitation: proteome undersampling. It has been estimated that 30% of known genes have never been observed at the protein level.[20] This shortcoming can be attributed mainly to two reasons. First, the ionization efficiency of peptides is low and is dependent on their biochemical characteristics. Therefore, the majority of peptides do not get ionized and cannot be detected. Secondly, due to the inherent nature of MS operation, there is a bias towards detecting the most-abundant peptides, leaving low-abundance species undetected.[21] The latter technological limitation can be overcome either by decreasing sample complexity using upstream fractionation methods or by depleting the most-abundant proteins. However, during depletion, low-abundance proteins that interact with those undergoing depletion may be removed as well.[22] Additionally, both approaches will significantly decrease the throughput of the analysis.[21] An additional factor that may contribute to proteome under-sampling is that protein identification by MS is dependent on the existence of genome sequence databases. Proteins or isoforms with unknown sequences will not be identified even if they exist in the sample. In an effort to overcome the shortcoming of undersampling, Human Proteome Project launched the Chromosome-centric Human Proteome Project (C-HPP) in Geneva, Switzerland in September 2011. The mission of C-HPP is to identify and characterize at least one representative protein isoform for every human gene in a systematic chromosome-based manner.[23] Extra attention will be paid to map the "missing proteins" (genes that lack experimental evidence at protein level) by employing a variety of approaches such as RNA-sequencing, recombinant protein standards and heavy-labeled peptides.[24] It is worth mentioning that C-HPP is an international collaboration consisting of 25 groups coming from 17 countries, with each group adopting one of the 24 human chromosomes and one group adopting mitochondrial DNA (www.c-hpp.org).

Quantitative MS-based proteomics has also been widely utilized during protein biomarker discovery studies. Numerous methodologies exist (Table 12.2) and can be described under two main categories: label-free and stable-isotope labeling.[25] Two broadly used label-free quantitative strategies exist and are based on (a) spectral count and (b) peak intensity. In the former strategy, the number of spectra originating from peptides of a given protein is used as a proxy of the protein abundance. In the latter methodology, the signal intensity of precursor peptide ions originating from a given proteins is measured and compared across multiple liquid chromatography (LC) runs. One criticism of spectral counting is that it can be biased by properties such as protein size, peptide length and peptide chromatographic behavior, all factors that are irrelevant to protein abundance. For that reason, statistical tools such as a normalized spectral abundance factor (NSAF) that accounts for the effect of protein length on spectral count, have been developed and should be implemented during data analysis.[26,27] In the case of measuring peak intensity, the major analytical caveat is that the peptide chromatographic profile need be highly reproducible because any changes in the elution or retention times would encumber LC-MS peak alignment.[28] Special software has also been developed to overcome this potential limitation.[29–32] Although label-free methodologies may offer higher dynamic range of quantification (2–3 orders of magnitude) in

Table 12.2   Quantitative mass-spectrometry-based methodologies.

Label-free	Spectral counting
	Ion intensity
Stable-Isotope Labelling	Metabolic protein labeling
	stable-isotope labeling with amino acids in tissue culture, SILAC
	Chemical protein labeling
	isotope-coded protein label, ICPL
	Chemical peptide labeling
	Isotope-coding affinity tags, ICAT
	Isotope tags for relative and absolute quantification, iTRAQ
	Tandem mass tags, TMT
	Enzymatic labeling
	incorporation of ^{18}O in the C-termini of peptides during digestion

comparison to stable-isotope labeling (1–2 orders of magnitude), they exhibit lower accuracy and precision.[25]

Stable-isotope labeling can be performed at the protein or peptide level in various ways: metabolically, enzymatically or chemically.[25] Metabolic labeling (stable-isotope labeling by amino acids in cell culture, SILAC) is the most accurate method because samples to be compared can be combined at very early stages of the experiment resulting in the least variation due to sample preparation. However, it is costly and time consuming and can be applied only in growing cells. Enzymatic labeling (incorporation of O_{18} during protein digestion) is a very specific way of introducing isotope labels into peptides but suffers from low labeling efficiency and peptide-dependant incorporation rates.[33,34] Numerous methodologies exist to perform chemical labeling (also in Table 12.2). Among these, isobaric tag for relative and absolute quantification (iTRAQ) offers the advantage of directly comparing up to 8 different conditions. However, labeling and combination of the samples occurs after protein digestion, causing results to be affected by variations in sample preparation.[35] Additionally, the combining of multiple samples into a single sample significantly increases the complexity of the analytical run and can increase undersampling.

During the last decade the research community has witnessed tremendous technological advancements in the field of mass spectrometry with respect to protein identification and quantification. Although limitations exist, technological advances in MS and proteomics are ongoing and will soon result in more sensitive and powerful instruments that will be able to cope with sample complexity and increase proteomic coverage.

12.2.3.2 Protein Biomarker Verification/Prioritization Technologies

Although currently available proteomic technologies may not be able to perform global proteomic profile of complex samples, they can still be effectively utilized for biomarker verification and/or prioritization studies.[21] Contrary to discovery-driven MS-based proteomics, targeted MS (Box 12.1) provides a good combination of sensitivity, selectivity and throughput.[35] As implied by the name, targeted MS requires *a priori* knowledge of the analytes to be detected and enables relevant or even absolute quantification of multiple peptides, and therefore proteins, in a biological sample.[36] Small-sample requirements, multiplexing capability, high selectivity and cost- and time-efficient development of assays are the major advantages of targeted MS assays. A major drawback of targeted MS is that direct analysis of complex specimens, such as plasma, allows quantification of proteins to only the low μg/ml level.[37] Most protein biomarker concentrations range from the ng/ml to pg/ml level. This inadequate sensitivity can be circumvented by adding a separation or enrichment step prior to MS analysis.[38–40] Nonetheless, similar to discovery studies, increased sample handling lowers the throughput and increases the preanalytical variables. As MS technologies develop, the sensitivities of instrumentation continue to improve and may soon reach the levels required to detect these low-abundance proteins.

Box 12.1   Description of Targeted Proteomics

Selective reaction monitoring (SRM) is used for targeted quantitative proteomics. SRM enables targeted detection and quantification of selected proteotypic peptides (PTP) and the PTPs are used as surrogate markers of the protein of interest. The selected PTPs should be unique to the protein of interest, should have known fragmentation properties, and should be detectable by mass spectrometry. The most widely used platform for SRM experiments is a triple quadrupole mass spectrometer. During an SRM experiment, the selected precursor ion of a PTP is detected based on the mass-to-charge (m/z) ratio in the first quadrupole (Q1), and is fragmented by collision-induced dissociation (CID) with an inert gas in the second quadrupole (Q2). Selected fragment ions are then monitored in the third quadrupole (Q3) and their signal intensity is recorded over time. SRM transitions combined with the retention time (RT) and machine parameters (ion source, collision energy) serve as unique quantitative identifiers of the selected PTP. At least two peptides and two to four transitions should be monitored per protein to ensure a reliable quantification. Absolute quantification of the protein of interest can be achieved by adding known amounts of isotopically labeled peptides to the sample prior to analysis.

12.2.3.3   Protein Biomarker Validation Technologies

Antibody-based detection techniques such as enzyme-linked immunosorbent assay (ELISA) and immunohistochemistry (IHC) have been the cornerstone for protein quantification in diagnostics. Based on the principle of antigen–antibody interaction and enzyme-conjugated antibody detection, immunoassays can be highly sensitive and specific, with the ability to detect analytes at the pg/mL level in complex biological fluids. The major drawback of immunoassays is the lack of high-quality immunoaffinity reagents for most proteins of interest. The need for highly specific antibodies was one of the motives for the creation of Human Protein Atlas.[41] The ultimate goal of the Human Protein Atlas project is the generation of multiple antibodies against all human proteins and the systematic exploration of the human tissue and organ proteome to determine protein expression profiles. By November 2011, 15 598 antibodies corresponding to 12 238 protein-coding genes were included in Human Protein Atlas portal (www.proteinatlas.org) covering >40% of the 19 559 human entries as defined by UniProt.[42] Generation and validation of the highly sensitive and specific antibodies required is a long and tedious procedure and it is only the very first step towards developing an immunoassay, many more steps must be taken to ensure that the analytical assay developed is appropriate for use in future studies.

12.2.3.4 Analytical Requirements

Regardless the technological platform of choice, developing reliable quantitative assays is essential for biomarker development. Poor assay choice or improper assay validation may result in biomarker failure, irrespective of the biomarker's true performance.[7,43,44] Although required assay characteristics depend on the intended application of individual biomarkers,[6] analytical assay validation should provide information regarding the parameters listed in Table 12.3. High analytical specificity is essential to ensure quantification of the analyte of interest versus structurally similar molecules, thus avoiding overestimation.[45] High analytical precision (within and between assay runs) is vital as highlighted by the results of Prostate Lung Colorectal Ovarian (PLCO) Cancer Screening Trial Specimens for ovarian cancer.[46] In this study, multiple candidate biomarkers were tested in phase II and III trials and clinical performance was strongly correlated to assay performance: only those markers with analytical assays achieving coefficient of variation (CV) less than 30% demonstrated adequate diagnostic sensitivity. Proper quality-control (QC) samples should be used to confirm specificity, monitor accuracy and precision, and determine run acceptance.[45]

As mentioned earlier biomarker discovery and validation can take place in different environments including academia, industry and government. Not all the laboratories that participate in biomarker research comply with federal regulations, known as Clinical Laboratory Improvement Amendments (CLIA). However, operating under CLIA-like practices will increase the reliability and quality of the data.[6]

12.2.4 Statistical Analysis

Statistics are an integral part of the biomarker development pipeline. A smooth collaboration between biomedical researchers and statisticians is pivotal for the

Table 12.3 Minimum requirements for biomarker assay validation.

Parameters for analytical assay validation
Reference standard of known purity
Standard/calibration curve
Recovery studies
Proof of specificity
Within and between assay precision
Assay dynamic range
Lower limit of quantification (LOQ)[1]/Lower limit of detection (LOD)[2]
Parallelism and dilution linearity
Reference range for pertinent population
Quality controls independent of calibrators

[1]LOQ: the lowest amount of an analyte in a sample that can be quantified with a predetermined degree of confidence.
[2]LOD: the lowest amount of an analyte in a sample that can be detected, but not necessarily quantified as an exact value.

success of a biomarker study. Biomedical researchers should provide biostatisticians with an extensive and well-defined "study problem" so that the latter
can comprehend the concept of the study. Biostatisticians, on the other hand,
should walk the researchers through the statistical methods and thoroughly
explain the statistical analysis so that the researchers are able to critically
interpret the results. Biostatistics should not be a black box for researchers.
Close collaboration should occur at the earliest stages of the study, giving the
statisticians the opportunity to participate in the design of the study. Given that
resources are finite, biostatisticians will assist in designing the most robust
experimental approach possible with the resources available.

Hunting down a *p*-value of <0.05 is not always the path to statistically
significant results. A simple but usually neglected rule in statistics is that
multiple hypothesis testing in the same population increases the possibility that
one of the hypothesis tests will reach significance solely by chance;[47] therefore,
a *p*-value of <0.05 is no longer significant. Multiple hypothesis testing would
refer to studying multiple biomarkers, different cut-off points or multiple
endpoints(for example disease-free survival and overall survival) in one cohort
of samples. Hidden multiple hypothesis can be a threat given that several
individual groups may perform biomarker studies using the same sample
collection. Another point for consideration, which could be also related to
multiple testing, is that the statistical test is selected *a priori* to ensure that the
selected statistical tool tests the hypothesis rather that fits the acquired data. By
defining the statistical approach before data analysis, researchers avoid testing
multiple associations until one of them shows statistical significance.

Development of biomarker panels, rather than single biomarkers, has
emerged as an attractive approach based on the biological heterogeneity of
human disease and the multiple molecular pathways involved during disease
progression. The most well-known examples of biomarker panels stem from
gene-expression studies, but similar panels should be expected at the proteome
level. The analysis of such multidimensional data poses a challenge as the
existing guidance and standards are directed at tests based on single signal
output.[48,49] Although, new methodologies for analyzing multisignal readings
are available, researchers should be very careful utilizing such approaches in
order to avoid overfitting.[50,51] Overfitting occurs when multivariable models
demonstrate discrimination between two conditions by chance.[52] A model is
prone to overfitting when the number of parameters tested is large and the
number of samples is small. Several ways exist to avoid overfitting. The most
direct way includes the existence of two distinct sample cohorts: training and
test. The training cohort is used to develop a classification algorithm that is
independently validated in the test cohort consisting of samples that are not
part of the training set.[52]

Statistical analyses hold the key to the success of a biomarker study.
Although studies designed with the help of a trained biostatistician still risk the
possibility of reporting false findings, the risk is drastically reduced compared
to findings coming from a poorly designed study.

12.2.5   Clinical Validation

Validation is perhaps the most ambiguously used term in biomarker development.[52] Clinical validation of a biomarker refers to establishing the clinical utility of the marker. Proper clinical validation should be performed in both a retrospective and prospective setting and by multiple independent investigators.[7] The assay used for clinical validation should have undergone a rigorous analytical validation and the specimens collected and handled strictly according to SOPs so as to minimize all sources of variation, as previously discussed. The number of patients required for the study, estimated using formulas to achieve sufficient statistical power, is usually high. Additionally, specimens should be accompanied with relevant clinical information.

Given the need of high-quality specimens for validation studies, sample availability is one of the main obstacles of biomarker development, and the need for biological specimens surpasses the available supply. One way to overcome this limitation is to develop central registries for biological specimens and generate universal regulations that will allow access to these valuable resources.[53] Sample collection, processing and storage during biobanking should follow prespecified guidelines such as those published by the National Cancer Institute (http://biospecimens.cancer.gov/practices/). Validation studies requesting biospecimens can then be evaluated and prioritized to ensure that well-designed studies, with high potential for clinical implementation, have priority access to biological resources.[53] This central registry could also maintain the data generated from these studies, providing both a comprehensive picture of the biomarker field and information on where further resources should be invested.[54] The creation of this type of registry would also address two more drawbacks of clinical validation: hidden multiple hypothesis and nonpublication bias.[54] The hidden multiple hypothesis threat (see statistical considerations) would be reduced because researchers would be able to review the number of studies that a particular cohort of samples has been utilized. Secondly, nonpublication bias would be controlled. Nonpublication bias refers to the higher likelihood for studies with positive results, as opposed to negative, to be published. As mentioned earlier, independent validation studies should be performed for a biomarker to achieve a high level of evidence and the consistency among the results for these studies is crucial for the final assessment of the biomarker. All validation studies should be published, regardless of the findings. If negative findings do not reach publication then the validity of published positive results remains questionable. Even in the case of peer-reviewed and published validation studies, the issue of inaccurate and/or incomplete reporting presents challenges to biomarker development. To overcome this potential challenge, major journals have adopted reporting guidelines such as the Standards for Reporting of Diagnostic Accuracy (STARD)[55] and Reporting Recommendations for Tumor Marker Prognostic Studies (REMARK)[56] for diagnostic and prognostic biomarkers respectively. Similar guidelines should be developed and adopted for other biomarker types (screening, predictive, monitoring) to ensure increased transparency and easy assessment of the study quality.

12.2.6 Financial

So far, we have been discussing barriers encountered during biomarker development that are related, in one way or another, to the biomarker performance. However, issues related to financial profit may also pose challenges in biomarker development. As mentioned previously, biomarker development is an arduous process with high financial burden that no single organization, company or academic group can afford to undertake alone.[4] The participating private sectors, such as industry, usually invest a considerable amount of money on the development of *in vitro* diagnostic (IVD) assays with the ultimate aim of commercializing the final product. Therefore, financial investment on a biomarker will be considered not only when it demonstrates remarkable clinical utility, but also when there is no other conflict of interest. A company would be reluctant to invest in a product with unclear intellectual property or patented status, regardless of the clinical performance of the marker.

In general, diagnostics have been regarded by industry as a less-attractive investment opportunity than therapeutics, and consequently are not as well funded as drug development.[57] This reluctance to invest in diagnostics poses an extra challenge for biomarker development and can be attributed, at least partly, to inadequate reimbursement. Obtaining reimbursement by social or private insurers allows for broader use of the test and is one of the main reasons to seek regulatory approval of a biomarker. The importance of inadequate reimbursement is highlighted in two government-commissioned reports recommending the re-evaluation of reimbursement rates for diagnostics.[58,59]

12.3   Concluding Remarks

The journey of a protein biomarker from the bench to the clinic is long and arduous. It encompasses multiple contiguous phases, requires collaboration among different stakeholders and carries a significant financial burden. Challenges are present at every phase and therefore every step needs to be meticulously planned and executed to warrant any possibility of success. Researchers need to be aware of the challenges that they may face during biomarkers development. *A priori* knowledge of the potential limitations will improve study design and enhance critical interpretation of results. Independent validation and unbiased publication of the findings holds the key for the discovery of novel biomarkers. Furthermore, scientific discussions and debates on published biomarker studies through commentaries, correspondence and scientific forums such as the Biomed Critical Commentary (www.bm-cc.org), SEQanswers.com and sharedproteomics.com, may pinpoint potential limitations and drawbacks of the study resulting in a more objective evaluation of the biomarker's clinical utility.

The paucity of novel biomarkers, in spite of the intensified research, has become a cause for alarm, as can be seen in documents such as the National Institutes of Health (NIH) Roadmap initiative and the US FDA's Critical Path Initiative (CPI) for drugs and diagnostics.[60] In an effort to improve success, organizations such as the Early Detection Research Network (EDRN) of the National Cancer Institute (http://edrn.nci.nih.gov/) and the National Heart, Lung and Blood Institute's (NHLBI) Clinical Proteomics Programs (www.nhlbi.nih.gov/) have been established to promote collaborative efforts for discovery and validation of new disease biomarkers.

The existence of highly useful biomarkers utilized currently in clinical practice, such as serum levels of cardiac troponin as an indicator of a cardiovascular damage,[61] indicate that the discovery of novel biomarkers is not impossible. The candidate markers are out there and it is up to the scientific community to properly utilize all the available resources in order to identify them.

References

1. A. S. Ptolemy and N. Rifai, *Scand. J. Clin. Lab Invest Suppl*, 2010, **242**, 6.
2. N. L. Anderson and N. G. Anderson, *Mol. Cell Proteomics.*, 2002, **1**, 845.
3. S. Gutman and L. G. Kessler, *Nat. Rev. Cancer*, 2006, **6**, 565.
4. T. A. Metcalfe, *Scand. J. Clin. Lab Invest Suppl.*, 2010, **242**, 23.
5. S. Narayanan, *Am. J. Clin. Pathol.*, 2000, **113**, 429.
6. B. N. Swanson, *Dis. Markers*, 2002, **18**, 47.
7. M. S. Pepe, R. Etzioni, Z. Feng, J. D. Potter, M. L. Thompson, M. Thornquist, M. Winget and Y. Yasui, *J. Natl. Cancer Inst.*, 2001, **93**, 1054.
8. O. S. Miettinen, *Am. J. Epidemiol.*, 1976, **104**, 609.
9. D. L. Baker, P. Morrison, B. Miller, C. A. Riely, B. Tolley, A. M. Westermann, J. M. Bonfrer, E. Bais, W. H. Moolenaar and G. Tigyi, *JAMA*, 2002, **287**, 3081.
10. A. F. Lomholt, C. B. Frederiksen, I. J. Christensen, N. Brunner and H. J. Nielsen, *Clin. Chim. Acta*, 2007, **380**, 128.
11. V. Espina, C. Mueller, K. Edmiston, M. Sciro, E. F. Petricoin and L. A. Liotta, *Proteomics. Clin. Appl.*, 2009, **3**, 874.
12. A. Babic, I. R. Loftin, S. Stanislaw, M. Wang, R. Miller, S. M. Warren, W. Zhang, A. Lau, M. Miller, P. Wu, M. Padilla, T. M. Grogan, L. Pestic-Dragovich and A. S. McElhinny, *Methods*, 2010, **52**, 287.
13. D. F. Ransohoff, *Nat. Rev. Cancer*, 2005, **5**, 142.
14. N. Rifai, M. A. Gillette and S. A. Carr, *Nat. Biotechnol.*, 2006, **24**, 971.
15. J. P. Ioannidis, *Epidemiology*, 2008, **19**, 640.
16. D. M. Thomson, J. Krupey, S. O. Freedman and P. Gold, *Proc. Natl. Acad. Sci. U. S. A*, 1969, **64**, 161.
17. D. F. Ransohoff and A. R. Feinstein, *N. Engl. J. Med.*, 1978, **299**, 926.
18. D. L. Sackett, *J. Gen. Intern. Med.*, 1987, **2**, 40.
19. A. M. Soto and C. Sonnenschein, *J. Biosci.*, 2005, **30**, 103.

20. P. Legrain, R. Aebersold, A. Archakov, A. Bairoch, K. Bala, L. Beretta, J. Bergeron, C. H. Borchers, G. L. Corthals, C. E. Costello, E. W. Deutsch, B. Domon, W. Hancock, F. He, D. Hochstrasser, G. Marko-Varga, G. H. Salekdeh, S. Sechi, M. Snyder, S. Srivastava, M. Uhlen, C. H. Wu, T. Yamamoto, Y. K. Paik and G. S. Omenn, *Mol. Cell Proteomics.*, 2011, **10**, M111.

21. P. Wang, J. R. Whiteaker and A. G. Paulovich, *Cancer Biol. Ther.*, 2009, **8**, 1083.

22. J. Granger, J. Siddiqui, S. Copeland and D. Remick, *Proteomics*, 2005, **5**, 4713.

23. Y. K. Paik, S. K. Jeong, G. S. Omenn, M. Uhlen, S. Hanash, S. Y. Cho, H. J. Lee, K. Na, E. Y. Choi, F. Yan, F. Zhang, Y. Zhang, M. Snyder, Y. Cheng, R. Chen, G. Marko-Varga, E. W. Deutsch, H. Kim, J. Y. Kwon, R. Aebersold, A. Bairoch, A. D. Taylor, K. Y. Kim, E. Y. Lee, D. Hochstrasser, P. Legrain and W. S. Hancock, *Nat. Biotechnol.*, 2012, **30**, 221.

24. Y. K. Paik, G. S. Omenn, M. Uhlen, S. Hanash, G. Marko-Varga, R. Aebersold, A. Bairoch, T. Yamamoto, P. Legrain, H. J. Lee, K. Na, S. K. Jeong, F. He, P. A. Binz, T. Nishimura, P. Keown, M. S. Baker, J. S. Yoo, J. Garin, A. Archakov, J. Bergeron, G. H. Salekdeh and W. S. Hancock, *J. Proteome. Res.*, 2012, **11**, 2005.

25. M. Bantscheff, M. Schirle, G. Sweetman, J. Rick and B. Kuster, *Anal. Bioanal. Chem.*, 2007, **389**, 1017.

26. B. Zybailov, A. L. Mosley, M. E. Sardiu, M. K. Coleman, L. Florens and M. P. Washburn, *J. Proteome. Res.*, 2006, **5**, 2339.

27. L. Florens, M. J. Carozza, S. K. Swanson, M. Fournier, M. K. Coleman, J. L. Workman and M. P. Washburn, *Methods*, 2006, **40**, 303.

28. W. Zhu, J. W. Smith and C.-M. Huang, *Journal of Biomedicine and Biotechnology*, 2009, **2010**, 1.

29. D. Bylund, R. Danielsson, G. Malmquist and K. E. Markides, *J. Chromatogr. A*, 2002, **961**, 237.

30. P. Wang, H. Tang, M. P. Fitzgibbon, M. McIntosh, M. Coram, H. Zhang, E. Yi and R. Aebersold, *Biostatistics*, 2007, **8**, 357.

31. A. V. Tolmachev, M. E. Monroe, N. Jaitly, V. A. Petyuk, J. N. Adkins and R. D. Smith, *Anal. Chem.*, 2006, **78**, 8374.

32. E. F. Strittmatter, P. L. Ferguson, K. Tang and R. D. Smith, *J. Am. Soc. Mass Spectrom*, 2003, **14**, 980.

33. K. L. Johnson and D. C. Muddiman, *J. Am. Soc. Mass Spectrom.*, 2004, **15**, 437.

34. A. Ramos-Fernandez, D. Lopez-Ferrer and J. Vazquez, *Mol. Cell Proteomics.*, 2007, **6**, 1274.

35. K. G. Kline, G. L. Finney and C. C. Wu, *Brief. Funct. Genomic. Proteomic*, 2009, **8**, 114.

36. V. Lange, P. Picotti, B. Domon and R. Aebersold, *Mol. Syst. Biol.*, 2008, **4**, 222.

37. T. A. Addona, S. E. Abbatiello, B. Schilling, S. J. Skates, D. R. Mani, D. M. Bunk, C. H. Spiegelman, L. J. Zimmerman, A. J. Ham, H. Keshishian, S. C. Hall, S. Allen, R. K. Blackman, C. H. Borchers, C. Buck, H. L. Cardasis, M. P. Cusack, N. G. Dodderxx, B. W. Gibson, J. M. Held, T. Hiltke, A. Jackson, E. B. Johansen, C. R. Kinsinger, J. Li, M. Mesri, T. A. Neubert, R. K. Niles, T. C. Pulsipher, D. Ransohoff, H. Rodriguez, P. A. Rudnick, D. Smith, D. L. Tabb, T. J. Tegeler, A. M. Variyath, L. J. Vega-Montoto, A. Wahlander, S. Waldemarson, M. Wang, J. R. Whiteaker, L. Zhao, N. L. Anderson, S. J. Fisher, D. C. Liebler, A. G. Paulovich, F. E. Regnier, P. Tempst and S. A. Carr, *Nat. Biotechnol.*, 2009, **27**, 633.
38. G. R. Nicol, M. Han, J. Kim, C. E. Birse, E. Brand, A. Nguyen, M. Mesri, W. FitzHugh, P. Kaminker, P. A. Moore, S. M. Ruben and T. He, *Mol. Cell Proteomics.*, 2008, **7**, 1974.
39. V. Kulasingam, C. R. Smith, I. Batruch, A. Buckler, D. A. Jeffery and E. P. Diamandis, *J. Proteome. Res.*, 2008, **7**, 640.
40. N. L. Anderson, N. G. Anderson, L. R. Haines, D. B. Hardie, R. W. Olafson and T. W. Pearson, *J. Proteome. Res.*, 2004, **3**, 235.
41. M. Uhlen, E. Bjorling, C. Agaton, C. A. Szigyarto, B. Amini, E. Andersen, A. C. Andersson, P. Angelidou, A. Asplund, C. Asplund, L. Berglund, K. Bergstrom, H. Brumer, D. Cerjan, M. Ekstrom, A. Elobeid, C. Eriksson, L. Fagerberg, R. Falk, J. Fall, M. Forsberg, M. G. Bjorklund, K. Gumbel, A. Halimi, I. Hallin, C. Hamsten, M. Hansson, M. Hedhammar, G. Hercules, C. Kampf, K. Larsson, M. Lindskog, W. Lodewyckx, J. Lund, J. Lundeberg, K. Magnusson, E. Malm, P. Nilsson, J. Odling, P. Oksvold, I. Olsson, E. Oster, J. Ottosson, L. Paavilainen, A. Persson, R. Riminixx, J. Rockberg, M. Runeson, A. Sivertsson, A. Skollermo, J. Steen, M. Stenvall, F. Sterky, S. Stromberg, M. Sundberg, H. Tegel, S. Tourle, E. Wahlund, A. Walden, J. Wan, H. Wernerus, J. Westberg, K. Wester, U. Wrethagen, L. L. Xu, S. Hober and F. Ponten, *Mol. Cell Proteomics.*, 2005, **4**, 1920.
42. M. Uhlen, P. Oksvold, L. Fagerberg, E. Lundberg, K. Jonasson, M. Forsberg, M. Zwahlen, C. Kampf, K. Wester, S. Hober, H. Wernerus, L. Bjorling and F. Ponten, *Nat. Biotechnol.*, 2010, **28**, 1248.
43. R. C. Bast, Jr., H. Lilja, N. Urban, D. L. Rimm, H. Fritsche, J. Gray, R. Veltri, G. Klee, A. Allen, N. Kim, S. Gutman, M. A. Rubin and A. Hruszkewycz, *Clin. Cancer Res.*, 2005, **11**, 6103.
44. J. A. Wagner, S. A. Williams and C. J. Webster, *Clin. Pharmacol. Ther.*, 2007, **81**, 104.
45. J. W. Lee, *Bioanalysis*, 2009, **1**, 1461.
46. D. W. Cramer, R. C. Bast, Jr., C. D. Berg, E. P. Diamandis, A. K. Godwin, P. Hartge, A. E. Lokshin, K. H. Lu, M. W. McIntosh, G. Mor, C. Patriotis, P. F. Pinsky, M. D. Thornquist, N. Scholler, S. J. Skates, P. M. Sluss, S. Srivastava, D. C. Ward, Z. Zhang, C. S. Zhu and N. Urban, *Cancer Prev. Res. (Phila)*, 2011, **4**, 365.
47. J. P. Ioannidis, *PLoS. Med.*, 2005, **2**, e124.

48. US Food and Drugs Administration, 2003.
49. Clinical Laboratory Standards Institute, 1995.
50. K. Dobbin and R. Simon, *Biostatistics*, 2005, **6**, 27.
51. G. Campbell, *J. Biopharm. Stat.*, 2004, **14**, 539.
52. D. F. Ransohoff, *Nat. Rev. Cancer*, 2004, **4**, 309.
53. M. C. Hinestrosa, K. Dickersin, P. Klein, M. Mayer, K. Noss, D. Slamon, G. Sledge and F. M. Visco, *Nat. Rev. Cancer*, 2007, **7**, 309.
54. F. Andre, L. M. McShane, S. Michiels, D. F. Ransohoff, D. G. Altman, J. S. Reis-Filho, D. F. Hayes and L. Pusztai, *Nat. Rev. Clin. Oncol.*, 2011, **8**, 171.
55. P. M. Bossuyt, J. B. Reitsma, D. E. Bruns, C. A. Gatsonis, P. P. Glasziou, L. M. Irwig, J. G. Lijmer, D. Moher, D. Rennie and H. C. de Vet, *Fam. Pract.*, 2004, **21**, 4.
56. L. M. McShane, D. G. Altman, W. Sauerbrei, S. E. Taube, M. Gion and G. M. Clark, *Nat. Clin. Pract. Urol.*, 2005, **2**, 416.
57. K. A. Phillips, B. S. Van and A. M. Issa, *Nat. Rev. Drug Discov.*, 2006, **5**, 463.
58. S. A. C. o. G. H. a. S. Department of Health and Human Services, 2005.
59. L. Kessler, S. D. Ramsey, S. Tunis and S. D. Sullivan, *Health Aff. (Millwood.)*, 2004, **23**, 200.
60. F. a. D. A. US Department of Health and Human Services, 2011.
61. L. Babuin and A. S. Jaffe, *CMAJ*, 2005, **173**, 1191.

Subject Index